"Czerski aims to greatly expand and even revolutionise the reader's understanding of what is going on in seven tenths of the planet that is not covered in land." —*Financial Times*

"Czerski is a wonderful writer. . . . a compelling and elegantly written story. . . . *The Blue Machine* really does change the way you see the world."

—Christopher Hart, *Daily Mail*

"Helen Czerski, urging us to see the ocean as a presence, not an absence, has done a remarkable job of shoehorning an overview of the whole shebang into a single, very readable volume."

—Jon Turney, *Arts Desk*

"In Helen Czerski's hands, the mechanical becomes magical. An instant classic."

—Tristan Gooley, author of *How to Read Water: Clues and Patterns from Puddles to the Sea*

"Awash with fascinating facts. Helen Czerski writes with authority, passion, and an easy conversational style. You will want to be out there on the ice and ocean with her. I loved it."

—Hugh Aldersey-Williams, author of *The Tide: The Science and Stories Behind the Greatest Force on Earth*

"*The Blue Machine* is quite simply one of the best books I have ever read. Helen Czerski is a consummate storyteller. . . . In places you'll drift serenely among corals or dense kelp forests, in others you'll ride Atlantic breakers or fear for your life in a tropical storm. . . . When you resurface, you will be bursting with enthusiasm and wonder and you'll understand how the ocean works and more besides."

—Dr. George McGavin, zoologist, entomologist, and broadcaster

THE BLUE MACHINE

Also by Helen Czerski

STORM IN A TEACUP: THE PHYSICS OF EVERYDAY LIFE

BUBBLES: A LADYBIRD EXPERT BOOK

THE BLUE MACHINE

How the Ocean Works

HELEN CZERSKI

W. W. NORTON & COMPANY
Independent Publishers Since 1923

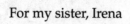
For my sister, Irena

Contents

Ocean Currents

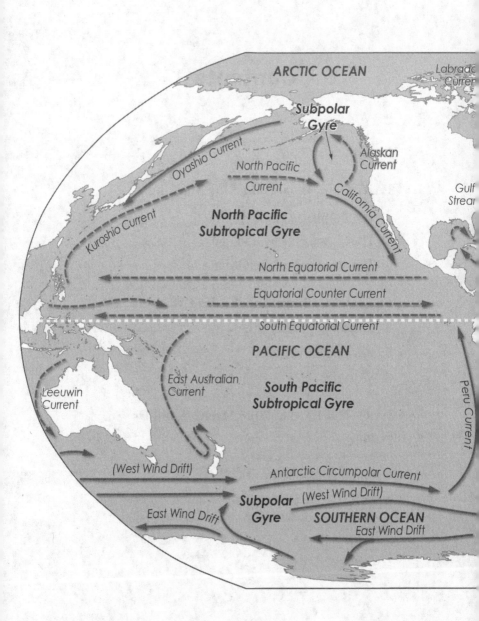

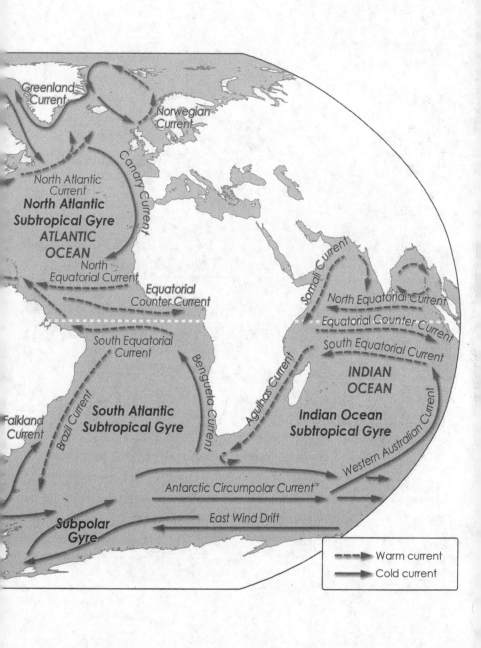

Ocean Winds

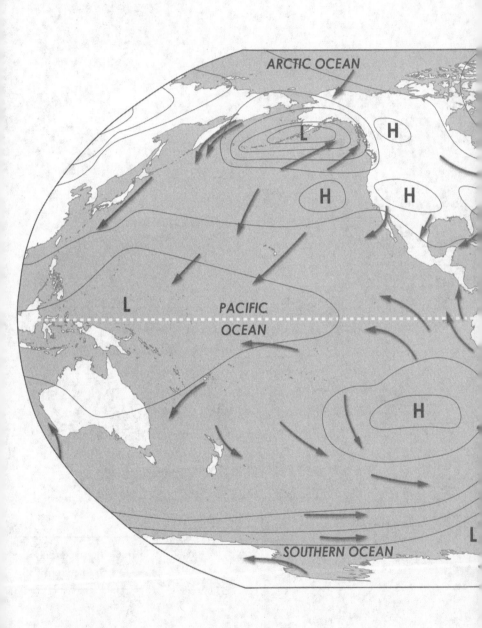

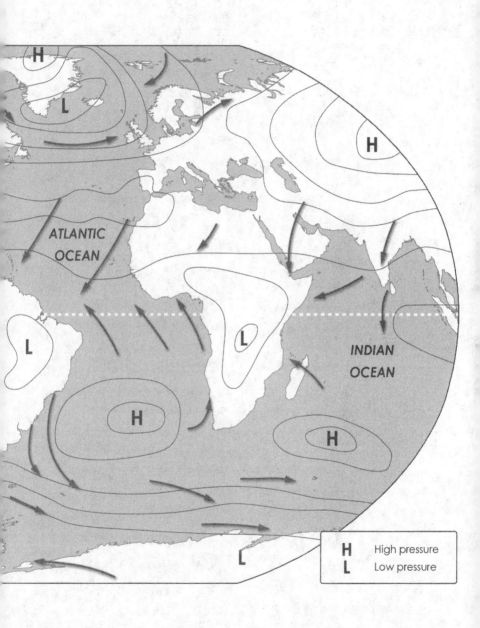

ATLANTIC
OCEAN

INDIAN
OCEAN

| H | High pressure |
| L | Low pressure |

Introduction

Equator

OUR CANOE FLOATS QUIETLY IN the darkness, a tiny drifting speck dwarfed by the starry sky. We wait while deep inky blue crawls upwards from the horizon, gently extinguishing the stars as it reveals the gigantic silhouette of Haleakalā, the larger of Maui's two dormant volcanoes. The support boat bobs a few metres away, between us and the eastern horizon. I'm in the fifth seat from the front in a six-person outrigger canoe, and the *ama*, the second floating hull that connects these canoes to the ocean like no other craft, rides the water to my left. The water is flat calm and the peace is immense. We wait.

Just before the first hints of pink light wash around Haleakalā, it is time. Kimokeo Kapahulehua stands up on the side of the support boat, facing the canoe, wearing shorts and a T-shirt and a ti-leaf *lei*. He calls out to us in the canoe, and to our teammates on the support boat. I know very few Hawaiian words, but I can follow along because this is a familiar expression of the deep Hawaiian connection to the ocean. The canoe isn't just a physical object, designed for a practical purpose. Every aspect of a canoe is a symbol of teamwork – to make it, transport it, paddle it and maintain it – and it is teamwork that holds these island nations together. This is about *'ohana*, the extended family, and about taking care of the people in your canoe. The ocean is as much a part of home as the land. It's changeable, and it can be hazardous, but if you show humility, and you

1

observe and learn, the ocean will support you and provide for you.

Today, we are setting out on a voyage around Maui's second volcano, Mauna Kahālāwai, from here in Kahului Bay all the way around to Kīhei, but we need skill and weather and luck with ocean conditions to complete the trip. Whether or not we succeed, it's the time spent together and what we learn on the way that's important. This ritual reminder lasts only a few minutes, because here at the equator the sun rises quickly. As bright lilac and pink fill the sky behind Haleakalā, Kimokeo finishes with a chant and we all join in. *E ala e* is the first and most significant chant that any Hawaiian child learns, intended for this vital moment, when sunlight first touches the ocean, and everything can begin.

Under the clear sky, we can almost feel the day's flood of intense sunlight on its way. Cam, steering the canoe from the sixth seat, just behind me, calls out: 'Ho'omākaukau', and we raise our paddles. They hang in the air for one last moment of stillness, as the first shaft of bright sunlight touches the ocean ahead of us. Then, 'Imua!', and six paddles cut into the water simultaneously. The voyage is under way.

Pole

Five months later, I'm lying on my tummy on the edge of a giant ice floe.[1] My colleague Matt is standing on a 3 metre square wooden platform floating on the ocean, and I'm attaching a rope to the side of it. The water is −1.8°C, technically much warmer than the air (which today is −8°C) but far more effective at stealing heat, so I'm focused on keeping my fingers and the rope dry. The platform has a squat round metal dome rising from its

---

[1] An ice floe is a sheet of floating ice. At 2 km across, this one was unusually large – most are much smaller.

centre, and the rest of the platform is a hive of mechanical servants tending to the dome's needs: large metal boxes and batteries,
exchanging data, electrons and air with the queen bee in the centre via chunky cables. This is Matt's experiment, designed to
capture and count any tiny particles that the ocean is spitting
into the atmosphere. I tug on my new knot and stand up, check-
ing that nothing has fallen out of the pockets of my heavy
flotation suit. When I nod, Matt strides off across the ice floe to
fetch some helping hands, the cheerful bouncing of the bobble
on his hat taking the edge off any 'serious polar scientist' look he
might have aspired to.

I spend a few minutes just looking at the view, a rare treat.
Just over a mile away on the other side of the ice floe, I can see
the Swedish icebreaker *Oden*, our home for these two months.
Towards the middle of the ice floe, a large tethered red and white
balloon, the size of a small camper van, bobs in the sky with
today's scientific experiment dangling beneath it. In the other
direction, the white sea ice stretches away for hundreds of miles.
We're only a few nautical miles from the North Pole, and the
summer sea ice here is about 2 metres thick. The ice floes around
ours shunt and grumble, moving slowly but just enough that
our work site looks different every morning. This afternoon, the
clouds are lifting to reveal rare blue sky directly above, but sunlight will never beam down from that space. The sun here slinks
around the horizon but never drops below it, casting astonishingly long shadows when the clouds allow. Even though this
spot is in the middle of six months of constant daylight, the sunlight feels purely decorative, a cascade of soft, subtle illumination
which will never keep you warm. The energy flow here is mostly
invisible, directed upwards, not downwards. Around me, the
Earth itself is glowing, radiating infrared light as what little heat
it has seeps away into the sky. Without the clouds as an obstacle,
this heat energy will just keep going out into space. A clear day
today means a cold day tomorrow, as the ice and the ocean lose

their energy to the universe. This process is a crucial entry in the Earth's energy budget, but we still can't predict it exactly. On this ice floe, we're in the middle of these mechanisms, surrounded by liquid clockwork, and we want to *know*. The cohort of scientists on *Oden* is here to observe, measure and analyse this spectacular environment, to use the latest equipment and logic to deduce the inner workings of the ocean and atmosphere.

Matt returns with three of our colleagues. Out in the natural environment, data are hard won, and that often means very physical effort. We want the platform to be floating away from the ice edge, so it can take measurements in open water. Hauling on long ropes in teams, and with much grunting and shoving, we manoeuvre the platform away from the edge, and watch to see whether it will stay put. Once we're happy, we anchor it in place, and prepare to return to the ship. Beneath the ice we're standing on, the ocean is deep and dark, quiet and cold, always there but only rarely the centre of attention.

*

The Earth's oceans are vast, and yet they often seem invisible. We had to go into space to really appreciate that the defining feature of our planet is not land but water. The Apollo programme sent men to the moon, but I think that its most significant achievement was to let all of us see the Earth. Two of the most influential photos ever taken shifted our perspective for ever: 'Earthrise', taken in 1968 during Apollo 8, and 'The Blue Marble', taken in 1972 during Apollo 17. Once you've seen them, you can't un-see that view and its significance, the fragile blue sphere floating in the cosmos, with everything we've ever known on board. But even then, the blue was seen as the canvas on which the land was drawn, the emptiness between the great continents, and a mystery that could probably wait until after all the important stuff was sorted out. Fifty years on from Apollo, humans are finally starting to pay proper attention to what's inside those vast blue expanses. It's overdue. But those forays into space gave us the necessary starting point: a blank map to fill in.

Maps of the Earth are rich and astonishing treasure troves, and globes are even better. The anatomy of our planet is full of fascinating detail: coastlines, mountain ranges, rivers and island chains, full of patterns and yet still so varied. It seems endless because it is; the more closely you look, the more there is to see. Lumpy continents give shape to the blue oceans, and we split our home planet into land and sea. It's natural to label land with fixed features, and the labels can last for decades or centuries without change. But every wonderful map that I've seen is guilty of misleading us in one fundamental way. They make it easy to forget one of the most important and breathtaking features of our planet: the ocean *moves*.

The blue of Earth is a gigantic engine, a dynamic liquid powerhouse that stretches around our planet and is connected to every part of our lives. It has components on every scale, from the mighty Gulf Stream gliding across the Atlantic to the tiny bubbles bursting at the top of a breaking wave. This is a beautiful, elegant, tightly woven system, full of surprising connections and profound consequences. The complexity can seem overwhelming, but at the largest scale, the logic is straightforward. Those seeking to understand the cynical side of politics are advised to 'follow the money', but planetary physics is immune to human cynicism. Our task is simpler, and more fulfilling. The key to unravelling the internal logic of the oceans is to bow to the physicist's instinct and follow the energy.

Our planet intercepts a tiny fraction of the mighty energy output of the sun, preventing it from flowing onward into the universe, and diverting it on to a much slower path through the mechanisms of the Earth: ocean, atmosphere, ice, life and rocks. On its way through the planetary system, this energy is carried by atmospheric winds and ocean currents, builds both mighty oak trees and the delicate lichen on stone walls, lifts a trillion tons of water into the sky every day, fuels every human and every owl and every ant on Earth, and powers the laptop I'm typing on. The oceanic engine is the heart of this system, hosting the majority of this flowing energy either as heat or as movement. Oceans are deep and broad, home to vast currents moving in different directions at different depths as water circulates around the globe, heating and cooling its surroundings as it goes. But energy is transient, only ever a temporary house guest. Eventually, after much recycling, that energy leaks away from the Earth as heat and resumes its journey through the universe. The first law of thermodynamics states that energy can neither be created nor destroyed, so the giant flows in and out are balanced. The Earth is just a cascade of diversions, unable to stop the flood but tapping into it as it trickles past; and the ocean is an engine for converting sunlight into movement and life and complexity, before the universe reclaims the loan.

Sunlight reaches everywhere on Earth, but it's more intense close to the equator because there the sun is high overhead, so the equatorial regions receive far more of the sun's energy than the poles. However, the Earth loses heat much more evenly, and plenty of energy is lost from the polar regions. That means that over a year, the equatorial regions experience a net gain of energy from the universe, and the poles experience a net loss. That contrast leads to a very profound conclusion: the atmosphere and oceans aren't just storing the energy that flows through the system; they're redistributing it. This is the dominant pattern behind the ocean engine: the overall shunting of energy from equator to poles. All aspects of the ocean and its influence on our lives fit somewhere into this mosaic: currents and stormy seas, the evaporated ocean water that later falls as rain over the Amazon, coastal erosion, migrating fish and airborne whale snot, expelled from the blowhole of the largest mammal on Earth and temporarily meandering through the atmosphere. They all have their role to play.

Describing the ocean as an engine is not metaphorical embellishment. The definition of an engine is something that converts other forms of energy (usually heat) into movement. We're used to the sort that has solid metal pistons driving ingeniously interlinked cogs and levers, all mobilized by temperatures high enough to fry an egg. The industrial revolution is long past, and yet a small army of enthusiasts keeps the world of steam alive – how could you abandon completely a technology that oozes that much character? The steam-powered steel dragons are beautiful and satisfying in a way that's rare in the modern world, because you can see exactly how they work. This piston drives that wheel which spins this little widget and so on down the chain – the elegant sequence of cause and effect is mesmerizing. But an engine doesn't have to be made out of solid materials.

As land, ocean and atmosphere absorb the sun's energy, they heat up. Some of that heat will almost immediately drive movement through convection – warm water may heat the air above it, making it buoyant, so newly warmed air often gets pushed upwards as cooler air slides in beneath it. As the wind moves across the ocean surface, the air pushes on the water, transferring energy back into the ocean as waves, and this energy eventually dissipates back into ocean heat. But this is only one route that the incoming energy could take through this endlessly fascinating mosaic. Earth's blue is closely connected to the other global components: atmosphere, the ice, life and the land, and all five work together as a single system. But the ocean is the big beast in Earth's planetary machinery. The engine that is Earth's ocean takes sunlight and converts it into giant underwater currents and waterfalls, hauling around the ingredients for life: nutrients, oxygen and trace metals like potassium and iron, shaping our coasts and transporting heat. This isn't just another engine, it's the grandest one of all: an engine the size of a planet.

It's got all the elegance of the most ingenious human-built engines but the mechanics here are more subtle and intricate. Instead of a nice tidy piston, we're faced with a flow of water that merges into the water on either side of it; it's definitely up to something, but it's hard to say where *this* pushes on *that*. But it is absolutely still an engine, converting light and heat into movement in myriad different ways.

The most frustrating thing about this engine is that it's so hard to watch directly. I was once asked what impossible invention I would most love to have, and there could only be one answer: a pair of binoculars that lets us see into the ocean the way we can see into the sky. Imagine standing on the prow of a ship and peering downward at stately currents sliding over vast mountain ranges beneath, giant plumes of tiny ocean animals on their daily vertical migration from the lower layers to the surface, and perhaps catching a glimpse of the great ocean voyagers: 4-metre-long tuna fish, turtles or a blue shark. But even though those binoculars won't be available any time soon, you can still see the engine at work if you know where to look. We humans don't live inside it, but we are affected by almost everything it does. For years, we thought we were independent observers looking out over the choppy surface out of curiosity, but we're actually tiny ants living on the shores of this great blue fluid mechanism, completely dependent on its output. It's the sort of perspective shift that can give you vertigo.

Humans and the ocean

As citizens of the Earth, we cannot escape the influence of the oceans, and we shouldn't want to. Humans have piggybacked on this deep blue engine for generations, our tiny vulnerable vessels trading and exploring where the surface took us, without regard to the inner workings of the deep. Battles have been won and lost on the basis of what the ocean threw at us, and

whole societies have grown up around its fertile places, respond-
ing to unseen ocean mechanics without any knowledge of why
there were fish *here* and not *there*. Even on land, the most suitable
regions for agriculture are often dictated by the nearby seas. The
ocean is deeply woven into human culture, and the threads
always track back to the engine and, ultimately, to the energy
flow. But even though they couldn't see the whole engine, intel-
ligent and observant humans from many cultures have seen
parts of the pattern, and have gained deep expertise in their own
waters, more than enough to navigate, fish, explore, trade and
rely on the oceans for their livelihood. Knowledge fed into culture, and myths and stories were used to explain the patterns
and to provide a foundation for thinking about the ocean: what
it was, why it mattered and how humans should behave towards
it. Attitudes to the ocean also fed back into the culture on land,
influencing even those who never went to sea. And every
culture's attitude to the ocean is partly geographical accident.

Science and culture are far more intertwined than most scientists would like to admit, and it's possible that one of the reasons
why ocean science hasn't really been visible is that many cultures consider the ocean to be a bit of a nuisance on a good day,
and very dangerous indeed on a bad day. In Britain, for example, a trip to the local seaside is widely regarded as a necessary
ritual of childhood, although sometimes viewed by the children
involved as more of a duty than a pleasure. In the north-west of
England, where I grew up, visits to the beach were often associated with being forced to paddle in freezing cold water, and
then competing to see who could lean furthest into the wind
without falling flat on their face. Looking under the sea surface
didn't really occur to anyone when I was at school, partly
because the water was cold, and partly because British coastal
waters are often too full of sediment for anything (even your
own toes) to be visible. Artists like J. M. W. Turner sometimes
painted tranquil seas and idyllic coastlines, but the clear

implication was that the sea was for looking at, not going into. Turner is better known for his paintings of ships being tossed about by violent seas under dark, angry clouds, an image reinforced by the British seafarers of the nineteenth and twentieth centuries recounting their adventures. As just one example, the polar explorer Ernest Shackleton wrote this of his extraordinary and heroic 1916 journey in a small boat to fetch help for his stranded crew: 'The tale of the next sixteen days is one of supreme strife amid heaving waters. The sub-Antarctic Ocean lived up to its evil winter reputation.'[2] It's not really the sort of description that encourages casual bystanders to pop along and have a look for themselves.

And it's not just the British. Iceland, perched at the top of the North Atlantic Ocean, is a nation built on the fishing industry and boasts a proud maritime heritage that has lasted for centuries. But if you walk along the harbour in Reykjavik, you meet a series of large display boards, each with a map of Iceland on it. Dotted around each coastline are black symbols designating shipwrecks, each one marked with the ship's name, the year, the type of boat and the number of men lost. Each map has 30–40 wrecks on it, and covers one decade. The maps go back 200 years, and you can't step on a boat without passing these memorials. There is no mistaking the message here: the ocean can and will kill you. I spent a while trying to ask Icelanders whether they ever went out on boats for pleasure, but the response was almost always a blank look. Up here, you go to sea to get fish, not to play. The seas around Iceland can be fierce, and fishing in those conditions is undoubtedly a dangerous pursuit. The clear

[2] The full story of this voyage, which was supposed to be the 'Imperial Trans-Antarctic Expedition', has become famous for Shackleton's extraordinary leadership and his daring voyage for help to rescue his men after their ship was crushed in the Antarctic ice.

lesson from their local history is to think very carefully before going anywhere near such peril.

On the other side of the world, Hawaiians who live their lives enveloped in the vast Pacific Ocean see things very differently. Close to the equator, squalls and gales are relatively rare, but storms thousands of miles to the north initiate the smooth ocean swells that make Hawai'i such a perfect place for surfing. Mastery of the ocean waves was a recognized royal pursuit, with kings and queens having their own special surfboards. Surfing was a ritual and a right, and central to Hawaiian society. The ocean was a part of life, and being in it and on it came naturally.[3] The ocean is a vital part of Hawaiian culture, partly because these small islands are completely enveloped in it and partly because their local ocean is much more benign than that around Iceland. Our human relationships with the ocean are as rich and varied as the ocean itself.

The accidental ocean scientist

My route into ocean physics wasn't planned or expected. I grew up in Manchester in the north of England, where 'ocean' was considered a very exotic concept because what we had was the sea. Two seas, to be precise: the freezing cold North Sea to the east, and the blustery grey Irish Sea to the west. Neither seemed particularly appealing. I studied physics because I wanted to know how things worked, and occasionally thought about geology, because I was interested in how the Earth worked, but the two never seemed to overlap. When I finished my PhD in experimental explosives physics, I spent six months writing up papers

---

[3] When I think of all the drab formal pictures of British kings and queens in my school history books, generally looking thoroughly miserable (and very self-consciously important), I can't help but think that a bit of surfing would probably have done them good.

and looking for another research topic, hopefully one that would let me continue to build interesting experiments, but without quite so much cleaning up after the experiment had blown itself to bits. Bubbles seemed to fit the bill, and Dr Grant Deane at the Scripps Institution of Oceanography took a chance on me and invited me to work as his postdoc for a year. I loved Scripps, and for the first three weeks in Grant's lab I felt completely at home. It was full of oscilloscopes and other familiar electronics, a giant Lego set of a lab, with drawers and cupboards housing all the bits needed to build any experiment you wanted. And then one day a giant frame turned up near the door, with buoys on the corners and waterproof boxes of sensors in the middle. It was designed to make measurements in the ocean, and it was solid and serious and lurked near the wall like a giant spider as my colleagues fussed around it. This was a type of beast I had never seen, had never imagined could be necessary. It was the absolute focus of their attention and it took me a while to realize why: it was their gateway to another world, their access to the alien realm beneath the waves. And so I stepped over the edge into the abyss of ocean knowledge that I had never known existed.

At first, I just listened and absorbed and tried to contain my astonishment. But very soon afterwards, a mystified indignation took over. How was it that no one had ever told me about any of this? How had I managed to get through three physics degrees and hundreds of books and articles and talks and *no one had ever mentioned the ocean*? This was easily the biggest scientific story I had ever heard. And so I read everything I could and learned to scuba dive and walked around ocean conferences with wide eyes and flapping ears, soaking it all up.

I'm still baffled that we don't talk about the ocean more, that this vast and crucial engine manages to be almost invisible. The more I found out about the ocean, the more jarring the invisibility became. The giant flows of the ocean engine are fascinating by themselves, but they also directly influence the parts of the

Earth that we breathe, walk about on, eat and use as raw materials; they are a huge part of the fabric of this rich and varied planet. This isn't merely a diverting tale about some salty water. This is the story that defines planet Earth.

My arrival in the world of outrigger canoes was also unplanned and unexpected. I was new to London and heard someone mention a local Pacific canoe club. I didn't know what an outrigger was, but I reckoned that if there was a group of people bonkers enough to scoot around the opaque and chilly waters of the Thames estuary in a canoe designed for the equatorial Pacific, we'd probably get on. I was right. But it was only after a year or so, when I had spent more time paddling on the ocean, that I made the deeper connection. It was unlike any other sport I'd tried because Hawaiian culture was folded into everything we did – and while that was perhaps hard for an outsider to spot, it was very obvious when you knew where to look. It was open and welcoming, respectful of everyone, social, tactile, enveloping you in the canoe 'ohana (family). These were people who stepped up when you needed help and tolerated difference. And then I learned about the history of voyaging across the Pacific, and the astonishing skill and observation that made this possible – including observation of waves and bubbles, my scientific research topic. The Hawaiians looked at the ocean, and I looked at the ocean, but we saw different things. The connection was the canoe. The ocean became like an optical illusion – by blinking, I could switch my view from one perspective to the other. But I was sure that they weren't as separate as they seemed. I'm a scientist, but before that, I'm human. How did the people of the greatest ocean civilization on Earth, the Pacific Islanders, see all the things I measured and scrutinized and reduced to parameters in a computer program? When I saw the ocean through new eyes at Scripps, I had only seen the physical engine, not the cultural housing. When I first understood the mindset of the canoe, it felt as though I had entirely missed the point for the second time. The canoe came to symbolize the complementary perspective that my scientific education hadn't provided: the tradition and culture of the ocean.

The development of a new scientific discipline

When you splash through the waves at the beach, you're connected via seawater to every drop of water in the global ocean. You might have to go the long way round, but you can get to exotic parrotfish, hydrothermal vents, icebergs and aquatic deserts without leaving the water you're paddling in. Many of us have been to a beach at least once, so it's not as if the ocean is completely inaccessible (although I'll admit that getting to large parts of it involves overcoming a few practical problems on the way). And yet even the basic principles of how the ocean works were obscure until a few decades ago. The first dedicated global oceanographic research expedition (considered by many to mark the start of oceanography as a discipline) was carried out by HMS *Challenger* between 1872 and 1876, and it returned with a vast collection of samples and measurements and observations. But although *Challenger* travelled nearly 70,000 nautical miles around the globe, mapping temperatures and currents, and fishing up all manner of ocean beasties from below, the scientists on board could only really pluck at the edges of their subject. It was like taking a huge work of art – say the ceiling of the Sistine Chapel – and mapping 360 pinpoints of colour while looping once around the ceiling. And works of art don't generally change with the seasons and shift from one pattern to another on cycles that last decades. Those scientists came back with only the tiniest sprinkle of samples from the ocean expanses they had crossed, but this haul was already rich and fascinating. For example, the expedition discovered the Challenger Deep, the deepest part of the ocean at around 11 kilometres below sea level, but all they could say about it was that it was there. Acquiring this new scientific knowledge was still a significant achievement, and the substantial report written by the *Challenger* scientists for the Royal Society was hailed as a huge advance in humanity's knowledge of the Earth. A start had been made.

But the expense and difficulty of mounting these expeditions made for slow progress until the Second World War, when submarine warfare ruled the waves. Suddenly, the military was interested in understanding this new fighting space, and once the war was over, oceanography finally had its golden age. In the 1950s and 1960s, every expedition came back with new ideas and unexpected discoveries, and by the mid-1970s the outline of the ocean engine could finally be seen. Then came satellite data, with the potential to reveal large-scale surface patterns that connected the dots drawn by many individual ships. Now we are entering the age of autonomous buoys and vehicles, which can drift for years or dive for days in the innards of the ocean, colouring in far more detail than any cohort of oceanographers could. And we're still learning. At every stage, we discover new mechanisms, new subtleties and new links. The ocean is intimately connected to the atmosphere, ice, geology and life; and although the ocean is the great energy reservoir, its mechanics depend on these other components of the Earth system. Every scale is important – from milliseconds to decades, and from the microscopic to ocean basins thousands of miles across. This is a story that isn't finished, and our oceans continue to astonish even the most experienced ocean scientists.

In the past decade, the focus has shifted slightly. As we explore this liquid engine and get to grips with its inner workings, it's been impossible not to notice how much the world up here on land depends on it. It regulates our weather and climate. El Niño, the combined atmospheric and ocean waltz that spans the Pacific Ocean at the equator, has a measurable effect on the GDP of the countries surrounding it. And the ocean soaks up around 30 per cent of the additional carbon dioxide that we're putting into the atmosphere, slowing the march of global warming but with significant consequences for the ocean itself. Thinking about the ocean isn't a luxury for the curious any more. It's a major part of our life-support system and we'd better take it seriously.

We're also learning that even this vast body of water isn't big enough to shrug off the influence of humankind. An increasing awareness of our effect on the ocean is slowly seeping into the public agenda, dragging behind it a conversation that is decades overdue. But this conversation faces a massive obstacle. It's almost impossible to discuss what to do about something changing if you don't initially know how it works. If a doctor tells a patient that they have a problem with their kidneys, the patient probably already has at least a vague idea about where their kidneys are and what they're up to. They learned about that part of their own personal life-support system at school. But that's not the case for the oceans. When we see a news story about the long-term decline in the numbers of krill in the Southern Ocean, it sounds generally like a bad thing. But there's far more to it than the risk of whales going hungry. Krill are a part of the ocean engine, life that is woven into the fluid machine, and we need to understand at least some of the context before we can discuss the change and take appropriate action.

To look more deeply at the ocean is also to look more closely at our own identity, and at what it means to be a citizen of an ocean planet. Zoom out far enough, and our story starts with sunlight arriving at the Earth, and it ends with the light that leaves the Earth, light that has been altered by its passage through the planetary engine, by being reflected, scattered, absorbed, emitted and transformed into many different types of energy before becoming light again. The light that leaves us to restart its journey out into the cosmos bears the unmistakable imprint of a dynamic planet: invisible infrared, green from forests, brown from rocks, white light reflected from clouds and ice, and the blue of the water. Sharp, simple sunlight has been converted into this beautiful palette: our signature on the universe, scrawled in the handwriting of a dynamic and living planet. And that signature is dominated by blue; our message to the rest of the universe is *We are ocean*.

More than any other scientific subject I've studied, the ocean reminds me that I'm human. I have had some of the greatest adventures of my life at sea, made enduring friendships, been scared and exhilarated and bored and tired and more content than I've ever been on land. Put this scientist in a canoe and the personal and professional merge, because really, they were never separate.

Our many connections to the ocean are written all over our history and our culture and our lives, hidden in plain sight. But now it is time to highlight and talk explicitly about those connections. Our attitude to the Earth's ocean determines our behaviour towards it, and we are the first generations who can really see the consequences of our actions. We need to decide as a society how we think about the blue of our planet, which will guide how we act towards it. Scientific knowledge is essential, but this is a cultural decision. Traditional knowledge about and attitudes to the ocean will be necessary to help us with the big decisions we face. How do we reconcile our many competing interests and attitudes with our increasing awareness of what's at stake? Individual societies have faced this question before, but now we need to refresh and share our cultural attitudes to come to a global consensus. The ocean engine is critically important, and humanity's rich cultural treasure-chest can't be ignored. We humans need to address both directly. We have one planet and one global ocean, and if we want future generations to experience the best a blue planet has to offer, we cannot afford to delay.

The voyage

Understanding the ocean better sounds relatively straightforward – but then we meet the beautiful swirling nuanced reality of a shell of water surrounding a rotating planet. There are big patterns and small patterns – currents and krill, sea ice and sediments – all overlapping without clear boundaries. Like any other engine, there is an underlying structure: there are components, and there are links between those components. You need both to make things happen; even the most ingenious piston is useless unless it's connected to something else. In a steam engine, a piston usually has two connections. In the ocean, one current could have dozens, perhaps hundreds. This is an essential feature of the system and it's what makes oceans endlessly fascinating. But the system obeys some basic principles, and we can hold on to those while we explore the maze of connections and components that keep the blue machine ticking along.

This book will take you on a voyage through the global ocean, hopping between stories of history and culture, natural history and geography, animals and people, to reveal the basic shape of the blue machine. Our adventure will traverse the ocean's innards, both the physical mechanisms and the life that is woven into the liquid engine. We will see how the ocean events we observe and notice are not the results of randomness, bad luck or the whims of the gods, but are actually just the surface expression of the deep engine which is always turning beneath. The full complexity of the global ocean can't be squeezed into a single book; but we can draw its outline and set out the fundamental principles underlying how it works – enough to serve as a map for further exploration. I hope that it will change your perspective on the ocean, and perhaps on yourself as well. And it is definitely true of the ocean that the more you know, the better it gets. So let's begin.

PART ONE

WHAT IS THE BLUE MACHINE?

1

The Nature of the Sea

THE EARTH'S OCEAN IS FICKLE when it comes to its appearance. Seawater can present itself as the startling turquoise of a shallow tropical bay, the ruffled grim grey of a blustery northern coast, a calm royal blue that stretches for a thousand miles – or perhaps temporarily cloaked in the cheeky orange of a sunset. When you look at the ocean, there's a wealth of detail to notice, and the most obvious visual characteristics often change by the minute. In contrast, when you walk up to the water's edge and touch the source of all that variety, the same three observations are always at the top of the list: it's noticeably warm or cold, it's salty, and it's liquid. These three fundamental features – temperature, salinity, and the weird and wonderful wetness of water – provide the foundation for everything the ocean engine does. But they are also as accessible to a ten-year-old child as they are to an experienced Polynesian navigator or an Atlantic fisherman. Each of them has a direct influence on the world we take for granted. And then once you step back from the seashore and broaden your perspective, this trio has a deeper beauty and influence to offer, as facets of a single interlinked system.

But as you zoom out, from a single breaking wave at a beach to a hundred-mile-wide bay eating into the coastline, and then further still to a whole sea, another contributor becomes apparent. We live on a rotating planet, and although no human can feel our daily pirouette around the Earth's axis, the shifting fluid ocean cannot ignore it. The effect of our globe's spin is to sculpt this liquid machine into beautiful loops and curves, giant whorls and vast underwater undulations.

It's tempting to dive straight into the patterns and the astonishing intricacies. But before we reach for the grandest perspective of the Earth's blue machine, we need to understand the major physical influences that make it work: temperature, salinity, density and spin. Since every machine needs energy to run, we will start with the way energy is stored in the ocean and the way that ocean scientists track it: temperature.

Temperature

I'm peering down through a horizontal metal grille set into a broad concrete base, and all I can see is the reflection of the sky and my face in the still water beneath. Keith Olson's face pops into the circle next to mine as he leans over the sump. 'This is probably the first time that sunlight has touched this water in a thousand years,' he says. We're in the middle of a small fenced-off section of a lava field next to the ocean in Kona, on the Big Island of Hawai'i. Inside the fence, four huge pipes each half a metre across crawl out of the ground, merge into two, and snake across the lava away from the ocean. There's a brilliantly obvious clue to what differentiates them in the paintwork on the pumps which are pushing water down those pipes: two are red and two are blue. This is the Natural Energy Laboratory of Hawai'i Authority (NELHA), and here in the middle of the Pacific Ocean they've got pristine ocean water on tap, or two taps if you're being precise: one hot and one cold. The combination of this water supply and a barren corner of an active volcano is generating an astonishing flood of fertility. And that temperature difference is the key to it all.

NELHA began as the response to a shock, back in the winter of 1973–4, when crude oil prices shot up from $24 per barrel to

$56 per barrel pretty much overnight. The state of Hawai'i, thousands of miles from the mainland and almost entirely dependent on oil arriving by ship for its energy, decided it had better try to wean itself off fossil fuels. If you're going to do that, being a chain of volcanic islands with abundant equatorial sunshine is a very good start. But Hawai'i's engineers realized that they had something else up their sleeve: access to the deep ocean. The Hawaiian Islands are shield volcanoes parked in a region of ocean that's 4–5 kilometres deep,[1] and as you sail outwards from their shoreline the volcanoes continue to drop away beneath you. If you run a pipe far enough down that slope, you can tap directly into something surprising: cold water. Today, NELHA's longest pipe reaches a depth of a thousand metres, and those blue pumps are bringing 5°C ocean water to the surface. The red pumps are hauling in surface water at 25°C. Between them, they pump an astonishing 113 million litres or 116,000 tonnes of ocean water into the technology park every day. But the point of all this isn't to supply water. The valuable commodity here is energy.

Water can store a surprising amount of energy as heat. Imagine two grapefruit-sized globes of water side by side,[2] one full of North Pole ocean at about −1.8°C, and one full of Persian Gulf water at about 30°C. The warmer water has more heat energy, and if you could channel all that extra energy into doing something mechanical, you could lift an SUV (weighing nearly two tons) up by about 7 metres, high enough to be level with the top

[1] Shield volcanoes have relatively gentle and smooth slopes, due to the runny lava coming out of them. The lava spills out over the landscape gradually, creating broad, shallow-sided cones, which look like shields laid on the ground.
[2] This isn't a random choice of fruit. The average grapefruit has a diameter of about 12 cm, which means that it's got about the same volume as one kilogram of water.

of a two-storey house. This is an astonishing amount of energy, and that's just what's held in a single kilogram of water. It's really hard to heat water up, so you have to add a huge amount of energy to make even a small change to the temperature.[3] But once you've warmed up the water, you haven't lost anything – the energy is just stored as heat until the water gives it away and cools back down. This makes water astonishingly effective at storing energy. The millions of litres of warm water being pumped by NELHA are carrying many thousands of gigajoules of energy every day, just because of their temperature. There's no doubt that the energy is there. The challenge for NELHA is to extract it.

The original idea was to use the temperature difference to run a heat engine which can generate electricity. This process is known as ocean thermal energy conversion, or OTEC.[4] There's heat energy in water at any temperature, but in order to extract it, you need a contrast – the warm water has to be paired with something cold. Then you need a technology to sit between the hot and cold water and to extract some of the energy flowing between the two, which in this case is a heat engine built by Makai Ocean Engineering that runs by continuously cycling ammonia.[5] The current version, opened in

---

[3] This is why your kettle is one of the most energy-hungry devices in your home.

[4] In this context, it's helpful to think of a fridge running backwards. Your fridge takes electrical energy and uses it to create a temperature difference between the inside of your fridge and the outside. What OTEC does runs the other way and achieves the opposite, converting a temperature difference into electrical energy.

[5] The way it works is beautifully simple, but of course there are lots of subtleties in the execution. Ammonia is a fluid that boils at very low temperatures and although it's a gas at room temperature, if you put it under a bit of pressure, you can move its boiling point to sit between the hot and cold water temperatures. As it passes the cold water, the ammonia cools and becomes

2015, is a demonstration plant connected to the electricity grid, and it's capable of generating 100 kilowatts of electricity. For that technology to work, you need a temperature difference of 20°C, and there are plenty of tropical islands where that's available. This energy source is an ideal baseload for the grid – it's very easy to control and you can turn it up and down as you need. The biggest stumbling block (apart from easy access to the deep ocean in the tropics) is that to make it cost-effective, you would need to scale it up considerably, and there are still questions about how you could do that, whether it's worth it and whether it would have any unintended consequences. But the concept of extracting solar energy that's currently being stored in the surface ocean undoubtedly works, if you're in the right place to do it.

Most local residents refer to the whole site as OTEC, but the demonstration energy plant is just the start. This whole corner of Kona is an incubator for small businesses that can use the hot-plus-cold ocean water supply, all perched on top of the lava field deposited by the Hualālai volcano in 1801. Keith drives me around the site, pointing out the astonishing variety of businesses hidden in unassuming buildings nestled on the black rock. There are huge tanks growing spirulina, kept at the right temperature by ocean water heating and cooling. There's a business that grows 50–60 per cent of the broodstock for one of the giants of the aquaculture world: Litopenaeus vannamei, the whiteleg shrimp. There's another growing abalone, a marine mollusc that is a favourite of chefs all over the world. There are hydrogen production and clam-breeding facilities, more farmed algae and a monk-seal hospital. And to top it all off, the NELHA offices are air-conditioned using ocean water. The cold salt water cools fresh water from

---

liquid. Once it reaches the warmer water, it boils to form a very high-pressure gas. That can be used to drive a turbine and generate electricity, and then the cold water condenses the ammonia ready to start all over again.

the normal water supply, which is then pumped around the buildings using solar energy. I'm not generally a fan of over-air-conditioned buildings, but I can easily forgive NELHA theirs.

NELHA is only a small-scale operation, and this type of energy extraction isn't going to solve the world's energy problems. Yet it demonstrates that warm water – just warm, not even hot – is a vast store of energy. It might be tricky for humans to extract that energy, but on a planetary scale the global ocean is a gigantic heat reservoir. Even a small increase in temperature represents a huge amount of stored heat, and so the temperature of the water is the critical measure of stored energy. But that heat energy is not evenly distributed around the planet. To understand where it is and the huge influence that it has, we need to go back to where it comes from and why it ends up in some places but not others.

Why warmth?

It starts with a star. When we think about our solar system, we are often distracted by the variety and mystery of the other planets that share our sun. These wanderers through our night sky provide rich fuel for humanity's imagination, conveniently packaged into seven giant and mostly inaccessible spheres. But the sun makes up 99.86 per cent of the total mass of the solar system, and although you could argue that 99 per cent of the variety is found in the planets, our solar system is really one massive nuclear reactor surrounded by spherical specks of dust that it can't shake off. At the sun's heart, 4 million tonnes of matter is converted into energy every single second as colossal temperatures and pressures force hydrogen atoms to fuse into helium atoms. The newly released energy rattles around inside the sun for tens of thousands of years before reaching the surface, but once free of its plasma shackles it pours out into the universe in all directions in the form of infrared, visible and ultra-violet light. The fraction of this flood that the Earth receives is just less than one-billionth of the total. The laws of physics make this energy the Earth's essential currency, setting a fixed budget for everything that happens on this planet.[6]

As with all budgets, the total is quickly split up by circumstances. A third of it is reflected straight back out into the universe, barely touching the Earth system. A small chunk is intercepted by the atmosphere. Nearly two-thirds makes its way down to the surface, and if that surface is seawater then raw energy from the sun finally touches the Earth's ocean.

But reaching the surface does not guarantee entry. Physics is the gatekeeper, selecting which light rays to admit on the basis of strict criteria. Imagine looking out across the ocean to the rising sun. On a still day, the reflection of a beautiful orange sunrise can be almost perfect, as if the ocean surface was a mirror. That's because when light arrives at the water surface at a shallow angle, water does indeed behave exactly like a mirror and the light is all reflected back into the sky (or, in the case of the sunrise, into your waiting camera lens). But as the day goes on and the angle between the sun and the water surface increases, sunlight is more likely to cross the air–water boundary. This means that the best place to appreciate the effect of sunlight on the ocean is in the tropics. Close to the equator in the middle of the day, the sun is high in the sky and therefore sunlight has the highest probability of piercing the ocean surface. And if

---

[6] Nuclear fission (and hopefully nuclear fusion in the future) sit outside this finite budget. But the world's total supply of nuclear energy is less than one-millionth of all the energy from the sun that arrives at the Earth. Geothermal energy contributes a bit too, along with solar and lunar tides, cosmic radiation, magnetic storms, and a few other bits and pieces. But the total of all of these is only one-fortieth of 1% of what we get from the sun (and almost all of that is geothermal).

you're near to the coast, there's a good chance that this light will illuminate the ocean's best-loved feature: the tropical coral reef.

A healthy coral reef is one of nature's most diverse and most extraordinary environments. The coral itself forms a seascape of colourful clustered bulges, large mounds, delicate fronds and branching antler-like protrusions. Brightly painted parrotfish dart about, gnawing at the coral whenever they pause. Red squirrelfish lurk in dark crevices, watching the world go by. Striking butterfly fish patrol their patch, ready to defend their territory at the slightest insult.[7] And the resident population of nudibranchs, worms, shrimps, clams, sponges and more creates a busy and diverse ocean city. Up near the surface, it looks to us as though there is oodles of light down here, generously illuminating the stripes, spots, camouflage, iridescence, sand and rock that the natural world has on display. But water isn't nearly as transparent to light as we generally think, and as you descend the surroundings get bluer and then blacker as the water absorbs all the sunlight on offer. As humans looking at the world, our focus tends to be on this visual loss.

But the laws of physics offer a different perspective: the light may disappear, but the energy can't. Visible light is converted into ocean heat, and so the photographer's loss is the thermometer's gain. Sunlight bathes the pretty reef fish twice: in light until the water has extracted its energy tax, and then in warmth. And so the ocean is heated by the sun.

This visible light – the colours of the rainbow that make up a human's visual world – is only half of what arrives from the sun. The other half of the sunlight reaching the Earth's surface is

---

[7] I have always wondered why the common western names for so many reef fish are based on land animals. Goatfish, sheepshead wrasse, hawkfish . . . the list seems endless. It is true that a frogfish really can look like a lost toad, but of course a toadfish is an entirely different species.

infrared light: the colours that sit beyond the red part of the rain-bow, unseen by human eyes. If you put your hand near a warm object in the dark, you can feel its warmth because it is sending out infrared light that contains energy. You know there is some-thing hot near by even though your eyes can't detect anything. Infrared light floods downwards from the sun and it does touch the ocean, just, but water is so opaque to these wavelengths that within a fraction of a millimetre this part of the light is absorbed completely and then reradiated straight back out to the atmos-phere. Counter-intuitively, the visible light is responsible for heating the ocean, and what we think of as travelling warmth – the infrared – makes no contribution at all.

Sunlight heats up ocean water wherever they touch, pouring energy into the reservoir. Less than ten minutes after leaving the sun, the energy is stashed away in long-term storage in the ocean, keeping the vigilant butterfly fish nice and warm. It takes a long time for water to heat up and cool down, so tropical waters remain pretty much the same temperature throughout day and night alike.

The heating is most dramatic at the equator, giving our planet a balmy belt inhabited by tropical fish. But as the Earth curves away from the sun in the northern and southern hemispheres, the direct heat contribution from sunlight decreases. Closer to the poles, the ocean is a very different place indeed.

The chill beneath the tropics

Just beyond the northern limit of the Atlantic Ocean, the Arctic Circle crosses Greenland and swoops eastwards across a 1,200-mile stretch of open ocean, barely skimming Iceland as it crosses grey choppy water towards northern Norway. At noon on the winter solstice each year the sun scrapes the horizon only briefly, and then buries itself behind the curve of the Earth, leaving the Arctic to face the unforgiving darkness of the universe for another day. Even six months later, when sunlight slants across the ocean surface for twenty-four hours continuously, the light rays are weedy, tightly rationed by the tilt of the Earth and weakened by long transit through the atmosphere. The consequence is a very different face of the global ocean: it's cold here, and relatively dark. There's still plenty of life, but the details have been tweaked by evolution to match the local challenges. Every creature carries the imprint of its environment.

Halfway between Iceland and Greenland, 400 metres below the surface, a patch of the gloom is moving. It glides slowly, taking four seconds to cover 1 metre, and its grey mottled skin hangs loosely on its long body. The sharp corners and sleek silhouette of closely related species are replaced by a softer look, reminiscent of a baggy old jumper which won't ever be thrown out because it's too comfortable. This lumpy-looking creature is 4.5 metres long and weighs a gigantic 400 kilograms. Speed is not its game. The Greenland shark never goes anywhere in a hurry. But then it has no need to. An individual of this size was probably born before the earliest glimmerings of the industrial revolution, and has been gliding around these waters for 240 years. It's thought that this species can live to be at least 300 years old (possibly more), doesn't reach sexual maturity until it's about 150 and keeps on growing throughout its life at about one centimetre each year. As far as we know, it's the world's longest-living vertebrate. This exceptional lifespan seems to be directly connected to the cold, which slows down the processes of life,[8] stretching out the shark's existence by a factor of ten. This slow giant will spend its leisurely life in water that's around 0°C, mostly hidden several hundred metres below the surface. One of the greatest mysteries carried by that baggy body is how it catches prey – adult Greenland sharks are found with stomachs full of fish like flounder and skate, and sometimes even

---

[8] The name of a recent international collaboration to study this species was brilliantly to the point: 'Old and Cold – the Biology of the Greenland Shark'.

with freshly killed seal. Any of those species could easily outswim this creeping predator. Adding to the mystery, almost all Greenland sharks are at least partially blinded by a parasite that sits in one or both of their eyes. Such a hindrance won't matter in the inky blackness of the ocean depths, but it will limit their hunting nearer the surface. We often assume that the predatory thrones at the top of the food chain must be occupied by the fastest, most observant and most aggressive animals, endowed with nature's most dangerous weapons in the form of teeth, claws and heft. But in the deep dark cold, the Greenland shark shows us that there is another way to live, and another way to hunt.

The contrast between a tropical coral reef and the normal Arctic habitat of the sluggish Greenland shark shows us the broad pattern of ocean temperature in near-surface waters. It's set by exposure to sunlight, which is tightly tied to latitude and season. The sunny equatorial seas delight visitors with temperatures of around 30°C, but in the central Arctic, water temperature can drop as low as –1.8°C. The global ocean energy reservoir is bursting at the equator, but offers only slim pickings nearer the poles. And yet every detailed map of sea surface temperature is a beautiful study of minor exceptions to this rule, each one opening a new door on the complexity of the ocean beneath. There are cool patches and warm patches parked incongruously next to coastlines, elegant swirls of balmy water far from the shore, and striking, snaking intrusions of cold water into friendlier surroundings. We'll explore some of those later in this book, but the foundation that sits beneath all the beautiful complexity of these patterns is that ocean temperature varies between –2°C and 30°C and is broadly correlated with how close to the equator you are. But this pattern of energy storage isn't just about the surface. The depths have a tale to tell too.

In August 2013, the research vessel R/V *Apalachee* was working in the Gulf of Mexico, close to the site of the blowout that had killed eleven crew on the Deepwater Horizon oil rig three years earlier. The latitude of this site is close to 29°N, far closer to the equator than to the poles, and the typical surface water temperature at that time of year is around 30°C. The researchers on board, led by a team from Florida State University, were sampling deep-sea fish communities to assess the impact of the oil spill. One of the fish they hooked was brownish-grey, 3.7 metres long and quite clearly a juvenile Greenland shark. It was brought out into the hot summer sunshine in a green net stretcher and parked on deck, far warmer in death than it had ever been in life. It was the first of its kind to be captured in the Gulf of Mexico – noteworthy enough to earn a write-up on a few online news sites, but the researchers weren't that surprised. So what was a polar shark doing in the Gulf of Mexico in the height of summer?

The hook that snagged the shark had been hanging 1,749 metres below the sea surface. And down there, the water temperature was 4°C, well within the comfort zone of a Greenland shark. The upper ocean might have been warm enough to take a bath in, but the depths were not, and this is normal. The shark would have been at home at that depth anywhere between the North Pole and where it met the hook, because the deep ocean is almost all that cold. That bright colourful map of ocean surface temperatures, with the nice broad red stripe at the equator, is just that: a map of *surface* temperatures. The temperature difference between the surface water, which basks in bright sunlight, and the deeper waters that haven't been touched by sunlight for decades can be dramatic. In the Gulf of Mexico, that warm surface layer is only 100–200 metres thick, but a huge section of the ocean basin inside the Gulf is almost 4,000 metres deep. Below 1,000 metres, this basin is filled with cold salty water very similar to the cold salty water that fills much of the depths of the North Atlantic basin. It flows in from the Caribbean, continually refreshing the pool. There are also other, more subtle, layers above that.

The ocean that we humans see – the sunlit water that you can comfortably wade into, full of life and character and food – forms only a small fraction of the whole ocean. Tropical coral reefs are an exception, albeit one that is vital for life in the ocean, and one that is an enormous joy with which to share a planet. But it is essential that you don't give up on the deep ocean at this point, on the basis that it sounds best left alone. That cold dark water beneath is just as interesting and necessary and important as anything that happens at the surface. Looking down at a flat map of ocean temperature seen from above isn't enough. We also have to pay attention to how things change with depth.

Stacked water

The thin skin of salt water that encapsulates our planet, 4 kilometres deep and 12,740 kilometres in diameter, has a dramatic internal anatomy. It's written in temperature and salinity, and it's built of horizontal layers. The details change as you cross the planet: currents eventually merge with their surroundings, water sinks in some places and rises in others, and slabs of water are disrupted as they slide across giant sea-floor mountain ranges. But the overall pattern is clear. In the deep ocean basins, there is generally a stack of three or four major layers which are known to oceanographers as water masses. Each one has a distinct character and history, and these are the biggest components of the ocean engine. Lesson one about the structure of the ocean is that it's layered, and those layers generally do not mix with each other.

The most striking internal boundary in the ocean is defined by temperature, and it's known as the thermocline. A thermocline is defined as a thin layer of the ocean that undergoes a rapid temperature change with depth, marking the transition between layers which have different temperatures. But when oceanographers refer to 'the thermocline', they're generally talking about the starkest transition of all: the boundary between warm sunlit surface water and the much cooler dark water below. This distinct warm upper layer, known as the 'mixed layer', exists over most of the global ocean, and it's the drive-train of the planet, connecting the powerhouse of the sun to the ocean engine which runs on that heat energy. The vast ocean basins are filled with much colder water, often untouched by the sun for centuries. In the Pacific Ocean the depth of the thermocline is typically between 60 and 200 metres, and beneath that, the character of the ocean water changes dramatically. This is why NELHA can pump cold water out of the ocean near Hawai'i, as long as it's pumping from a depth of 1,000 metres or more, well below the thermocline.

The ocean we humans see is only the lid on the whole system; but it's absolutely critically important to the whole engine. Let's stick with the lid – this surface mixed layer of warmed water – for now. Like everything else in the ocean, it moves. And so it's time to look at some of the exceptions to the general rule that temperature is highest at the equator and decreases uniformly towards the poles. The consequences are bountiful and beautiful and awesome, but as so often in history, when humans come along, the story has a tendency to end with a bang.

A work of art, drawn in temperature

Many of the ocean's temperature idiosyncrasies have been known to mariners for centuries. One of the earliest people to start weaving a global pattern from these local oddities was an endlessly enthusiastic German naturalist and scientist, notable for describing broad swathes of our planet with both the skill of a scientist and the heart of a poet.

Friedrich Wilhelm Heinrich Alexander von Humboldt was

an intellectual fidget with a deep passion for the natural world.[9] Born in 1769 in Berlin (then part of Prussia), he chose a path of global exploration which would have been a great disappointment to his mother, had she lived to see it. Her hope was that he would become a high-flying Prussian civil servant. Instead, he bubbled over with curiosity throughout years of travel to exotic locations, studying and sketching and seeking answers to everything. Later in life, he came to view nature (including its human components) as one giant web of connections,[10] in stark contrast to most scientists of the time who were very busy categorizing everything into neat pigeonholes.

In 1802 von Humboldt saw the Pacific Ocean for the first time, and set sail from Lima in Peru to follow the South American coastline northwards until he got to Mexico. Much later, in a giant work called *Cosmos: A Sketch of the Physical Description of the Universe* (no one could accuse him of lacking ambition), he described the oceanic current that they rode on that journey:

> At certain seasons of the year the temperature of this cold oceanic current is, in the tropics, only 60 degrees, while the undisturbed adjacent water exhibits a temperature of 81.5 degrees and 83.7 degrees. On that part of the shore of South America south of Payta, which inclines furthest westward, the current is suddenly deflected in the same direction from the shore, turning so sharply to the west that a ship sailing northward passes suddenly from cold into warm water.

[9] Exceptional in his ability to connect the scientific, cultural and artistic aspects of the world he saw, he inspired both Charles Darwin and Ralph Waldo Emerson. For the full story of his life, I recommend the excellent book *The Invention of Nature* by Andrea Wulf. Among his many notable achievements, he was the first person to recognize human-induced climate change. In 1800.

[10] Thereby predating by 170 years some of the ideas that James Lovelock would later incorporate into the Gaia hypothesis.

On the Celsius temperature scale (and even back then, he included a conversion table in his scientific writing, because both Celsius and Fahrenheit were commonly used), he was recording that the ocean water was 16°C instead of the 28°C you might expect at that latitude. Not only that, but there were sharp boundaries between warm and cold water. Lima is only 12 degrees of latitude south of the equator, well within the tropics by anyone's standards. On a modern map of global sea surface temperature, the anomaly is even more apparent. A tongue of cool water crawls up the western side of South America, stretching half the length of Chile and up into Peru. It stops just short of the equator, around 4°S. This odd feature has had a huge influence on global geopolitics, the Atacama desert, a lot of fishermen and a surprising number of pigs. Von Humboldt would have been delighted.

The most notable inhabitant of this cold current is the Peruvian anchoveta, or anchovy, a small fish of large consequence. It's got a slim, silvery body, lighter on the bottom and darker on top, and big round eyes that seem to take up most of its head. At most, it can reach around 20 centimetres in length, but it makes up for its diminutive size by hanging out with millions of others in vast dense schools. Those schools share the cold surface water in this region with plenty of others: mackerel, sardines, hake and mullet, which in turn feed bigger predators like bonito, sea-lions and various species of seabirds. This narrow stretch of ocean is a mobile feast, a luxurious exception when compared to much of the open ocean. To get to the root cause of that biological bounty, we need to follow back down the chain of full stomachs. Big predators like bonito are fuelled by the smaller fish like the anchoveta, who devour krill, tiny shrimp-like creatures that dine off phytoplankton. These are the most critical ocean inhabitants, because they spend their time busily converting the sun's energy into something the rest of the food chain can use. So far, so good. Once enough solar energy has been converted, the rest of the food chain is fat and happy. The astonishing density of this ecosystem must be fed by an equally astonishing population of phytoplankton. But what is this unusual deluge of Earth's tiniest powerhouses doing here, in the cold current rather than the warm water next door?

This brings us to one of the most significant consequences of the layered ocean. Phytoplankton are a diverse, complex and beautiful group of organisms, but their major needs are quite simple. They need sunlight, nutrients, carbon dioxide and water, although this last one has never yet been considered a limiting factor in the ocean. It's the first two that hold things back. The upper (generally warm) layer of the ocean has plenty of sunlight. But nutrients there can get used up quite quickly. Down in the deeper, colder water hidden beneath, there tends to be plenty of nutrients, but no sunlight. Tiny single-celled phytoplankton are capable of a truly monumental task: building the foundation of the entire ocean food chain. But they can only do this when *both* light and nutrients are present – both the energy and the raw material – and our layered ocean tends to keep these critical ingredients frustratingly separate: light up top, nutrition down below. The cold current noted by von Humboldt (and later named after him) is the site of a very effective solution to this fundamental problem.

A single anchoveta, scooting around and scooping up krill, generally stays within the top 50 metres of the ocean. The critical action happens both above and below, with the anchoveta sandwiched in the middle and vacuuming up the consequences. At the ocean surface, the wind drives water westwards, out into the Pacific Ocean. All along this coastline, the warm surface layer is constantly pushed offshore, and the push is so strong that the warm water moves away from the coastline entirely. You might think that would leave a hole, but down below the anchoveta, at a depth of about 300 metres, water from the ocean's cool lower layer moves in to fill the gap, drifting eastward until it meets the coast and then moving upwards. Our anchoveta is swimming around in sunlit water from the deep ocean. Literal-minded ocean scientists call this upward flow an upwelling. Cold, nutrient-rich water has escaped from underneath the warm lid, and as it comes up to meet the sunshine, all the ingredients for life are there in huge quantities. It doesn't happen along every coastline, but it happens here in style. The phytoplankton can gorge themselves silly on sunlight, stashing away solar energy on a monumental scale.

Breaking the normal ocean rules like this matters. This narrow ocean region, covering perhaps 0.05 per cent of the global ocean surface, is responsible for 15–20 per cent of the entire global fish catch. Until very recently, the annual catch of the Peruvian anchoveta has consistently been the largest of any single wild fish species, reaching a maximum of 13.1 million tonnes, or about 200 billion individual fish, in 1971 (although it should be noted that this extraordinary harvest did cause a population crash the following year). In 2018, the Peruvian anchoveta harvest was approximately 5 million tonnes, out of a total global catch of around 90 million tonnes. At this point, you could be forgiven for wondering why Peruvian anchovies haven't made it to your local fish'n'chip shop. It turns out that although the sea-lions are perfectly happy with this small oily mouthful, humans haven't been so keen on it.[11] Even fans of anchoveta-based dishes describe the flavour using words like 'distinct' and 'bold'. Back in the 1950s, almost no one was eating them. But humans were reluctant to leave this plentiful ocean bounty alone, and centuries of animal husbandry had taught them exactly what to do with food they couldn't face: feed it to the pigs.

After the Second World War, food was scarce and prices of

[11] Although there are now quite a few campaigns by conservation charities and celebrity chefs trying to change this.

basic commodities could swing dramatically. In Britain, the wartime government had enthusiastically nudged people to produce their own food. This led to a lot of household chickens and the existence of 'pig clubs': groups of individuals who clubbed together to raise pigs, half of which they kept and half of which went to the government for the war effort. Pigs make good use of household leftovers because they need protein – they can't survive on grass alone as cows can. After the war, scaling up pig production sounded great, but household leftovers weren't going to supply industrial farms. Pig farmers had a problem.

Five thousand miles away, Californian sardine fishermen had more than a problem. Their entire industry had just fallen off a cliff. The famous sardine canning factories that formed the backdrop of John Steinbeck's novel *Cannery Row* were empty after four decades of explosive growth in the sardine fishery and ruinous shrinkage in the sardine population. From 1934 to 1946, around 500,000 tons of sardines were harvested each season as the protests of fishery biologists were swept aside. But by 1947, it was over. The fishery had collapsed. The Californian industrialists were not to be deterred, and took their newly available equipment, their expertise and their money to Peru in order to develop the anchoveta fishery. But instead of canning the fish for human consumption, the focus was on producing fishmeal.

A fish is an exquisite and fascinating creature: a stunning answer to the question of how to live in the sea, one which has been honed by evolution over hundreds of millions of years. Fishmeal is all of that, but dried, squashed and ground up into powder.[12] It's also astonishingly rich in protein: between 50 and 70 per cent by weight. And in 1950, farmers were just waking up to its potential.

---

[12] A reasonable fraction of fishmeal is made from bycatch and trimmings which would otherwise go to waste. But plenty of it comes from 'food grade' fish even now.

And so the world bought fishmeal as fast as Chile and Peru could haul anchovetas out of the Humboldt Current, while the lessons from California about the long-term consequences were completely ignored. Between 1950 and 1973, world fish harvests tripled, but the amount of fish directly consumed by humans stayed the same. The rest went to fishmeal, as a supplemental food for livestock, and this became an essential ingredient for modern industrial farming. Britain imported all that it could get, and by 1960 half of all fishmeal was being used as pig food. With the addition of industrial farming methods and antibiotics, farmers could grow more pigs more quickly, in less space and for less money. By 1960, Peru was the world's top producer of fishmeal, and in 1964 it caught 40 per cent of the entire global fish harvest. When overfishing and environmental conditions caused Peru's fish harvest to collapse in 1972, shutting down the fishmeal supply, the price of British bacon doubled almost immediately.

And so the extraordinary consequence of upwelling water along the coast of South America isn't just that it has produced a huge marine ecosystem in a relatively tiny area. It's that it has provided the biological bounty to feed pigs and chickens (and, increasingly, farmed fish grown in other countries) all over the world. Those animals were raised to feed humans, who were probably blissfully unaware of the marine source of their protein, and also its colossal cost to the natural environment. The tiny anchoveta was just a link in the chain, the vehicle for nutrients. And the origin of that incredibly productive fishery is written in the surface temperature map of the ocean, because of the disturbance to the layered structure that it implies.

Under the weather

The ocean touches life on land in many ways, and food that comes directly from beneath the waves is only one of them. The most noticeable, for most humans, is the weather. This takes us back to temperature as a marker for stored energy, because although the ocean engine is fuelled from above, when sunlight hits its surface the atmospheric engine is fuelled from below. The ocean heat reservoir acts as a hotplate, powering the weather.

Two and a half thousand miles to the north of Peru, one of the most committed land animals on Earth is 5 metres up a tree and happily wallowing in gifts from the ocean. The tree is 6 miles inland, hidden from the nearby Caribbean Sea by a carpet of dense rainforest. A heavy downpour has just finished, and trickles of water are still dripping through the lush leafy canopy. Some of them splat on to a clump of bedraggled greenish-beige fur which is considering going for a walk. There's a promising sprig of leaves, 2 metres further along the branch, and this brown-throated sloth is eyeing it up with unhurried enthusiasm. In the sloth world, tummy fur with a belly-parting is the norm, providing an upside-down animal with a million tiny gutters to guide the rainfall off to the sides. As the sloth starts to move, the dripping speeds up, freeing the water to continue to the forest floor. The annual rainfall at this particular spot is more than 4 metres per year, nearly twice the national average here in Costa Rica, and it all comes from the ocean.

Each raindrop exists because warm water evaporated from the ocean surface to form water vapour, which was then carried high into the atmosphere until it condensed to form cloud droplets. Wind moved those clouds sideways until the droplets grew large enough to fall as rain, watering the rainforest by dumping vast amounts of borrowed ocean water on to land. But this huge distillation and distribution system is only half the story. Its invisible twin is the huge amount of ocean energy that this process deposits directly into the atmosphere, fuelling the weather at the same time as it propels the water cycle. The sloth and its leafy home rely on both.

As we have seen, seawater forms a liquid storage system for the sun's energy. That energy is held in the jostling of individual water molecules, which bump and swirl around each other faster and faster as the temperature increases. At the surface, some of those molecules get up enough speed to escape from the masses, freeing themselves from the liquid to zoom up into the atmosphere. That's the process of evaporation. But only the most energetic molecules can escape, and the transition comes with an energy price tag (known to physicists as latent heat). An escaping molecule must carry this booty whenever it's gas rather than liquid. This means that the process of evaporation cools the ocean surface, because the ocean loses the additional energy that must travel with the evaporating water. But it also means that high up in the atmosphere, when the molecule condenses to join a cloud droplet and become liquid again, this extra energy must be dumped. The cloud droplet carries on, growing and travelling and eventually falling back to Earth. But the energy stays in the sky, powering convection and winds, driving our dynamic weather.

Although the ocean is a very effective store for heat energy, it's extremely sluggish when it comes to moving that heat around. In contrast, our atmosphere is pretty poor at storing heat, but it can shunt energy around very quickly. A warm ocean can provide a steady supply of energy to the skies, keeping our atmosphere ticking along even when clouds block the sun for days or weeks. The vehicle for the majority of that energy supply is evaporating water, but the ocean surface also radiates heat directly, and transfers a little bit via conduction right at the surface. All this makes the ocean a vital energy reservoir, buffering the sun's irregular supply and smoothing the flow that gets distributed around the globe. It's also the source of the vast majority of water in our skies. Sea surface temperature is constantly sculpting global weather patterns, and that makes it critical information for modern weather forecasts.

As the sloth reaches its lunch, the rain starts again, and it licks water off the nearest leaf as an aperitif before attending to the leaf itself. This rainforest and the sloth are here because of the reliable supply of warm rain provided by the ocean. Palm trees grow in Cornwall in the UK because of the warm Gulf Stream waters to the west. Hurricanes that batter the east coast of the United States are fuelled by warm Atlantic water, and a recent study suggested that increased rainfall over the Amazonian basin is caused by warming water in the Atlantic. The Earth's weather is inextricably linked to its ocean.

This brings us back to that cold ribbon of water that cosies up to the west coast of South America. It's a rare feature of nature that can claim to form a direct link between South American politics and British turnips via the medium of bird poo, but then the world is full of surprises. The cooler waters of the Humboldt Current have a direct effect on the local weather: they prevent rain. This doesn't bother the Guanay cormorant or the Peruvian booby, two of the bird species that feast on the fishy abundance of Peruvian anchoveta swimming in those waters. But what goes in must come out, and when the birds return to the local islands to nest, their droppings stack up. Long before the anchoveta boom, in the early 1800s, some of those stacks of guano were 30 metres thick: towering grey mounds of dried poo that were exceptionally rich in nitrogen, phosphorus and other trace minerals. The local Inca communities valued this ecological treasure trove because it was an exceptionally effective fertilizer for their crops. Our name for this agricultural white gold, 'guano', comes from the Quechua word for fertilizer, *wanu*. The Inca protected the guano piles and punished anyone who disturbed the birds, correctly recognizing a healthy bird population as the key to a sustainable supply of poo. But the western world wasn't nearly as disciplined. It was von Humboldt himself who noticed the unfamiliar substance and brought it back to England in 1804, little suspecting how important this acrid white powder would turn out to be. It took a few years for European farmers to establish that a sprinkle of guano dramatically boosted the yield of their crops, but by the 1840s the guano boom had begun.

The Peruvian government got to work on harvesting their substantial stinky white islands with enthusiasm. The grim labour of digging the noxious guano out was mostly done by slaves, and the disruption caused considerable damage to the seabird populations that had converted anchoveta into fertilizer in the first place. But that didn't stop the traders, and guano exports expanded rapidly. By the 1850s, half of Peru's guano exports (200,000 tons of the stuff) ended up in Britain; it was the only manure British farmers bought in significant quantities. And what Britain appears to have done with it, more than anything else, was fertilize turnips.[13] A century before Peruvian anchoveta were feeding Britain's pigs, Peruvian birds were fertilizing our root vegetables.[14] *Monty Python*-style debates posing the question 'But what has the Humboldt Current ever done for us?' could prove time-consuming.

Why was it necessary to import bird poo from South America when Europe was full of birds that were presumably producing a similar quantity of poo per bird? The critical difference was the cool waters of the Humboldt Current. Since the cold ocean surface suppressed rain, South American guano dried quickly and was chemically undisturbed by rainwater, leaving pristine deposits. The wet and humid conditions prevailing in Europe meant that bird poo was either washed away or chemically changed by the rainwater. The cool ocean water not only generated the conditions for the anchoveta and therefore the birds, but it also preserved the nitrogen-rich consequences in a uniquely effective way.

The profits from the guano trade dominated Peru's economy,

---

[13] This is one of history's oddities, because even at the time it was apparent that turnips need a fertilizer which is richer in phosphates than nitrates. Plenty of other crops would have benefited more than turnips. But from 1840 to 1860, turnips it was.
[14] Turnips were often fed to livestock, so you could also argue that the pigs were getting the benefit either way.

giving the country a new-found financial stability for the duration of the 'Guano Era'.[15] But such a valuable resource was the source of considerable international envy. The United States of America passed the 'Guano Islands Act' in 1856, giving its citizens the right to take possession of guano islands on behalf of the United States, in a move widely regarded as the country's first experiment with imperialism. The Chincha Islands War was fought over some of the most valuable islands from 1865 to 1879, and the precious guano was one of the resources under dispute in the War of the Pacific in the early 1880s, when Bolivia lost its coastline to Chile. The fact that Chile had guano wars and Florida didn't is no accident. It happened because the structure of the ocean engine directly caused a bonanza in one place and not in another. The patterns that influence civilizations – weather, resources, culture – are often a consequence of the patterns that the ocean engine generates. Humans are usually just scooting about on the surface, dealing with the problems right in front of their noses and paying no attention to the turning of the ocean engine beneath. But the engine is always there, working as a servant of the laws of physics, not of humans. We have seen that the temperature itself and the energy it carries matter to human civilizations. But their influence on planet Earth is far larger.

When we think of the temperature of the whole Earth, we land-based humans tend to focus on the temperature of the atmosphere. But as scientists started to search for the Earth's heat energy, it became apparent that the atmosphere itself was no more than a distraction. Each square metre of the Earth's surface has ten tons of atmosphere above it, and our atmosphere extends hundreds of kilometres above ground level (although 75 per cent of its mass is in the lowest 10 kilometres). Imagine picking 1 square metre of

---

[15] In Yorkshire, in the north of England, there's a saying 'Where there's muck, there's brass', meaning that where there's a dirty job, someone can make a lot of money from it. This story is surely the ultimate global endorsement of that phrase.

ocean surface, and adding up the heat needed to raise the temperature of the entire atmosphere above that square by 1°C. If you put that same amount of energy into raising ocean temperature by 1°C, you would only be able to heat the top 2.5 metres of the seawater below your square, a piddling sliver when you've got 4 kilometres of ocean depth to play with. So the total heat energy in the atmosphere is tiny compared with that in the ocean. The atmosphere moves much more quickly than the ocean, and so can help shunt energy around, but it's the transport system, not the warehouse. And what about land, the other 30 per cent of Earth's surface? It's true that solid land absorbs heat quickly and that 1 cubic metre of rock can store a fair amount of heat energy. But rock doesn't move, and so the heat stays near the surface. The vast depth of rock below doesn't contribute to the land's storage capacity, because the surface heat never gets there. So neither the land nor the atmosphere contributes much to heat storage, a task that is left almost entirely to the seawater.

The ocean is Earth's thermometer. The vast swirling blue acts as an energy reservoir for the whole planet, and so its average temperature is a measure of the amount of heat energy stored on the Earth at any moment in time. It provides a giant buffer that smooths our climate's spiky energy supply, evening out the difference between day and night, and between summer and winter. The stability that this provides is critical for life on Earth, saving us from huge swings in temperature that could easily freeze or fry the delicate biochemical machinery upon which we all rely. But temperature is only one of the big influences that drive the ocean engine. The next one is inconspicuous, common and utterly mundane in everything except its effect on our world: salt.

Salt

One of the great absurdities of the ocean is that you can be floating on the sea, directly connected to 97 per cent of all the water on Earth, and yet die of dehydration.[16] The culprit, of course, is salt. When it's dissolved in water, the components of salt split apart and sneak into the gaps between water molecules, becoming invisible in the process. But these tiny intruders have a gigantic effect on the way the ocean engine functions. Planet Earth without its salt would be a very different place indeed.

Salt is essential for life as we know it. An adult human body contains around 200 grams of salt, and this isn't a passive cargo. It's constantly at work, helping signals flow through our nerves and muscles and regulating our blood. If we lose too much, through sweating or urination, it needs to be replaced. Animals that predominantly eat meat can top up their salt supplies from their food, but plant-eaters need to find a separate source in order to stay alive,[17] and that includes humans in agricultural societies. Salt is only mundane if you have enough of it, and away from the coast it can be hard to find. So the history of human civilization is peppered with salt: control of salt supplies by the British Empire, the Chinese and the Aztecs, among many others, ingenious technologies for salt extraction and transport, and plenty of trade and tension and skulduggery,[18] served with a hefty side order of salted herring.[19]

And yet our planet is not short of salt. A bath filled with sea-water contains 5 kilograms of it, enough to fill a medium bucket. That's a *lot* of salt. If an asteroid large enough to vaporize the ocean's water were to hit the Earth tomorrow, the layer of salt left behind would be about 65 metres thick across the whole sea

---

[16] As famously stated in *The Rime of the Ancient Mariner*: 'Water, water, everywhere, / And all the boards did shrink; / Water, water, everywhere, / Nor any drop to drink.'

[17] This is why elephants, goats and deer gather at salt licks for the salt, but the carnivores only come along for the prey.

[18] Mark Kurlansky's excellent book *Salt: A World History* tells the story in detail.

[19] See chapter 6, 'Voyagers'.

floor: 49 billion billion metric tons. It's reasonably evenly distributed throughout the world's seawater, but like everything else in the ocean, there are patterns that tell stories if you know how to look. Deciphering those clues is an operation so valuable that today both NASA and the European Space Agency have expensive satellites in orbit constantly watching the concentration of salt at the sea surface. But it wasn't always obvious that it mattered, or even why it was there. This edible rock has puzzled many of history's great thinkers.

On the basis of his portraits, the famous seventeenth-century scientist Robert Boyle is perhaps the last person you might expect to be sniffing rancid seawater. Ostentatiously long and luxurious curls cascade down over his shoulders, and rich backdrops and fountains of lace label him clearly as wealthy and powerful, a man of elevated social and financial standing. But he built for himself an even greater scientific stature, founded on robust and careful probing of the nature of reality and a refusal to believe anything unless someone (preferably him) had actually done an experiment to check. This may not sound unusual today, but this was trailblazing stubbornness in the seventeenth century, a period when superstition and the pronouncements of history were generally considered to be all anyone needed to understand reality. Boyle inspected the world around him and wrote chatty but rigorous tracts about his experiments: the formation of ice, the workings of air, the meaning of colour and plenty more besides.[20] He was one of the founding members of the Royal Society (and its precursor the 'Invisible College'), completely embodying its motto *Nullius in verba*, usually translated as 'Take nobody's word for it'. And in 1674 he published *Observations and Experiments about the Saltness of the Sea*.

---

[20] One of his more memorable but incidental experiments was to put a viper in a vacuum, an early foray into hyperbaric medicine (the study of what pressure changes do to the body).

He starts by taking aim at Aristotle's suggestion that seawater is salty because sunlight somehow imparts saltiness to it, having left fresh water in the sunlight to see if it became salty (it didn't). Recognizing the limitations of his own single data point led him to an ocean-wide test: if Aristotle was right, the ocean would be fresh at the bottom and saltier at the top, and Boyle declared that no credible sources can be found to suggest that this is the case. He collected accounts of pearl fishers, famous for free diving to great depths to harvest their treasure, who stated that the ocean is just as salty at the bottom as it is as the top,[21] then carried on to explore alternative explanations with enthusiasm. He certainly wasn't above a diversion or three along the way, either. He tested and found wanting the assertion that seawater could not become putrid by leaving a bucketful in sunlight and declaring that 'it did, in a few weeks, acquire a strongly stinking smell'. And just a few years after his assistant Robert Hooke used a microscope to shock everyone with the world of the small, Boyle coyly noted that many seafarers thought that 'vast tracts of it [the sea] are imbued with stupendous multitudes of adventitious Corpuscles, which by several ways diversifying its parts, keep it from being a simple solution of salt'. But he saved his most modern thoughts for last, as he speculated on whether the ocean is becoming saltier over time and whether it has the same salinity everywhere. He was confident about asking big questions, the sort that are so fundamental that it takes a brilliant mind to even see that they exist, but he also knew the limits of his own answers. Ever reluctant to speculate beyond the evidence he had in front of him, Boyle stated that 'a great number of observations, in different climates and different parts of the Ocean' would be necessary to answer the question of whether the ocean has the same salinity everywhere. In the 350 years

---

[21] Although no one actually knew where the bottom of the ocean really was at that point.

since he wrote that (and mostly in the last thirty years), ocean scientists have made billions of those observations. Boyle the nit-picker would have loved the detail, but I think that the ocean machinery thereby revealed would have delighted him far more. But before you can get to how the salt is distributed, you have to take a step back: what exactly *is* this stuff that we call sea salt?

There is a small metal pot of salt perched high up on my bookshelf, next to a rock from Antarctica and a water sample from the North Pole. I bought it in Kona, Hawai'i, and it proudly boasts that it's salt from the cold deep water piped up at NELHA. If I were to eat it, the deep sea would become a part of me, helping my nerves send signals and my muscles contract. But is this exotic salt any different from the rock salt mined in Northwich or Nantwich near where I grew up?[22] The answer was hidden in the waves until the *Challenger* expedition of 1872–6. Although there had been ocean research expeditions before, *Challenger* set new standards and new ambitions for oceanography. It had a dedicated scientific crew and truly global reach, looping around the Atlantic, Pacific and Southern Oceans over four years. But for all the amazing new and unexpected science that came out of it, it certainly wasn't funded out of the goodness of anyone's heart for the purpose of satisfying human curiosity.

The *Challenger* expedition sat at the crossroads where scientific questions met both commercial and naval interest in laying telegraph cables across the deep sea to improve communication networks. The Royal Society had the pure science questions, the Hydrographic Office had the surveying experience, and the Navy had ports, coal, protective power and the ship itself: HMS *Challenger*.[23] The 'scientifics' on board were most excited about weird deep-sea life and unseen sediments, while the Navy was

---

[22] The ending -wich was associated with salt-producing towns in England.

[23] This was the height of the British Empire, when Britain controlled a huge amount of infrastructure across the world. The reason that the *Challenger*

definitely most interested in the shape of the bottom of the ocean. But along with the critical surveying work, the 'scientifics' collected data on every aspect of the ocean that was available to them. They also found time for a considerable number of parties at the various colonial ports they called at, making up for long months at sea with short sojourns on land where they were treated as guests of honour at balls, dinners and the local tourist sites.

The mountain of data that *Challenger* brought back was staggering. It took almost twenty years after they returned for it all to be analysed and published, and so although the scientists on board were collecting the raw material, they often couldn't immediately see its significance. This gap is common even today (although it's usually between one and four years these days rather than twenty), because analysis is slow and painstaking, and your focus at sea is on collecting as much high-quality data as possible. If you ask a scientist what they learned as soon as they step off a ship, the answer is frequently 'we don't know yet'. The best surprises are often yet to come, as the data repaint your mental image of the ocean you thought you were in.

The answer to the question of whether sea salt is the same the world over sat quietly in seventy-seven water samples brought back from the *Challenger* expedition, parked among pickled fish, boxes of sediment and endless mollusc shells until they were sent off to William Dittmar in Glasgow. He spent several years coaxing chemical clues out of seawater from every region and many depths, all detailed in the long and impressive report he published in 1884. It had one overall conclusion: the *amount* of salt in seawater might vary from place to place, but the *ratio* of all the major components (sodium, chloride, magnesium,

expedition had global reach was that the British Empire had global reach, and was determined to keep it.

potassium, etc.) was always the same,[24] in every ocean and at every depth. My prized salt from tropical Kona is exactly the same as salt from the grey and gloomy Irish Sea.[25]

Aristotle may have been wrong about how salt got into the ocean, but he provided a decent starting definition: salt is what you get when you remove all the water from seawater. If you let seawater evaporate from a large shallow bowl with gently sloping sides, white salt is left behind in distinct rings, and that's the first clue that 'sea salt' isn't just one thing. The constituents of sea salt are all mobile fragments of planet Earth that just happened to end up dissolved in the ocean. It all started on the early Earth: a harsh and hot place, blanketed by an atmosphere rich in carbon dioxide. Any exposed land was just bare rock regularly rinsed by acid rain. When the acid reacted with the rock, it liberated a select group of atoms: sodium, potassium, magnesium and calcium. These are all small molecules that reside in the upper left-hand corner of the periodic table, which indicates that, given the chance, they'll surrender one or two electrons and hang around as ions with a positive charge. This trait means that they dissolve easily in water.[26] But our young planet

[24] Dittmar wasn't the first person to have this idea: Johan Forchhammer had suggested it back in 1865, just before he died. But Dittmar was the first person to conduct the really thorough chemical survey that put the principle beyond doubt.

[25] I did know this when I bought it. I paid $10 for 99g of salt precisely because it was such a ridiculous thing to do. Also, I liked the idea of salt that had not seen sunshine for a few hundred years.

[26] The water molecule is considered 'polar' by chemists, which means that within the molecule itself, clusters of negatively charged electrons tend to congregate in particular places on the molecule. The places with electron clusters become slightly negative and the areas left with a lower electron density become slightly positive. This is why water is liquid at room temperature – the positive parts strongly attract the negative regions of nearby molecules, and it takes a lot of energy to pull them apart (i.e. to evaporate the water). But all those strong positives and negatives also mean that it's easy for atoms that

was also extremely volcanically active, constantly huffing and puffing as its innards settled down. The volcanoes spat out enormous quantities of hydrochloric acid, the highly reactive partnership between hydrogen and chloride ions, and also sulphur compounds. This process added chloride and sulphate to the ocean, ions with a negative charge. All these ions hide in the gaps between the ocean's water molecules, each held in place with its negative or positive charge. One litre of seawater contains 18.9 grams of chloride, 10.6 grams of sodium, 2.6 grams of sulphate, 1.3 grams of magnesium, 0.4 grams of calcium and 0.4 grams of potassium, along with tiny traces of lots of other elements.[27] In the ocean, these different elements have very little to do with each other. But when you remove the water, the positive ions have to pair up with the negative ones to form a solid salt, and this happens in sequence, starting with the least soluble combinations. This is what makes the rings in the bowl as brine evaporates. The dominant pairing is sodium chloride: 'common salt'. You could view ocean water as an ionic dating agency: it provides a place for the sodium and chloride to mingle, but it's only when the water itself quietly gets out of the way that the familiar pairing appears. Sea salt really is just what's left, and it's the same all over the world.

Sea salt snobs need not be too worried. Although the salt itself doesn't vary, nature may have added a few other things to fashionable cooking salts. All of our salt comes from the ocean, but most of it has been hanging around on land for a few million years before we get to it. We could use lots of energy to heat up seawater to evaporate the water until only salt is left. But it's far

---

have gained or lost an electron or two (like the positively charged sodium ion, Na^+) to fit in.

[27] The next one on the list: 0.14g of bicarbonate, is the hidden enabler in the entire global ocean carbon cycle, as we'll see in chapter 3, 'Anatomy'. This is about the same in volume as seven grains of rice per litre of water.

easier simply to go to places where the sun did all the hard work millennia ago. Rock salt is formed when shallow seas dry up, dumping their salty cargo in deep layers on to newly relabelled land. Irrespective of the method used to remove the water, an unpurified salt is likely to carry traces of algae, bacteria and (in the case of rock salt) other minerals. These do produce some variation in colour and taste. This is what you pay for if you're into posh salt. The pink or black salt carries the chemical imprint of its temporary home on land. Boring old 'table salt' has been processed to remove everything except the sodium chloride.

On the *Challenger* back in the 1870s, John Buchanan didn't know much about where the salt had come from. But as the expedition's chemist, it was his job to work out how much of it there was, and especially *where* it was – was it the same at all depths and in all locations? Chemical analysis would have required a very precise daily dance of delicate laboratory glassware – on a rolling ship, a recipe for disaster if ever there was one. So Buchanan constructed a brilliantly simple and robust device. He knew that adding salt made seawater more dense, so measuring density gave you a measure of salinity. His elegant solution was a cylinder that had almost exactly the same density as seawater, with a tall thin section sticking up out of the top. If you put this into a seawater sample, it would float almost entirely submerged, but slightly lower or higher in the water, depending on the density of the water.[28] The exact measurement could be read off a scale on the thin top section. The whole thing was mounted on a suspended tray in the middle of the ship, so as the ship rolled around it, the water sample stayed unperturbed. Over the four years of the expedition Buchanan did this two thousand times, in the process becoming the first human

---

[28] This method wasn't entirely new; in 1674 Robert Boyle had described sending a captain out to sea with a very similar apparatus.

ever to map the whole ocean's salinity.[29] He found that it was pretty similar everywhere *Challenger* went, varying by only 10 per cent, although there were clear patterns within that narrow range. He discovered that different layers of the ocean had different salinities, and that the Atlantic is saltier than the Pacific. Oceanographers now measure salinity on the Practical Salinity Scale, which is typically equivalent to grams per kilogram.[30] The average ocean salinity is around 35 on this scale, while the surface of the Atlantic is around 37 and the Southern Ocean is around 34. The next big question was why there should be such variation. The details are still debated today, but part of that answer was hiding in plain sight even back then, in the contrast between two substantial seas with very different characters, both of which had carried traders for centuries: the Mediterranean and the Baltic.

Although the global ocean is a single connected body of water, it includes a few hangers-on which barely touch the rest. One of the most well-known is the Mediterranean Sea, which marks the southern edge of Europe, kissing the Atlantic at the majestic Strait of Gibraltar. The connection is a channel only 14 kilometres wide, but the Mediterranean then opens out to stretch nearly 2,500 miles eastward from Spain to Syria, providing coastlines for twenty-one countries. On the northern side of Europe, the entrance to the Baltic snakes off the North Sea at the Kattegat, a twisting shallow channel between Denmark and

[29] His 1877 paper does include data from previous expeditions which had taken samples from more limited areas.

[30] There have been several ways of measuring ocean salinity, and this – the Practical Salinity Scale – is the most commonly used now. It's actually a ratio of electrical conductivities so it has no units, but the resulting numbers are very similar to g/kg. In 2010, a new measure for absolute salinity was developed, TEOS-10, which is more robust scientifically but complicated to calculate. It's much harder to measure salinity than it sounds as though it should be.

Sweden which narrows to a 15-mile bottleneck before opening up into 1,000 miles of open sea splitting Sweden from Finland and Russia. From a structural point of view, the Mediterranean and the Baltic look pretty similar: they're both largeish seas which are weakly connected to the Atlantic. But for the animals that swim under the surface they could not be more different. The Baltic is cold and almost fresh,[31] with annual average surface temperatures around 8°C, and an astonishingly low salinity of around 8. The Mediterranean is warm and salty, with average annual surface temperatures of 20°C, and a salinity of 38. This is oceanic chalk and cheese.

There are two processes at work here, neither of which directly involves the salt itself. Their effects are apparent because the entrances to these two seas are tiny, far too small for enough open ocean water to get in and average everything out. The first process removes water molecules: this is evaporation, where water molecules escape into the air from a warm ocean, leaving everything else (salt, algae, fish, dolphins) behind. The second process returns fresh water to the sea in the form of rainfall and river run-off. Both the Mediterranean and the Baltic have many rivers running into them, carrying the collected rainfall from a significant swathe of the surrounding land and dumping it all in the sea. What's different for each of these two seas is the balance between these two processes. The chilly northern Baltic receives floods of river water and rain, and is so cool that very little fresh water evaporates from the surface. The balance of this sea is strongly tipped towards fresher water – there just isn't enough salt to balance out the inflow from the rivers. On the other hand, the warm Mediterranean gives away a huge amount of fresh water to the atmosphere via evaporation but gets very little back from the shallow rivers after their journey over hot parched

---

[31] The technical name for water with a salinity midway between the salty ocean and salt-free fresh water is 'brackish'.

land. So it's very salty: constantly losing water molecules and getting very few back.

The point is that the salinity of the ocean doesn't depend on the salt itself. Salt is just a passenger, carried by the water around it. There aren't any significant salt inputs or outputs feeding or depleting the global ocean:[32] what is in there now is pretty much all there will ever be. If we see variations in the salinity of the ocean, those patterns are telling us not about what the salt has been doing, but about what the water has been doing.

The concentration of salt makes an enormous difference to the physical structure and machinery of the ocean. But for many of the living parts of the ocean engine, salt is also the difference between life and death, a constant challenge that requires constant vigilance. Too much salt will kill you very quickly, and that leaves us with a conundrum. A human will die if seawater is their only source of liquid, but the fish seem to manage just fine. So how does an ocean native drink?

Coping with salt

The cool water off Nova Scotia is a foggy turquoise, lit by diffuse sunlight above and fading into darkness below. The fog is made up of tiny fragments of drifting organic life, individually invisible but collectively cloaking every resident in fuzzy ignorance of everything more than 5 metres away. The ocean is quiet, disturbed only by an occasional breaking wave at the surface and the very distant deep hum of ship engines. A leatherback turtle emerges from the fog and glides slowly through the bright nothingness. From nose to tail she is nearly 2 metres long, a solid, mottled grey oval with huge flippers and a snub nose. She has travelled nearly 2,500 miles from her breeding ground in the Caribbean, and she is hungry.

---

[32] Rock salt formation is a very small fraction of the overall amount of salt.

At a molecular level, the turtle isn't too different from us. The average salinity of her body is around a third that of seawater, and her reptilian kidneys can't produce urine that has a higher salt concentration than her blood. Her body is a neat package of low-salinity life, and her cells will fail if her insides come anywhere near the salinity of the water in which she swims. Her leathery skin is the fortress that keeps the salt out.

From the gloom below comes a haunting call: the long, slow cry of a humpback whale. These whales feed on fish, and those fish are much less salty than the ocean. As they're digested, their carbohydrate and fat release water, and the fish themselves contain useful water in their cells. So if a whale is careful, squeezing out the seawater that comes with each mouthful of fish before it swallows, it can get enough water from its food without taking in too much extra salt. We don't yet know for certain, but it seems likely that whales don't need to drink.[33] The work of eliminating excess salt is largely done for them by their fishy prey, who are experts at drinking seawater and then pushing salt back out into the environment though their gills, urine and faeces. Few ocean vertebrates drink, but all of them face the challenge of keeping the water in and the salt out.

The leatherback turtle is the master of this game. The turquoise gloom in which she swims is home to a living buffet of jellyfish, which is what this turtle feeds on. Every minute or two, a dark, pulsing silhouette emerges from the fog, a messy cascade of orange tendrils hanging from a colourless dome. A slight twist of her flippers, and the turtle is bearing down on the hapless mass of jelly. One snap and a puff of debris is all that remains.

---

[33] This is a particularly astonishing achievement when you remember that whales evolved from land mammals that moved back into the ocean. The evolutionary challenge of evolving to survive long dives on deep breaths may have been dwarfed by the challenge of dealing with the complete lack of fresh water.

But the turtle's salt budget has just taken a hit. A jellyfish is really just a small bucketful of ocean masquerading as life. It's 96 per cent water, and most of the other 4 per cent is salt, making the jellyfish as salty as the ocean. Less than 1 per cent of the jellyfish is organic material and therefore useful food, and so the cost of dinner is that the turtle must accept three times as much salt as food in every mouthful. But our leatherback is huge: 450 kilograms and in need of putting on another 100 kilos before she returns to the tropics to breed. The paltry dose of nutrients from this one jellyfish is almost nothing. And so she must eat and eat and eat, stuffing herself with 80 per cent of her own body weight in nutrient-poor jelly every single day. And the cost adds up – she takes in more than 10 kilos of salt every single day. How could any animal survive that without shrivelling up from the inside out?

The solution is both ingenious and (to us) heartbreaking: this gentle giant weeps as she eats. A huge proportion of her head is taken up with salt glands, organs that remove salt and push it out of her tear ducts. Leatherback tears are thick and viscous and almost twice as salty as the ocean. To keep eating without killing herself with salt,[34] the turtle must cry around eight litres of tears *every hour*. But this is the cost of living in seawater. As the turtle slowly sculls onwards, fading into the turquoise, her body is sorting the ocean, scrimping and saving the nutrients, rejecting the salt, and flushing through the water.

Our society is obsessed with animals that live in 'extreme' environments: microbes that can tolerate temperatures of −20°C

---

[34] In order to avoid ballooning, the turtle also needs to digest everything very quickly, and can probably hold only 25% of its body weight in its gut at any one time. It needs to use up its own energy to stay warm as it gives away energy to the huge amount of seawater passing through. So as it eats, huge plumes of very liquid faeces emerge from the other end. All of this really isn't a tidy way to live.

or 120°C, or bacteria that can survive volcanic springs with the pH of stomach acid. But it's easy to forget that for many ocean animals, the ocean itself is an extremely harsh environment which can only be survived with strict marshalling of salt and water. Both are essential, but it's impossible to take in one of them without affecting the other. There is no way to avoid salt in the ocean, so life must walk that tightrope or accept an existence as salty as the sea.

We humans eat the ocean every day. The sodium bustling about in your synapses right now, sending signals around your body . . . it came from the sea. The chloride that helps your body regulate blood pressure and control the gateway between the inside and outside of your cells . . . that came from the sea too. When a drop of sweat falls on to your tongue and you taste the salt, you're tasting the ocean. But these ions are rare escapees. The salts in seawater do cycle very slowly between land and ocean, over millions of years. But on the scale of a few centuries, the amount of salt in the ocean is pretty much fixed. So why would NASA bother tracking it with satellites?

Essential character traits: temperature and salinity

In our everyday human lives, debates about combinations of temperature and salinity are generally focused on the quality of soup and the risk associated with icy roads. Up here on land, departures from what's desirable are relatively manageable since we are usually dealing with isolated systems (one pot of soup or one road). There is an outside to bring some salt or heat in from. But the ocean is huge, and only a tiny part of it is any-where near its edges. It also mixes incredibly slowly. Different parts of the ocean can have very different temperatures and salinities, even though there aren't necessarily any barriers between them. The combination of temperature and salinity tells you a lot about where water has been and what's happened to it recently, because those markers change only slowly and only for a handful of reasons. Heat and salt can't just vanish, so oceanographers use them to label water masses. For us, it's like asking about someone's character – it's the first thing you want to know about any part of the ocean someone is describing to you. And it's not just about what the water *is*, but about what it's likely to *do*. Because the temperature and salinity of water have a huge influence on where the water itself can go.

Density

The rippling deep blue skin of the equatorial Pacific sparkles in the sunlight, unbroken except for the occasional dark back of a whale rising to breathe. The water stretches outward for hundreds of miles, seemingly empty to the casual glance. But zoom in, and the smooth surface is interrupted by the occasional passenger: perhaps a fragment of seaweed or a twig washed off the land, or perhaps an unexpected clump of bubbly white foam about the size of a dime. On the underside of one particular bubbly clump, bobbing about on top of water that's 4 kilometres deep and suspended over the vast abyss by nothing more than snot and air, is an upside-down bright purple snail. This is *Janthina janthina*, the snail that tried a conventional mollusc lifestyle and then evolved to float free.

Living cells are deceptively tidy parcels of water and organic material, each one a tiny biochemical factory containing all the instructions, machinery and control systems needed for the business of life. Almost everything in the parcel that isn't water is a long chain molecule, which happens to be a very efficient way of packing atoms together. The consequence of this is that living cells are generally more dense than water, which means that they have more mass per unit of volume. If you live out your life stuck to a rock or munching on the innards of a dead sheep, this distinction will make very little difference to you. But if you live your life in the ocean you are likely to sink unless you take evasive action, because gravity will pull more strongly on you than on your surroundings. There are lots of marine snails,[35] and the majority crawl about on rocks and sediment at the bottom of the ocean. They are more dense than ocean water, and so they win the gravitational competition with their environment and stay at the bottom. But one of *Janthina janthina*'s ancestors broke the rule.

We now think that it was a female that went first, possibly freed from a rock by the bump of a gentle current before twirling gently on the slow journey up towards the sunlight, hanging from mucus-based egg sacs that had filled with gas. The great thing about density is that it's the average density of a whole object that counts, and so you can compensate for squeezing a lot of mass into a modest volume by sticking yourself to something that has lots of volume and very little mass. The egg sacs acted as puffed-up balloons, lifting the female snail all the way up into the all-you-can-eat buffet at the sea surface. Adaptations over many generations led to *Janthina janthina*: as the shell turned purple to hide against the blue sea and became paper-thin to minimize weight, both males and females evolved to create deliberate bubble rafts and the species left the sea floor behind.[36]

It's a lifestyle that comes with genuine jeopardy. If the snail ever becomes separated from its bubble raft, it will sink and die with no hope of return. It's safe as long as the bubble raft and the snail together are less dense than water, keeping them floating at the ocean surface. The snail has no internal bellows to inflate its bubbles, but instead uses a funnel-shaped part of its foot to

---

[35] They have gills instead of lungs, in case you were wondering.

[36] This isn't the sea snail Murex, the origin of Tyrian Purple, much beloved by the Roman Empire. Interestingly, they have chemically very different types of purple, but still, both purple.

trap air from the atmosphere, coat it with mucus, add it to the bubble raft and then let the seawater harden it in place. These snotty pockets are reasonably robust, but the snail must never run out of bubbles, so continual maintenance is required. With every bubble it folds, the snail becomes less dense overall and the bubble raft will poke higher out of the water. With every bubble that pops, it becomes more dense, sinking downward slightly. If enough bubbles pop that the snail starts to match the density of water, the terminal inky black of the deep awaits. Gravity sorts the density hierarchy so that everything is where it belongs: bottom-dwelling marine snails are most dense, so they settle at the bottom; seawater is intermediate and sits on top of them; and the least dense *Janthina janthina* with its bubble raft is parked at the top. But the rules of gravity don't just apply to snails. The physical structure of the ocean is entirely based around a strict gravitational hierarchy, and so density is critical to everything that happens in the ocean. Even tiny changes in density can have gigantic consequences.

The most extraordinary polar ship ever built

It's a day of grey and white, and several steep-sided triangular buildings squat sternly at the water's edge, their sharp lines slicing into the soft snowy sky. Stepping inside is a relief, followed by an eye-popping change of perspective. I'm confronted with a vast rounded bulk rising up above my head and out to either side. It's a wooden ship, masts stretching into the peak of the building, sitting on the floor and filling its home. Its sides are brightly painted in cream, black and red, but it's the shape that's distinctive: the hull has a deep, rounded belly and the ship is only three times as long as it is wide. It's a bowl masquerading as a ship, smooth, plump and curvaceous. I have never seen a ship this shape, or anything wooden that looks so solid and so comforting. Nothing I've read has really prepared me for the reality, and I've read a lot about this vessel. For anyone who has worked in the polar oceans, coming here is a pilgrimage. This ship is a unique expression of the challenge involved in delving into the way nature *is*, of deep commitment to testing a scientific hypothesis, and of meticulous preparation bundled up with humanity and humility and teamwork. Resting quietly inside the museum that was built around her after some of the greatest voyages in the history of polar science and exploration, this is *Fram*. She was the first ship to cross the Arctic Ocean, and the only ship ever built on the principle that the way to defeat the ice was to surrender to it completely.

In the late 1800s, western polar explorers were the astronauts of their day: venturing beyond the known world to face an unimaginably harsh environment, huge physical hardship and a significant risk of dying during their quest. The development of photography meant that an eager public could drink in visual evidence of alien landscapes and terrifying hazards, and the rapid development of scientific ideas made scientists even hungrier for measurements and observations which would enable them to understand the world. Most of all, there was an informal list of tantalizingly achievable 'firsts': to the North and South Poles, through the Northwest Passage, and the discovery of whatever it was that filled the significant blank areas on the map at the top and bottom of the world. Fridtjof Nansen stepped into this world sideways, from a youth spent outdoors skiing and fishing, university studies in zoology and a PhD in the nervous system of marine creatures. Already an experienced polar adventurer by his early twenties, Nansen read about a new and controversial idea: that there might be an ocean current crossing the Arctic Ocean from one side to the other. There was a smoking gun: pieces of the ship USS *Jeannette*, which had been found on the coast of Greenland even though the ship was known to have broken up on the opposite side of the Arctic. Most attempts to reach the North Pole had started from the Greenland side, but

Nansen suggested that perhaps the solution was to start on the Russian side and travel with this proposed transpolar current, rather than against it. But the problem wasn't really the current, if indeed it existed. The problem was the ice itself.

A water molecule, as shown in chemistry textbooks and on classroom whiteboards, appears pleasingly straightforward. A chunky round oxygen atom has two smaller hydrogen atoms stuck to it like the ears of a Disney mouse, at a nice tidy angle of 106°. But this apparent simplicity is barely even skin-deep. The oxygen and hydrogen share a cloud of negatively charged electrons, and different parts of the molecule have strikingly different electric charges: negative around the oxygen and positive around the hydrogen. The attraction between negative and positive sucks the 'ears' inwards, making the water molecule smaller than almost all other molecules. And the ears themselves are constantly moving around, wandering and adjusting to the surroundings, so the 106° angle is only an average over time. But the real complication comes when other water molecules are near by and the positive and negative electric charges on one molecule feel the pull of all the other negatives and positives around them. The outcome is that water molecules are intensely attracted to other water molecules, and this is the only reason why such a small molecule is a liquid at room temperature. Other similarly sized molecules – methane, nitrous oxides, carbon dioxide – easily fly free of each other to form gases at atmospheric temperature and pressure. But water molecules are locked in a crowded liquid dance with each other, constantly twirling and swapping partners, but unable to leave the dance floor. If you give them extra energy by heating the water up, the molecules speed up and spread out, and maybe a small number will have enough energy to escape and become a gas. But if you remove energy by cooling the water down, the molecular movement slows down and the liquid water draws in on itself, shrinking as the molecules get closer together. If even more energy is lost, the molecules must eventually lock together in a new regular way to form a solid: we call this ice. But here water throws one of its many curveballs: as energy flows out of the cooling water and the molecules begin to lock together in a regular pattern, they push themselves apart to create a new structure that's more open, so the new frozen solid takes up *more* space, not less. The consequence is that ice is less dense than the coldest liquid water, by about 8 per cent.[37] This is why ice floats in your drink: the denser liquid wins the gravitational race to the bottom, and the less dense solid is left floating at the top. It is hard to overstate how weird this is: almost every other substance shrinks when it freezes, and so the frozen parts will sink to the bottom. But frozen water easily floats on top of liquid water, in all circumstances.

This oddity is why the poles of our planet are white: we only have frozen polar oceans because water ice is less dense than the rest of the ocean and so floats on top of it. If our oceans were made of almost anything else, any part of it that froze at the surface would sink down until it mixed in with warmer water below, cooling the depths and leaving the surface unfrozen. The only way to freeze the surface would be to freeze the whole ocean, all the way down. But instead, as the surface cools, the water freezes and the ice stays at the top, forming an insulating lid on the rest of the ocean.

The problem for anyone wanting to reach the North Pole is

[37] The full story is even weirder than that. As liquid fresh water cools down towards zero it consistently becomes more dense until it reaches 4°C. As it cools past that point, little spaced-out clusters of water molecules start to form. They're not solid ice, but they're taking up just a bit more room than the rest of the water and the cooling liquid becomes less dense. And so fresh water has a maximum density at 4°C, although the density difference between 4°C and 0°C is less than 0.2%. Ocean water is different: it increases in density all the way down to the freezing point without this little blip, because the salt prevents those clusters forming.

that the ocean surface doesn't freeze in a single tidy sheet. As the surface of the ocean loses heat energy, ice crystals form and then get thicker and wider as more water freezes on to them, eventually growing into large ice floes. There are still tides and currents and wind shoving the ice around so the floes break apart, squeeze together and rotate as the forces of nature push on them. All of this leaves you with open water to cross between irregular ice floes, and treacherous edges and pile-ups. If Nansen wanted to reach the North Pole in a ship, the ship had to be able to survive this mosaic of shifting untrustworthy ice, the ocean's polar lid. During the Arctic summer, the ice would partially melt and during the winter it would freeze solid, all the while groaning and creaking as it was pushed around by the ocean and the atmosphere. But the biggest problem was that as ice floes responded to the push of wind or tides, possibly a push that originated miles away, the ice itself could become a shrinking trap. If your ship sat in the middle of freezing ice, or of ice floes being pushed together by the wind, the ice would act as a vice gripping the ship's hull, generating crushing forces that no ship could withstand. But Nansen and his chosen ship architect Colin Archer had a plan,[38] and the plan grew into *Fram*.

Stepping on to the deck, my first impression was of how comforting it is to be surrounded by wood like this. I love wood as a material, with its grain that tells stories of the lives of trees, and its tough and chunky bulk. *Fram* has a large open deck made of dark-stained wood, and it feels surprisingly spacious. At 39 metres long, 11 metres wide and around 5 metres depth from the main deck to the keel, this ship could hardly be called large. But she was home to thirteen men and a pack of dogs, all of whom left Norway equipped for a voyage of three to five years. Below

[38] Archer's parents were Scottish, but he was born and brought up in Norway. Apart from *Fram*, he is most famous for designing rescue boats, with a distinctive boat design named after him.

the main deck there is a cabin for each crew member, a galley (kitchen area), dining area and lounge, as well as a sewing machine, a carpentry area and a coffee grinder. A small windmill attached to the mast charged a battery so the ship had electric light even in the depths of the polar night. I was expecting it to feel cramped, and of course I was looking at it without the supplies, the dogs, the scientific equipment and the thirteen occupants. But it felt like a home, a comfortable, reassuring, wooden home. It wasn't until I stepped into the hold in the bow that I saw the real character of *Fram*. As the walls narrow towards the rounded nose of the ship, chunky dark oak buttresses start to criss-cross the space, getting closer and closer together, each one wider than the span of my hand. *Fram* is a floating fortress. Even the sides of the ship are 70–80 centimetres thick, and the bow has 1.25 metres of oak protecting the men inside from the grinding ice outside. In spite of that gigantic strength, the real trick of this ship is her smooth rounded sides and bowl-shaped belly. As the ice squeezed inwards, *Fram* was designed to slide upwards, escaping the vice by popping up like a seed squeezed between your thumb and forefinger. The aim was not to push through the ice, but to be carried by it – to freeze in, lifted by the squeezing, and to become part of the ice, starting on the Russian side and then hopefully being pushed across the Arctic Ocean by the transpolar current, passing right by the North Pole and emerging on the other side of the Arctic. On the way, its occupants would keep busy by making as many scientific observations as they could, using the latest techniques to measure the water and the ice to begin understanding the Earth's last great unexplored ocean.

The black-and-white photographic portraits of the thirteen team members, taken before they left in 1893, are striking. They stare out with intent and determination, none more than Nansen himself, whose pale eyes pierce the decades with a predator's watchfulness. This was a man prepared to put his life on the line to test an idea, backed by the best team he could assemble in the best ship the world could produce. And so they set out in the summer of 1893, reached the ice edge on the Russian side in September at 77° 44′ N, and settled in to go wherever the ice would take them.

As summer shifts into winter, the Arctic changes from a place of twenty-four-hour daylight to twenty-four-hour darkness. Heat leaches away from the ocean surface into the atmosphere, robbing the water of energy as the sun disappears. The freezing point of fresh water is 0°C, but the salt in seawater reduces the freezing point, and so the temperature of the surface liquid water in the Arctic can go down to −1.8°C, depending on the water salinity. Any temperature changes will affect the density of the water, but as Nansen and his men and their dogs started their first winter inside this wooden bubble, something else was changing the density of the water that *Fram* floated in. As the sea ice thickened around the ship and the sturdy oak structure held strong to pass its first test, a delicate process was under way, one that not only gave *Fram*'s occupants fresh water to drink but also gives the deepest parts of the ocean engine their impetus to turn. The salt was on the move.

Sea ice is one of Earth's most formidable features: weighty, fickle, shifted only by the Herculean forces of atmosphere and ocean. But its origin is fragile and delicate, and far more complicated than the formation of ice in your freezer. It starts with tiny needle-shaped crystals of ice called frazil ice, which form as a few water molecules lose so much energy that they are forced to give up their mobility and lock on to other water molecules to form a fixed, rigid network. This is a crystal, and other water molecules will gradually settle on top, converting the fluid dance floor into a strict and systematic array. But this rigid array can only accommodate water molecules. The salt is left out. As frazil ice crystals bump into each other and start to stick together to form bigger ice structures, they may trap pockets of salty seawater, but those pockets will only ever partially freeze. The remaining liquid gets increasingly salty as the water molecules abandon it to join the crystalline ice. These brine pockets are likely to escape back into the ocean as the ice grows, compresses and cracks. And so the formation of sea ice is not important just because it creates floating solid ice that acts as a lid on the ocean. It's important because it's a *sorting* process. Ice, largely free of salt, floats to the top of the ocean because it is less dense and therefore buoyant. And just underneath the growing sea ice, a slow stream of extra-salty seawater forms and sinks downwards, because the cold and the salt make it more dense than the rest of the seawater. In the regions where the ice grows, a conveyer of salt slumps downward during the ice's growth season.

As *Fram* drifted into the interior of the Arctic Ocean, Nansen and his team measured the ocean around them. They sent down water samplers to record the temperature and salinity of the water at depth. They drilled holes through the ice and sent down sounding lines to discover that the Arctic is nearly 4 kilometres deep in places – which came as a surprise, because it had been assumed that the Arctic was shallow. And in those depths, there was plenty of potential for the water to arrange itself according to its density. Sea-water density depends on temperature and salinity, and there are several processes that affect it: a temperature change, an injection of fresh water from rivers or ice melt which makes it less salty, removal of water by evaporation, or brine rejection during ice formation which makes it more salty. The temperature and salinity of ocean water aren't just labels; they're the critical factors that determine where water will sit within the ocean. The whole ocean is fluid, so there is no barrier to movement; density is all that matters. Between them, temperature and salinity generally alter the density of ocean water by less than 1 per cent, covering a range from 1,020–1,028 kilograms per cubic metre. It seems like a tiny difference, but it matters. The *Fram*'s occupants had no shortage of drinkable water because they could melt the sea ice and snow at the surface, benefiting from the low density of ice which forces it to float at the top. Most of the Arctic Ocean has a very cold and relatively fresh layer at the top (brine formation happens only at certain times of year and only in certain places), and the lack of salt outweighs the cold to make this float on top of what's below. The middle layers of the Arctic Ocean are taken up with salty and relatively dense water,[39] which is slightly warmer than the surface layer (at around 1°C instead of –1.7°C). The additional density due to the salt outweighs the buoyancy due to the warmth, keeping the warmer waters counter-intuitively trapped in the depths, away from the surface. Below that are the densest waters in the world ocean, the result of the seasonal ice formation above, which creates an annual dose of water that's just as salty but even colder. And even though this cold salty dense water may form in the Arctic, it doesn't have to stay there. The density itself can force water to move.

Scandinavia has a long history of northern ocean exploration, and just down the road from the Fram Museum in Oslo, the Viking Museum tells the story of Nansen's ancestors. Between the eighth and eleventh centuries, they ventured out across the seas and returned home with tales of foreign lands, riches to exploit, exotic goods to trade and adventures at sea. But possibly the most famous Norse explorer of all – Erik the Red – sailed right over one of this planet's most dramatic geographical features without ever noticing it. While the buoyancy of his ship kept Erik safely at the ocean surface, density was driving an enormous rearrangement of the ocean just a few hundred metres beneath him.

[39] The dominant picture in the Arctic (of relatively fresh cold water lying on top of salty slightly warmer water) is quite different from that in the rest of the global ocean, where the top layer – the mixed layer – is warmer than everything below.

The largest waterfall in the world

The Vikings had a sophisticated culture, with impressive craft skills and trading networks, but Erik the Red lived up to the more bloodthirsty Viking stereotypes that still pervade modern culture. According to the Icelandic Sagas, he was born in Norway in AD 950 but was forced to leave at the age of ten when his father was banished for manslaughter. The family moved to Iceland, and Erik grew up and lived on a farm of his own until a murderous tit-for-tat with a neighbour meant it was his turn to be punished by exile from the village. He moved to a new village, but was soon kicked out again following a dispute over pagan heirlooms which ended in manslaughter. Iceland apparently didn't want Erik and neither did Norway, but he was a capable seafarer and so took his boat westwards to investigate rumours of a large land far across the sea in that direction. He crossed what we know today as the Denmark Strait, a journey of around 440 miles, and found that there was indeed an enormous land mass that was apparently uninhabited. But it was covered with ice, and he had to sail around its southern tip before he found a coastline approachable enough to land on. After spending the three years of his exile exploring, he returned to Iceland, spread the word about this amazing new land (it's suggested that he named it 'Greenland' to make it sound more appealing), and persuaded enough people to join him to found the first European settlement there.

That first crossing from Iceland to Greenland must have been grim, even though modern scientific records suggest that the period around AD 1000 was unusually warm in northern Europe. To the north, on the right side of Erik's boat, lay the cold Arctic Ocean. To the south, on his left, lay the North Atlantic, a region of fierce storms and warmer water. And underneath him was a ledge, a reasonably flat region of shallower water that divided the deep Arctic Ocean and the deep Atlantic Ocean. That ledge sat 100–300 metres below the sea surface, invisible to Erik but still high enough to act as a barrier between the oceans on either side, each 2–3 kilometres deep in that region. But the water is not the same on either side of the ledge. The basin on the north side is full of dense, salty, cold Arctic water, and like water over-flowing from a bath this dense water slithers across the ledge to meet the Atlantic. There it encounters warmer water which is less dense, and so the huge cold overflow slides underneath the Atlantic waters, hugging the slope and tumbling downwards underneath the rest of the ocean until it reaches the bottom, 2.5 kilometres below the ledge. This is the Denmark Strait Over-flow, the largest waterfall in the world, plunging down a long underwater mountainside to join an underwater deep pool at the bottom. It's estimated that 3 million cubic metres of water flood down this cataract every single second, more than one thousand times the flow of Niagara Falls. The flood isn't smooth: as the cold water slides downwards, it drags along some of the warmer water above it, creating a huge plume of turbulence and warming slightly as it goes. But it is continuously supplying a huge quantity of cold polar water to the bottom of the Atlantic. This is why the deep ocean is cold – wherever dense water forms it will sink, and cold water is dense so it ends up at the bottom. The ice formation underneath *Fram* was creating new water masses that were later sent off on a journey into the global ocean. Density matters because it dictates the structure of the ocean.

Erik would have seen no sign of the drama below him, no indication that he was above a precipice separating two great ocean basins. He would have met a cold surface current, and then the ice-covered rock of the east Greenland coast rising in front of him. But had he chosen to turn south instead of west, to head down into the Atlantic, he would have met much warmer surface water, far warmer than the latitude would suggest. For the balmy Gulf Stream curves around the top of the Atlantic, flowing along the coast of North America and then out across the Atlantic towards Britain. It keeps northern Europe far warmer than it would otherwise be, and it's the reason that palm trees can grow in Cornwall. It's there because of another major influence on the ocean engine: the Earth's rotation.

Spin

There is a kind of vertigo that comes with imagining the Earth spinning in space as it trundles around the sun, whirling away the years, the centuries, the geological eras. Seventy million years ago, an Earth day was only 23.5 hours long, but life pulsed with every pirouette. Clams sitting on the shallow sea floor grew rapidly as they spun around the sunny side of the planet, but construction slowed daily when their journey took them around into the dark, facing outwards into the vast universe. Four and a half thousand years ago, Egypt rolled out of that darkness each day, and every sunrise illuminated newly added stone blocks as the Great Pyramid rose out of the plains. And on one morning in 1945, just as New Mexico rotated out of the Earth's shadow for perhaps the trillionth time, the first atomic weapon was detonated and humanity passed into the Atomic Age. The Earth's vast history has happened one day at a time, one revolution at a time, and the turning has never stopped.

In 1918 in Europe, the First World War was entering its final chapter. The trenches along the Western Front, a human-made wasteland of mud, misery and death, had been circling the Earth's axis for nearly four years. Irrespective of the vast human disaster unfolding on top of it, this patch of land continued to move eastward around the planet at 310 metres every second (700 mph), carrying millions of soldiers on the global carousel into the darkness and then the light and then the darkness again. Despite the huge sideways speed, there was no way of feeling the movement, because everything a person could see around them was also spinning at the same speed. And yet the cycle of sunrise and sunset was the evidence that the celestial clock was still ticking, that in spite of the gigantic wave of death and destruction, time had not stopped.

By March 1918 the Kaiser's army had advanced to within 75 miles of Paris, but life in the city carried on, with churches and theatres and factories keeping their doors open in spite of the rationing and the hardship. Although the fighting was relatively close, no weapon could reach the city from the front and so its citizens felt relatively safe. But at 7.18 a.m. on 23 March 1918, just as Paris emerged from the dark side of the Earth to face another day's flood of sunlight, that changed. A boom echoed across the city as a part of the Quai de la Seine exploded, followed by another fifteen minutes later and then another, destruction sprinkled randomly across Paris by the cruel lottery of fate. It quickly became apparent that the explosions hadn't been caused by bombs dropped from planes or Zeppelins, but by shells fired from a gun, one either far closer or far larger than anyone had imagined possible.

As the shells continued to fall for days and then weeks, the relative peace of Paris was shattered, along with the morale of the population, who fled the city in droves. Eventually the origin of the shells was identified: a small hill 75 miles to the north-east of Paris, behind enemy lines. This was a huge distance away, and the idea that shells had hit the city from that hill was staggering. But sitting up there were the three biggest guns in the world, each one a brand-new and ambitious behemoth. Each gun had a barrel 37 metres long and was capable of launching a 106 kilogram shell with a muzzle velocity of 1,640 metres per second, nearly five times the speed of sound. This was the first and only deployment of the 'Paris Gun';[40] and while it was

---

[40] This is not the same as 'Big Bertha', an earlier long-range gun used in the same war.

ultimately of limited military use,[41] it soared into the record books. The range and the size would have been enough by themselves to achieve this, but during the three minutes each projectile took to travel to its target, the shell was propelled to a height of 40 kilometres (five times the height of Everest), making these the first human-made objects to reach the stratosphere.

But as its tiny human inhabitants scuttled across its surface, wreaking death and destruction on themselves, the Earth never stopped turning. And for the three minutes between being fired and landing, the Paris Gun's shells were temporarily disconnected from the planet. There was no friction with the ground to drag them around on the circle drawn by Paris around the planetary axis. When fired, the shells matched the sideways speed of the city, 310 metres per second to the east. But then they flew according to Newton's laws, along the parabola that took them up into the stratosphere and down again. During those three minutes, Paris turned underneath them, moving a total of 34 miles to the east along its circular path. Each shell also covered those 34 miles, in addition to the extra 75 miles covered by its parabola. But Paris travelled along a circle, glued to the planet as all cities are, and the liberated shells travelled on a straight path. The outcome was that, compared to the situation on a perfectly still and flat surface, the shells appeared to be deflected 400 metres to the right by the time they reached their target. The engineers knew how to correct for this, by aiming the gun ahead of where Paris was going to be when the shell landed, knowing

---

[41] There were several reasons for this, but perhaps the most striking is that the process of propelling the shell along the barrel was so violent that it shaved a significant fraction of the barrel away on every shot. Each barrel could therefore only be used for sixty shots, and had to be used with a special series of shells, each one slightly wider than the last. Relatively little is known about the details, because all these guns were destroyed by the German army before the end of the war.

that physics, in the guise of the Coriolis Effect, would bring it back on target. A sideways deflection of 400 metres was huge, but unavoidable: the laws of the universe state that it must be so.

The Coriolis Effect

Trajectory calculations are different on a spinning planet, because anything that isn't firmly attached to it will travel on a simpler path, free of the constraints of rotation. To an observer standing on the ground, the travelling object will look as though it's being deflected, but that's only because we're standing on something that is spinning. If you ever played catch on a merry-go-round as a child, you will have seen this in action: if you stand on the edge of a roundabout and throw a ball directly across the middle, it appears to bend to the side, taking a curved path instead of a straight one. This apparent deflection is known as the Coriolis Effect, and its consequences depend on where you are: in the northern hemisphere, non-attached travelling objects will be deflected to the right, and in the southern hemisphere they'll be deflected to the left. Paris in 1918 was about 7.5 km across, so failing to account for the deflection wouldn't have saved the city. But the correction is notable because the shell travelled far enough for it to matter. Exactly the same physics applies to Olympic javelin throwing, a tennis serve in the Wimbledon Championships, and the spud guns that I used to play with as a kid. But because the javelin, tennis ball and potato fragment are only free of the Earth for short distances and times, the difference is so tiny it's unmeasurable.[42] But the ocean certainly isn't firmly attached to the planet, and it moves long distances over

---

[42] Anyone showing you whirlpools at the Equator which apparently rotate in different directions on the north and south sides is pulling your leg. The Coriolis Effect is far too small in that situation to matter, not least because it's effectively zero at the equator, whichever hemisphere you're standing in.

long periods of time, so the Coriolis Effect has a huge impact on how ocean water moves around. The resulting patterns are mostly too large for a seafarer to see with their own eyes, but certain situations can provide clear clues. Fridtjof Nansen, floating across the top of the world in *Fram* in the 1890s, was the right man in the right place.

Nansen found the first part of the *Fram*'s drift worrying. The ship and the ice that surrounded it were definitely drifting northwards, but progress was slow and the ship would sometimes circle around or go backwards for a while. It looked as though the transpolar drift was real, but rough calculations suggested that it might take *Fram* seven or eight years to dawdle across the Arctic and reach open ocean on the other side. For a dynamic get-up-and-go team of explorers, the slow pace of the days was incredibly wearing. On the other hand, the science was extremely productive. Along with the wealth of measurements tracking the obvious things – weather, ice thickness, ocean water – came a few observations that were a huge surprise. In retrospect, possibly the most significant was the observation that ice floes were not blown by the wind in the way you might predict. If you put a toy duck on the surface of a bath and then blow sideways on it, the duck will move directly away from you, following the direction of the 'wind'. But Nansen noticed that ice floes don't do this. His observations showed the ice drifting in a direction consistently 20–40 degrees to the right of the wind.[43] He was well aware that the whole Arctic Ocean was

[43] We know now that it can be anywhere between 0 and 90 degrees to the right of the wind, depending on the circumstances. I've seen this while working at the North Pole and it's disconcerting, because the wind, ice and water all appear to be moving independently. It doesn't help that during the polar summer, the sun is just skirting the horizon all the way around and the ice floe you're standing on is probably rotating as well as drifting in the wind, so there are no intuitive fixed directions.

twirling away at the top of the world, and he correctly identified the spin of the Earth and the Coriolis Effect as the likely
cause of the strange ice movement. But a full mathematical
explanation had to wait until these observations landed on
the desk of a Swedish graduate student named Vagn Walfrid
Ekman.

What Ekman uncovered, right at the start of his career, became
one of the pillars on which all of physical oceanography rests.
Most ocean surface currents are pushed along by the wind, but
that doesn't mean that they move in the same direction as the
wind. On a spinning planet, things aren't that simple. Ekman
gave a mathematical framework to what Nansen had suspected:
the way to think about this is to imagine the wind dragging
along a thin layer at the top of the ocean, which then drags along
a thin layer underneath that, and so on all the way down. It's
like starting with a stack of paper and pushing sideways with
your hand on the top sheet. Your hand only moves the top sheet,
but friction between the top sheet and the next one down forces
the lower sheet to move too, and then that sheet pushes on the
next one down. But in the ocean, the Coriolis Effect kicks in as
soon as the top layer is moving, appearing to push it to the right
(in the northern hemisphere) just as the shells from the Paris
Gun were deflected to the right. As it accelerates, that top layer
is forced to move to the right of the wind. But the next layer
down isn't pushed by the wind – it's pushed by the top layer
which is moving in the new direction. And so when the Coriolis
deflection is added, that next layer moves at an even greater
angle to the wind. As each layer pushes on the layer beneath, a
spiral is set up, known as the Ekman spiral. As you go down the
layers, the wind-driven current slows down and its direction
moves to the right. Some way down the spiral, there's actually a
slow current moving in the opposite direction to the wind, even
though it's caused by that wind. Add up the flow over all the
different directions, and the outcome is this: *when a steady wind is*

driving a current, the net effect is that water moves at 90 degrees to the right[44] *of the wind.* This is a direct consequence of the Coriolis Effect and the spin of the Earth. It's a really peculiar thing to think about, but that's because the Earth is so big that it's possible to completely forget that sunrises and sunsets are direct evidence that the whole thing is spinning.

This counter-intuitive idea has very practical consequences. The reason that there is cold water upwelling off the coast of Chile, producing the fertile ecosystem responsible for ancho-veta, seabirds and mountains of guano, is the wind. The winds that push the currents are flowing northwards, along the coast-line. The mechanism that Ekman identified means that even though the wind is blowing northwards, the Coriolis Effect drives the water westward, at 90 degrees to the left (because it's the southern hemisphere) of the wind direction. As the warm surface water is pushed away from the coastline, cooler water must flow into the space from underneath. Tropical seabirds were a long way from the thoughts of Nansen, as he watched the ice that was carrying his ship. But all of us live on a spinning planet, and the consequences are found throughout the global ocean, if you know how to look.

After eighteen months of drifting, *Fram* had reached 84° 4′ N, having crawled past the previous record for the furthest north two months earlier. On 14 March 1895, Nansen and Hjalmar Johansen set out with a dog team to try to ski the rest of the way to the pole. They reached 86° 13.6′ N after three weeks, but the way ahead was blocked by what Nansen described as 'a verita-ble chaos of iceblocks stretching as far as the horizon', so they chose to turn back. The North Pole would remain beyond the reach of humanity for a little longer. After a long trek, the pair reached Franz Josef Land, an improvement on the sea ice but

[44] That's in the northern hemisphere. It's 90 degrees to the left in the southern hemisphere.

still frozen and uninhabited. Meanwhile, the *Fram* drifted onwards, as the men on board tracked the ocean, the magnetic fields and the weather, did their ski practice and maintained the ship. *Fram* was now a floating laboratory, taking full advantage of her unique position. In November 1895 she reached a new ship record of 85° 55′ N and then drifted onwards and southwards as boredom kicked in on board. In the summer of 1896 *Fram* finally popped out of the ice and made her way to the north of Norway, where Nansen and Johansen met them only a week after they themselves had arrived back in civilization. The reunited crew sailed southwards, receiving a hero's welcome in every port until *Fram* docked in Oslo (then called Christiania) on 9 September 1896, three years, two months and sixteen days after they had left. Nansen had conclusively proved that the transpolar current was real, and brought back with him a huge pile of data that would form the foundation of polar oceanographic knowledge for decades. *Fram* had more adventures ahead: a four-year expedition to explore the ocean around north Greenland, and then carrying Roald Amundsen to Antarctica for his successful quest to be the first to reach the South Pole. And then in 1936, *Fram* was hauled up on to the shore in Oslo, into the shelter of its own museum.

On my visit to the museum, eighty-six years later, I find it hard to leave and after a while it occurs to me that this is because this ship is such a powerful expression of ocean science and exploration done well. That dedication to the cause is familiar to me even though I've never been involved with anything even a tenth as bold as Nansen's expedition. The tough wooden structure of *Fram* gives it a much more human feel than a steel ship could have, and the combination of working with natural materials while working with nature highlights the flexibility and resilience of the whole project. It's unlikely that anyone will ever build another ship like this. As modern research ships get bigger and more stable, they separate oceanographers more and more from the ocean itself. *Fram* became part of the ocean: a connection rather than a barrier. I think that modern ocean science needs more, not less, of that.

We can now see most of the fundamental principles which dictate how the ocean engine works: temperature represents the vast amounts of heat stored by the ocean engine, salt drives deep water into motion because it's important for density, and the fluid ocean is strongly influenced by the spin of the Earth. There is one other huge influence on the way that the ocean engine turns: the wind, and adding that will let us explain the large-scale patterns of the whole ocean engine. But before we look at the importance of the atmosphere, we need to consider something else. The ocean may be mighty, but it's constrained by its own shape. In this theatre of nature the ocean's drama fills the space available to it, and the effects of heat, salt, spin and wind are the fundamental motivations behind the play. It's time to set the stage.

2

The Shape of Seawater

LIQUIDS ARE SHAPE-SHIFTERS. PART of the definition of a liquid is that it takes the shape of whatever container you put it in (which is why there have been occasional investigations into whether cats qualify[1]). The Earth's gravity pulls liquid seawater towards the centre of the planet, and so the ocean has filled the regions of Earth's surface which are closest to the middle.[2] But for all its lumps and bumps when you're a puny human, the Earth's surface is actually very smooth. The distance from the bottom of the deepest ocean trench to the top of Mount Everest is just 0.3 per cent of the radius of the planet. The average ocean depth is a fifth of that: 3.68 kilometre, only 0.06 per cent of the Earth's radius. To put that into perspective, if you imagine the

---

[1] The 2017 Ig Nobel Prize for Physics was awarded to Marc-Antoine Fardin for a paper discussing whether cats are solid or liquid. Figures include cats taking on the shape of the various containers they've put themselves into (jars, sinks) and pouring themselves across smooth floors.

[2] The pedants out there will note that this isn't quite technically true because the Earth isn't a perfect sphere with uniform symmetrical gravity. The planet's spin makes it bulge at the equator, and rocks vary in their density and distribution and therefore in their gravitational pull. The average ocean surface actually follows the geoid, which takes account of gravity being very slightly stronger in some places because of how Earth's mass is distributed. The point remains that the ocean falls 'down', wherever 'down' is in the local gravitational field, until it can't fall down any further.

Earth as an inflated blue party balloon, the ocean depth is equivalent to the thickness of the stretched rubber. But if the Earth's watery shell is so thin, why does it matter if it's skinnier in some places and a bit thicker in others? The answer is that the ocean engine is dominated by horizontal motion, by layers of water sliding over each other but staying at the same depth. Anything that pushes a water mass up or down is causing an important shift in the workings of the engine, as we saw with the upwelling off the coast of South America. And of course, if there's a very large immovable obstacle blocking the way, like a continent, water flowing towards it will have to change direction completely.

If you were looking down on Earth from outer space and feeling mean, you might characterize the ocean as a thin spherical puddle with some bits missing. But the weird shape of the ocean container is what makes the ocean engine interesting: it puts constraints on how the water can move, and this produces glorious variety in the workings of the engine. The ocean shape also isn't static – the edges are constantly moving as tides and seasons come and go, and the surface wouldn't be perfectly smooth even if you removed the endless waves ruffling it. So before we can get to the ocean's internal anatomy, we need to investigate its boundaries – its shape. We'll start with what we see when we see the sea: the top.

Top

At 3 a.m., the camp started to wake up. I had slept under the stars on the beach, while others roosted on the picnic tables under the local paddling club's hale (a roof-only hut) or curled up on the ground next to the canoes. The canoes had been rigged the night before, the traditional ropework done by torchlight and then hidden under a modern waterproof cover. At 3.45 a.m. Kimokeo called all forty paddlers into the hale, where both familiar and unfamiliar faces were dimly lit by four electric bulbs. As always before a voyage, all of us joined hands in one big circle. As always, Kimokeo thanked the elders – the kūpuna – and talked about our place between the heavens, the ocean and the earth. He had talked the night before about the principle of a voyage – that it isn't fundamentally about paddling technique or technical excellence, but about the connections among the crew. A female Hawaiian elder had also spoken: 'The ocean can embrace you, and the ocean can break you. When you go into the channel tomorrow, you must go with *aloha*.' *Aloha* to the Hawaiians is more than just a greeting. It means love and peace and compassion, and it's an expression of the most important values of cooperative human existence. Our circle was the physical reminder of our group, our team, and the interdependence of all the paddlers in a canoe. There was a moment of stillness after Kimokeo finished speaking, then a squeeze of hands before the circle rearranged itself to move the canoes to the water and begin the voyage.

These voyages are a modern expression of old traditions. The focus is the canoe, the vessel that allows you to traverse the ocean using skill and cooperation and connection. It's the concept that matters, not the material details. So when Kimokeo stands on the pier to sound the conch shell in the darkness, and to receive an answering tone from the gloom, it's not at all unusual that he's wearing a bright orange modern lifejacket. He is seventy-two, after all, and his wife insists on it.

We are about to cross from the Big Island of Hawai'i to Maui, a 56-mile journey when you include following the coast around Maui to our destination, and three six-person canoes will make the crossing. Each canoe has its own support boat which will carry one crew of six while the other six paddle for an hour, and we'll switch every hour for the whole day. In the dark, I sit at the back of the support boat as it tows our canoe out to the tip of the Big Island, and we arrive there just before sunrise. It's a stunningly beautiful morning, and as the stars disappear and the ocean turns pink there's a sudden perspective shift that makes the hairs on the back of my neck stand up. Our familiar human-scale canoe shrinks without warning to become a tiny and insignificant dot against the giant silhouettes of the volcanoes Mauna Loa and Mauna Kea that now dominate the horizon. As the sun rises, the first crew take up their paddles and the voyage begins.

The water between the Big Island and Maui is called the 'Alenuihāhā Channel by the Hawaiians, a name that translates as 'great billows smashing'. The trade winds are funnelled though the gap between the islands, creating the fastest winds in Hawaiian coastal waters and a steady army of waves marching towards the west-south-west. The last thing I think before I jump off the support boat to start my first hour in the canoe is that this is an astonishing example of wind-driven breaking waves, the topic I've spent ten years studying. There's a 15 knot wind pushing up 1.5 metre waves which are breaking at the crests and everything is beautifully lined up. I don't think I've ever seen waves this perfect, so regularly spaced and arranged like a textbook diagram. After that, I don't have time to think about anything for a while.

Switching involves jumping off the back of the support boat into the ocean and then treading water in a line as the canoe comes up to meet the group. The paddlers in the canoe hop out, leaving their paddles behind, and the new crew haul themselves out of the water and into the canoe to replace them, a process which is rarely elegant in open seas. But as soon as everyone is in, you paddle. For this first part of the crossing we are paddling across the wave field, so the canoe is nearly parallel with the wave crests and long ridges of water are continually rising up on my right, temporarily filling my world on that side. The *ama*, the canoe's stabilizing second hull, sits out to my left, keeping the canoe upright as we paddle forward. I'm in the second seat from the front, and the paddler in front of me is often paddling in air while a wave rolls underneath, lifting the nose of the canoe from the surface. Then the splash from the nose hitting the water launches over both of us and we dig our paddles deeper. But I don't care about that. My overriding emotion is a deep, visceral admiration for the canoe. It *works*. The support boat bounces around, safe in spite of the waves. But this canoe feels far safer *because* of the waves – because it's a part of the ocean, glued to it by the six paddles pushing it forward and the skill of the steerer (at the back in the sixth seat) at navigating nature. The hour goes incredibly quickly, and when I get back on the support boat, Mauna Loa and Mauna Kea already seem very far away.

But it's my second stint in the canoe that brings the real revelation. The canoe has turned now so that it's running with the waves and the wind. The steerer and the stroke at the front of the canoe are both very experienced Hawaiian paddlers, and it becomes clear that they're in the canoe for one reason only: to feast on a full hour's adrenaline rush as the human-plus-canoe-plus-ocean combination hits the peak of its potential. I did not know that being in a canoe could ever be like this. As each wave tilts the canoe, the Hawaiians frantically call 'push, push, push', and the human engine races to stay with the wave as it propels us forward, accompanied by enthusiastic whoops of pure joy which rapidly turn back into calls to push to meet the next wave. There is total focus, no time to think, just another wave and another, and the punishing physical effort required to maintain this state is completely forgotten in the joy of riding with the ocean. I have heard coaches talk about the need for your paddle to connect with the water, but I've never felt it like this. My paddle and the other five are holding the canoe and the ocean together, gluing us to the ocean surface to share its energy. The reward for creating that connection is a flood of raw exhilaration and absolute joy in the beautiful things that nature and humans can do together. And it goes on and on . . . I can feel the ocean through my paddle, a connection that science has never given me, and I can't get enough of it. When Hawaiians talk about connection to the ocean, the heavens and the earth, they're not empty words. It's this experience, in words whose meaning is enriched for ever once you have experienced the raw potential of being part of nature, not opposed to it. When I climb back on to the support boat, after an hour of flying with the ocean, I'm astonished at how much closer Maui is, and how fast we must have been travelling.

The voyage takes nine hours in total, and the conditions do turn against us a bit at the end. But the memory of that downwind section keeps me going through all the rest.[3] When I got out after that second stint, my first words to my coach were: 'You never told me a canoe could do that.' He laughed and said: 'That's the drug.' Count me an addict.

Since waves are temporary and mobile, it may seem odd to treat them as an important part of the ocean's shape. A huge storm in the middle of the North Atlantic with waves 10 metres high is just lightly dancing on top of an ocean that's 4 kilometres deep. If you want to play at being Poseidon, pick up your trident, stand in a swimming pool that's around 1 metre deep and blow gently on the surface to get the tiniest ripple you can manage. That's approximately the vertical effect of a force ten gale on the shape of the sea surface out in the open ocean. But the shifting shape of waves is active, not passive, and it's directly connected to the workings of the ocean engine. Waves are a link between atmosphere and ocean. They have also had an outsized influence on human affairs. Every wave has come from somewhere and is on its way somewhere else. Waves are

[3] One of our paddlers had a GPS-enabled device that tracked our journey, and it proudly declared at the end of the day that we had undergone a 586 m elevation gain. I'm not convinced that it's a particularly accurate number, but it was a stark reminder of how many big waves had carried us up and down that day.

caused by the wind, although the link isn't always straightforward. That makes life tricky if you want to predict the waves tomorrow or next week, and so for most of human history the arrival of waves was just a lottery. No one really took wave prediction seriously until the Allied forces in the Second World War were on the brink of setting themselves up for catastrophe.

Waves and war

In 1943, a young oceanographer, Walter Munk, was aiding the war effort in Washington DC when he learned of Allied plans to land troops in North Africa. The troops were to be conveyed to the beach in small landing craft which would come into the shore so that the bow of the boat could be lowered, allowing troops to scramble off. The problem was the shifting ocean surface shape: as soon as the waves at the beach reached a couple of metres in height they would push the landing craft sideways and fill them with water. After a bit of research, it became obvious to Walter that waves at the proposed landing site were almost always higher than the safe limit. Unless the landings happened on the small number of calm days, they would be a failure. There were only two routes to a successful outcome: blind luck or a scheme for predicting future waves. No one had tried to predict waves before, and the US Navy wasn't convinced that it was necessary. Munk consulted with his previous boss Harald Sverdrup, the Director of the Scripps Institution of Oceanography, and Sverdrup thought it was possible. His authority was enough to convince the Navy that the attempt should be made. The only way to make a prediction was to follow the whole journey of the waves, so they began at the beginning.

When you blow on a cup of tea with a smooth, stationary surface, the stream of air pushes downwards on that surface, making a dimple. Surface tension makes the liquid's surface behave a bit like an elastic sheet, and if you push it rapidly downwards in one place a disturbance will ripple out across the rest of the surface. The very beginning of waves in the ocean is similar, though the downward push doesn't come from a continuous downward wind. It comes from the ever-shifting swirls of the turbulent air above, which deliver puffs of faster wind and changes in pressure that distort the surface. Once there are ripples, the sideways flow of wind has something to get a grip on, and it pushes on the upwind side of each ripple, forcing the ripple to get bigger and steeper as it grows into a longer-lasting ocean wave. After a while, the ocean surface is a jumble of waves of different sizes, all travelling at slightly different speeds in slightly different directions, and all carrying energy that has been given to them by the wind.

In 1943, Munk and Sverdrup didn't know any of the details, but they knew that the mixture of wave sizes and shapes depended on the speed of the wind and the distance over which it had been blowing on the ocean (known to oceanographers as the fetch). They could use a weather forecast to predict the waves caused by wind out in the open ocean. But this jumble of waves caused by local wind, called the wind sea, isn't what was causing problems for the landing craft. The important thing about waves is that they're a shape that must *travel*. Once the wind has died down the waves are still there, moving outward across the sea surface and carrying their energy with them. The next stage was to adjust for what happened to the waves between their creation in a storm (which could be hundreds of miles from the wave prediction area) and their arrival at the landing beach. Some of the waves, especially the shorter ones, lost their energy quite quickly. But the longer waves kept going, spreading out and becoming smoother as the small lumps and bumps fizzled out. These long, smooth, leftover waves are known as swell, and they're often most obvious on days with little wind, because then there's no wind sea sitting on top of them. They can easily keep going until they reach a coastline. The final stage of Munk's and Sverdrup's analysis was to consider what happened at the landing site itself, where the sloping beach makes the waves steeper again just before they break. This last calculation is where the wave height experienced by the landing craft became apparent. The wave prediction model considered each of these three stages separately, and although it was rough and ready, it got the big picture right.[4]

This new method was successfully used to predict calm days for the landings in North Africa, and since the US Navy was now convinced that this might actually be a useful exercise, Munk and Sverdrup established a school for Navy and Air Force officers in La Jolla in California. They kept refining their wave prediction methods as they taught, but the real test was still to come.

As the oceanographers did their calculations, the planet kept turning and the Second World War rumbled on. But the end was coming. Germany had occupied most of western Europe since 1940, and in the early summer of 1944, the Allied forces were ready to launch a huge invasion from the British coast in an attempt to retake that territory. This invasion was code-named Operation Overlord, and the plan depended on a surprise crossing of the English Channel to land 132,000 troops

[4] It's also a good example of how subject areas carry their history with them. Munk found that the standard mathematical descriptions of waves didn't match with the intuitive understanding of the landing-craft steerers. So he invented a quantity that could be calculated mathematically but also matched the intuition of someone who was in a boat experiencing those waves. It's called the 'significant wave height', calculated as the average of the highest third of the waves. This is the quantity that every wave forecast model and wave measurement at sea still uses today, although the technical definition has changed slightly.

in northern France by ship on a single day, with another 24,000 transported by air. As he watched the debates about the ships needed to do the job, the British Prime Minister Winston Churchill grumbled in his diary: 'The destinies of two great empires [are] seemingly tied up in some god damned things called LSTs' (LST stood for Landing Ship, Tank). But those destinies were also tied up in the ocean engine, which controlled the water those ships would have to cross and more importantly, the waves that would be rolling in with the ships as they tried to land. As in North Africa, too much swell would cause huge casualties before the troops ever set foot on land. The best available weather forecasts were sent to Munk and Sverdrup in California and also to the British Meteorological Office, and they made their predictions of the ocean surface shape and what was likely to be happening as the effects of storms from the previous days reached the coast of northern France. The landing also required a full moon, the right tides and favourable weather, so the window of opportunity was narrow. Wave predictions found that on the preferred landing day, 5 June, successful landings would be impossible. On the following day, conditions would be 'very difficult, but not impossible', but further delay would miss the window for the tides and the moon. Walter Munk later told colleagues that he understood that the swell forecasts were what persuaded General Eisenhower to delay the landings until 6 June, which we now know as D-Day. The ocean waves that could have caused one of the greatest wartime disasters for the Allies rolled in on 5 June to plunge over empty beaches, but after one more rotation of the planet the ocean shape had changed; and so the most decisive amphibious landing in human history had far milder waves to contend with. Progress on the first day was far from ideal, and even the lowered swell caused a fair amount of seasickness in the landing craft, but this invasion marked the beginning of the end of the war. The temporary shape of the ocean surface has real and very practical consequences for human affairs.[5]

But it also has consequences for the ocean engine. As the wind blows across the great expanses of the ocean surface, it is constantly donating energy to waves. The waves lose that energy either gradually over many miles or suddenly when they break. We associate breaking waves with the coast, but most of them are out in the open ocean, providing a route for wind energy that has been converted into waves to shift back again into a tiny amount of ocean heating. The shape of the ocean surface is a temporary energy store, one of many such reservoirs that are constantly being filled and emptied as energy trickles through the Earth system. Munk and Sverdrup's work formed the foundation of today's wave and swell predictions, although they both moved on to other fields of study.[6]

The top of the ocean is not flat

If you somehow switched the wind off all around the world so that there was no wave generation, the ocean surface would still have a very distinctive shape. These patterns are too subtle to be seen directly by the human eye, but measurements of the

[5] Tide predictions also played a critical role here, informing the military that only 5, 6 and 7 June would have suitable tides. That story is told in Hugh Aldersey-Williams's book *Tide*.

[6] During my first month or two at the Scripps Institution of Oceanography, I was invited to dinner by another British scientist who worked at Scripps. They'd also invited a neighbour who was very enthusiastic, and encouraged me to tell him all about my work on ocean bubbles even though I was only a beginner. The neighbour turned out to be Walter Munk, then nearly ninety, who had been thinking about ocean waves for more than twice my lifetime, and yet had still listened with respect, encouragement and interested questions when I told him about my ocean ABCs, without ever dropping a hint about his vast knowledge or making me feel uncomfortable.

surface shape from precise scientific instruments give us a whole new way to know how the ocean engine is turning. The biggest features are attributable to the variations in gravitational pull from different rocks underneath, and they're astonishing to contemplate. There's a 54-metre-high dome in the North Atlantic Ocean surface that's just over a thousand miles across. There's also a huge hole 94 metres deep just to the south of India. No one sailing across those areas would ever know – you need satellites to spot fixed features like these – but the ocean is full of them. Mapping of gravity strength and sea surface height can actually be used to work backwards to find the shape of the sea floor; if there's a big underwater mountain, there's likely to be a small bump in the ocean surface just above it, as water is gravitationally attracted towards its rocky bulk. This ocean surface shape is independent of waves and currents and weather, and it's called the geoid.

But when you switch the wind (and therefore also ocean currents) back on in our hypothetical situation, even more lumps and bumps and ridges will appear, mostly less than 1 metre high. These irregularities are far broader than waves – many miles across – and the waves ride on top of them. Some of them are directly caused by the weather. The centre of a big rotating storm in the North Atlantic might well have an atmospheric pressure at the centre which is around 4 per cent lower than the atmosphere outside the storm. That pressure drop is responsible for a significant reduction in the downward push on the ocean surface, which will bulge upwards, forming a dome which is 40 centimetres higher than its surroundings, and which travels along with the storm. It's not a wave or a tsunami, just a region of ocean which bulges upwards. A bulge of this kind can make the damage caused by waves at a coastline significantly worse, especially at high tide, because the higher sea can reach further inland. And as great currents flow, they also cause ridges in the ocean surface, as the forces tugging sideways on them are balanced by water piling up. This is incredibly useful to ocean scientists, because it means that if you make very detailed satellite measurements of sea surface shape,[7] you can work out where the great (and lesser) ocean currents are flowing. Specialized devices on satellites called altimeters monitor these changes to give us a critical insight into what the ocean engine is doing today or this week.

The surface matters, then; but the overall structure of the ocean engine is dictated by the shape of the sea floor. So let's leave the surface behind and fall downwards through the darkness, until we land with a gentle bump on the bottom of the ocean.

Bottom

Possibly the greatest tragedy of studying the ocean is that the physics of seawater forbids the grand panoramic views of the sea floor that would give our imaginations a framework to anchor our understanding. Light is absorbed or scattered within a few tens or hundreds of metres, so even if you could floodlight a large area, you could never gaze into the distance and *see* where you are. Astronauts experience the overview effect, the intense emotional experience that goes with seeing the Earth from space and themselves as tiny specks in orbit around it. We can imagine ourselves in their shoes because we have the photographs they take, and we can immerse ourselves in the visual feast of the curvature of the Earth with an International Space Station solar panel in the foreground. Hundreds of humans have made the trip to the deep sea and they too have brought back astonishing photographs. But the deep-sea images are close-ups: detailed and fascinating rather than grand and

[7] Modern satellites can routinely measure changes in sea surface height of around a centimetre.

perspective-shifting. So the voyage to the lower extremity of the ocean engine is blank in our minds, although this is a dramatic journey to an alien part of our world. What are we missing out on?

'It's like falling through the stars.' Professor Deb Kelley is sitting in her office at the University of Washington, describing being encapsulated in a submersible descending into the depths. 'You're in a small sphere, it's dark, and it's relatively uncomfortable. And it's super-intense. You don't have any sense of direction and you don't have any sense of speed. I spend a large part of the time on the descent with my face crammed up against the window, looking at the bioluminescent critters – I have no idea what they are. And then about 100 metres off the bottom, they turn the lights on. And out here [the coast near Seattle], you forget you're in the water because it's so clear. It's like flying. But it's really intense, and that makes it phenomenally exhausting.'

Deb is a marine geologist who has done more than fifty dives in the deep-sea submersible *Alvin* since the mid-1990s. Her main expertise is in ocean volcanoes and hydrothermal vents, and she has seen with her own eyes what the deep sea floor has to offer. She is softly spoken but enthusiastic, and her office is littered with pieces of rock, including an irregular black glassy chunk about the size of a shoebox with a perfectly circular hole drilled straight through it. She tells me that this is part of a lava flow, taken from the top of an underwater volcano. It sits on the desk in front of me while we talk, glinting in the sunlight, begging questions.

Even after all her decades of studying and visiting the sea floor, Deb gives the impression of still being amazed by what's down there. She describes a submersible pootling along huge plains that stretch away into the gloom, incredibly flat – 'It's like driving across the floor' – and then the shock of the searchlights meeting a sudden wall, tens of metres high, which marks the start of a volcanic subduction zone. In tectonically active regions, there are arches and columns jutting up from a jumbled and rough base. The remains of lava channels look like emptied rivers. Most dramatic of all are the crown jewels of the darkest depths of the ocean, the deep-sea vents. 'They have such a startling amount of colour,' Deb says. 'There's bright purples and blues and white, and chimneys that are so covered in animals that you can't even see the rock.' She keeps coming back to the speed of change in these volcanic areas, and how different the same site looks if you return to it even a few months later. It is the job of science to use quantitative evidence to develop understanding, but it is tales like these – the experience of the deep sea – that has really changed our opinion of it.

It's easy to forget how recently our mental image of the deep sea was of an empty canvas: a yawning hole of nothingness. Thrilling stories of sea monsters kept ocean storytellers in business, but the ocean itself was less of a place and more of a threat. It's less than one hundred years since the first voyagers returned from this mysterious realm. Their tales made the deep ocean a real place, and suddenly that empty canvas had a character and occupants and mysteries made of substance rather than fantasy.

The first human deep ocean explorer

Off the coast of Bermuda, 400 metres down, the ocean is dark. This isn't the complete absence of light that feels like sensory deprivation, but rather an inky blackness smudged with just enough of a glimmer to emphasize the darkness filling the space around it. The sunlight has been sucked away – but this spot is still only a tenth of the way to the sea floor below. The only sound is the occasional cry from a distant whale, and there is no way to tell what time of day it is. The dense water carries silent shreds of life, pulsing gelatinous bells that occasionally produce colourful flashes of light, and the minuscule detritus of surface death drifting downward through the darkness. A dragon fish slides by, a long dark thin ribbon with teeth. Days and years pass. But on 3 June 1930, a small steel sphere appears suddenly, suspended from a slender cable like a spider hanging from the ceiling of a cathedral. The sphere contains fragile snoopers from the sunlight above: William Beebe and Otis Barton, who would be crushed by the pressure were it not for their metal bubble.

Beebe himself understood exactly how small and vulnerable they were, emphasizing the complete isolation he felt when he later wrote that they 'dangled in a hollow pea on a swaying cobweb a quarter of a mile below the deck of a ship rolling in mid-ocean'. The popular image of frontier-busting explorers is one of strength and activity, of athletic figures straining every muscle and brain cell to overcome the obstacles in their way. But Beebe and Barton dangling in their cramped hollow pea were the first of our species to see deep-sea creatures in their own habitat, the first to see the deep ocean rather than just fish things out of it, and the first to describe how it felt to be a land mammal cast fully into this alien world of seawater. Their obvious vulnerability and their lack of the ability to do anything other than look is perhaps a more honest way to enter a new environment: not as masters with the tools to modify their surroundings, but as wide-eyed tourists soaking it all up and hoping for the best. In contrast to the analytical, scientifically justified expeditions that were filing away the vital statistics of the ocean with the backing of important institutions, Beebe and Barton were there just because they thought it was possible and they wanted to see for themselves. Beebe was already well known as a naturalist and writer who shared not only his enthusiasm for the natural world, but also the fun and adventure of learning about it. Barton was the engineer who had designed the steel bathysphere which would become famous the world over as the key character in Beebe's book

Half Mile Down.[8] It's a work full of wonder, punctuated by the twists and turns of a genuine adventure. Together, Beebe and Barton set a world record for the deepest dive in a submersible (923 metres in 1934) but they also generated huge interest in the deep sea and were credited by many ocean biologists of the 1950s and 1960s for providing the initial spark that led to incredibly satisfying careers. It wasn't the vastness of undersea mountain ranges that captured the public's imagination; it was the fleeting encounters with weird animals only a metre from human eyes, only partially illuminated by a searchlight before they swam off into the dark.

Half Mile Down starts with Beebe's vision of the future, one where people living on the coast will pop on a diving helmet and swim out to inspect submerged ocean gardens which are an extension of their land gardens as easily as they pop out to the shops. In this shiny imagined future, personal access to the ocean will become a normal part of life, with sea anemones growing in underwater allotments for competitions at the local show and artists waxing lyrical over the quality of underwater light for their underwater paintings. Modern scuba diving has made some of this possible for the lucky few, but ninety years later, the training, ocean access and equipment are still prohibitively expensive for most. Many early science-fiction writers would be disappointed with our progress – we're well into the 2020s and it's still the case that most of us will never have direct experience of either outer space or the deep sea. But what we do have is a completely different outlook on our planet from those

[8] *Half Mile Down* is well worth reading, partly for the adventure but mostly for the honest description of the realities of the bathysphere: 150 cm in diameter, and made of steel 3 cm thick. It had three windows (although only two were used) and the entrance was through a circular hole 36 cm across. This carried the two of them for several hours, frequently filled with condensation, and occasionally leaked a bit.

early ocean optimists, and it came from slow, painstaking collection and analysis of data. A large part of the process of science is the careful accumulation of measurements, and so it's often thought that the outcome is 'facts': things that are considered true.[9] But the real impact of science, often ignored and at best underestimated, is the *perspective* that comes from the interpretation of those measurements. In the mid-twentieth century, the shape of the deep sea was the subject of a huge debate, controversial because of the massive change in perspective that seemed to be on the cards.[10] With that new perspective came new questions, and a different way of thinking about sea-floor shape.

The ocean doesn't just rest in a few linked hollows where the rocks happen to sit a bit lower than those around them. The shape of the sea floor is created by what those rocks are doing and what they're made from, and over most of the ocean that's fundamentally different from what is going on underneath the land. Humanity's curiosity has long generated an insatiable desire to poke things to see what happens, and in the 1960s this led to an audacious attempt to poke a hole not just into the sea floor, but all the way through it. While NASA was gearing up to send the first Americans outward into space, geologists were planning a vast scientific exercise in planetary navel gazing: the Mohole.

[9] Of course, 'facts' aren't the point: what the scientific process gives you is the least wrong interpretation based on the best available data at the time. It's all subject to modification, although the scale of the necessary modifications tends to shrink over time as any given hypothesis is tested in more and more varied ways.

[10] An excellent account of the background to this debate and some of the reasons why it happened relatively late in the history of science is given in Naomi Oreskes' book *Science on a Mission*.

The deepest hole in the world

It was Walter Munk's idea. In 1957, geologists were edging ever closer to the evidence that would finally end the debate about whether continental drift was real. The idea that the vast and very solid continents might have spent their history wandering around the planet had been around since 1596, but had been relatively easy to dismiss because, frankly, it seemed bonkers. What could possibly move a continent? And yet the evidence had been accumulating for decades that perhaps this wasn't even an exceptional event, but just a normal part of what continents do. It had already been established that the Earth has layers, and that the thin top layer – the crust – and the next layer down – the mantle – were made of quite different things.[11] It was also known that the Earth's crust comes in two quite different flavours, and we now know these as the continental crust, which is on average 35 kilometres thick and made of rocks like granite, and the oceanic crust, which is only 5–10 kilometres thick and made of denser rocks like basalt. Both the continental crust and the oceanic crust are floating on top of the mantle, but because the continental crust is thicker and more buoyant, it sticks up further and also goes down deeper. By contrast, the oceanic crust is delicately parked on top of the mantle, and is far less conspicuous than the continents, partly because it's thin and reasonably flat, and partly because its surface is almost always a

[11] A tectonic plate is a rigid section of the Earth's surface which moves around. Confusingly, this isn't the same as 'crust', even though they're commonly mixed up. The 'crust' and 'mantle' are different because of their chemical characteristics: they're made from different types of rock. But a tectonic plate is normally made up of the crust in that area plus the solid part of the underlying mantle – so a tectonic plate includes all the solid bits. Beneath the tectonic plates, there is ductile mantle which can flow slowly. Basically, where you put the boundary depends on whether you're a chemist or a physicist. The aim of the Mohole was to get into the mantle – to get down to a different type of rock.

few kilometres lower than the surface of the continents and water fills up the lowest spaces first. The deep ocean is deep because it sits on top of oceanic crust, and these great ocean basins are called basins because they really are like a bathroom sink in a countertop: a significantly deeper area that keeps the water in it.

In 1957, Munk was in a meeting at the National Science Foundation, reviewing proposals for future scientific work. The ideas on the table were all perfectly competent and well thought-out, but he thought they seemed . . . rather unadventurous. There was a discussion about whether there were much bigger, bolder scientific ideas that could be funded, the sort of experiment that would divide the history of Earth science into 'before' and 'after'. Walter suggested a direct investigation of the innards of our planet by digging the deepest hole that had ever been dug, one that punched right through the Earth's crust so that samples of the mantle below could be retrieved. At that time, almost all knowledge of the interior of the Earth came from logical deduction conducted at a considerable distance from the business end of things, and the thought of being able actually to touch the rocks involved was intoxicating. It would be an amazing test of the new geological theories, and it was certain to attract a lot of public interest.

High-profile geologists and oceanographers liked the idea, and began to push for the necessary funding.[12] But no one could deny the scale of the technical challenge. Digging deep holes is difficult, mostly because it's hard to keep the hole straight and to move drill bits up and down from and to the cutting face. The easiest way to reach the mantle must be via the shortest hole, and that meant digging through thin oceanic crust rather than

---

[12] It didn't hurt that the Russians had also been discussing deep drilling, since US politicians certainly weren't immune to the temptation of 'winning' both the Space Race and the race to the depths of the Earth.

thick continental crust. This not only had to be the deepest hole ever dug, but it had to start in a place no human could touch. The entrance to the bowels of the planet would be on the deep sea floor. The name of the boundary between the crust and the mantle is the Moho,[13] and so, inevitably, the vertical tunnel that crossed that boundary would be the Mohole.

The project had a very encouraging start. Preliminary holes were successfully dug in 1961, using an experimental ship capable of drilling into the sea floor 3,800 metres beneath it. There was indeed great public interest in this first phase: John Steinbeck (who won the Nobel Prize in Literature the following year) was sent to cover the event by *Life* magazine.[14] Even to dig those early holes, it was necessary to find a way to keep the ship floating at precisely the same location while the wind, currents, waves and tide tried to move it around. To do this, engineers installed four thrusters and used them to make delicate adjustments to ship position, a system still used today and known as dynamic positioning. The drill bits munched through the thick layer of sediment that coated the deep ocean floor and reached into the rock below, bringing back fresh samples of oceanic crust for the first time. The researchers even got a telegram from President John F. Kennedy, lauding the 'historic landmark in scientific and engineering progress'. Phase One was a winner.

---

[13] Named after the scientist who discovered it in 1909, Andrija Mohorovičić. He realized that the seismic data from earthquakes showed that there was a reasonably sharp transition in density at the bottom of the crust, which was clear evidence for the existence of two layers of different material. The full name of this transition region is the Mohorovičić Discontinuity, but because no one wants an eleven-syllable name for a simple concept, it became the Moho.

[14] His article pulled no punches when describing the drill ship as having 'the sleek race lines of an outhouse standing on a garbage scow', but it's a beautiful description of the nerves and excitement on board a ship with a truly historic mission. A 'scow' is a flat-bottomed boat.

But the Mohole never became reality. Phase Two of the project was quickly mired in cost increases, project management problems and bickering about the priorities and who was in charge. Public interest morphed into derision. With the costs of the Vietnam War looming, the US Congress finally pulled the plug in 1966. At the time of writing, fifty-six years later, the Moho still remains unpunctured by humans, although several current projects are getting close.

On superficial inspection, Project Mohole may appear to be an expensive (although entertaining) boondoggle. But it spawned international and collaborative ocean drilling projects that continue to this day, working from the specialized drilling ships *Glomar Challenger*, *JOIDES Resolution* and others. Those vast flat expanses of the sea floor turned out to be some of Earth's most detailed history books, stretching back 'only' 200 million years (on a planet that's 4.5 billion years old), but holding undisturbed treasures for those who can reach them. Sea-floor drilling opened that door, and spawned whole new fields of science. The invention of dynamic positioning also gave the oil industry a kickstart when it came to offshore oil exploration in shallow seas, which would make up 30 per cent of total oil production by 2011.

A lumpy bottom

Half a century on from the Mohole attempt, we have a clear picture of the overall shape of the bottom of the ocean. Seeing the pattern starts with the continents, the ancient big beasts of the plate tectonics world. If you stand on the west coast of Ireland and look across the Atlantic, the rocks that make up the seabed a few hundred metres out to sea will be pretty much the same as the ones you're standing on. This is part of the continental crust, and it has a pleasing symmetry to it – where it thickens so that great mountain ranges reach up towards the sky on the top, the base also intrudes further downward into the mantle below, forming a bulge on both sides. Continents are what most humans spend most of their time standing on. Continental rock can be billions of years old, and although continents can collide, split apart and rearrange themselves, the continental crust itself is just a passenger. As you gaze out to sea from the coast, you're looking at an ocean that's so full that water is spilling over on to the continents, like a swollen river flooding on to the towpath. The depth of the sea near the coast is just the depth of that over-spill, usually not much more than a couple of hundred metres. These shallow flooded bits of continent are known as continental shelves, and at some coastlines these shelves can stretch many hundreds of miles away from the shore. They're important coastal regions, often home to lots of life, but they don't actually contain that much water.

To find most of the ocean, you have to sidle up to the edge of the continent, the edge of the shelf, and then the sea floor will start rolling away from you, deepening until you reach the deep flat floor of the ocean basin. Underneath you now is oceanic crust, the dense volcanic filler that seeps into the gaps left by the continents as they move apart, and the depth of water here is suddenly 3,000 metres or more. Now you're on the abyssal plain, which covers more than half of the Earth's whole surface. This is one of the ocean basins, the deep, flat-bottomed expanses which contain almost all the water on Earth. These basins fit between the continents, and so we've given them different names: the Atlantic, Pacific, Indian, Southern and Arctic; but there is only one connected ocean. It might look as though it's segregated, like the hollows in the trays that used to be used for airline food, but the water doesn't care. The Pacific basin alone contains 52 per cent of all the water on Earth, but there are no boundaries between the water that happens to be in that basin and the ocean anywhere else.

As the processes of plate tectonics move the continental

crust around, it is the thin, flattish oceanic crust that must accommodate most of the changes. The approximate shapes of the continents are fixed as they trundle around, but the ocean outlines are constantly changing. And so we come to the nature of the deep sea floor, not just a technical description of its shape, but what it's *like*. And what it's like depends on what it's doing. Down there in the depths, some parts of the floor of the great ocean basins are very busy indeed. And since 2014, ocean scientists have been able to spy on that activity in real time.

Drama in the depths

Pacific City is a small coastal town about 200 miles south of Seattle, on the west coast of North America. After Native Americans were driven from the area by intruders, imported disease and forest fire in the first half of the nineteenth century, the rich fish stocks offshore attracted non-native pioneers. Overfishing was followed by a shift to tourism, and today more than half of the properties in the area are second homes. Its defining feature above ground is the stunning rocky coastline, perfect for peaceful contemplation of the ocean waves rolling in from the Pacific. But underneath the sleepy beach, it's a different story.

Two buried cables are swarming with constant electronic activity: an 8 kilovolt power supply and fibre-optic connections capable of transmitting 240 gigabits of data per second. That's enough to transmit about 900 two-hour HD movies every single minute. The cables run under the beach and out to sea. One crawls 60 miles out across the continental shelf at a constant depth of around 300 metres, and then drops over the bumpy edge of the continent to reach the vast flat abyssal plain 2,900 metres beneath the sea surface. Computer commands accompanied by power keep rolling westward, interrupted briefly at a chunky yellow metal box and then onward until they reach an underwater mountain chain 300 miles offshore. The ultimate destination is two clusters of sensors, one at the foot of the mountain chain and one near the top. The second cable from the beach hugs the continental shelf near the coast, linking up three other sensor clusters. This web of more than 150 sensors forms the eyes and ears of the Regional Cabled Array,[15] a surveillance system in the deep, and it's here because this area is a microcosm of the whole deep ocean floor. Since all tectonic plates are moving, the boundaries between them are exceptions to the general rule that the oceanic crust is reasonably flat. There is no way to move the eggshell-like pieces of crust around a sphere without altering the shapes of those plates. And so these boundaries are places where vast geological forces insist on change. Plate edges grind against each other or pull apart, allowing mountain ranges made of underwater volcanoes to grunt and vomit lava into the gaps, or give up old sea floor to be tugged down into the mantle, losing its final confrontation with a thick continental plate. The Regional Cabled Array can watch almost all of these processes in real time, all within a few hundred miles of the North American coastline.

An ancient piece of oceanic crust has been gradually sliding eastwards underneath the western side of North America for 200 million years. The Earth's conveyor belt has almost finished recycling it, but a last tiny portion is still to be dragged down and gobbled up, and that piece is the Juan de Fuca tectonic plate. It's a triangular fragment, with an active spreading zone on one side and the subduction zone at the continental edge. The mountaintop sensor cluster of the Regional Cabled Array is sitting on top of the most active subsurface volcano site in the North

[15] Funded by the National Science Foundation, and part of the Ocean Observatories Initiative.

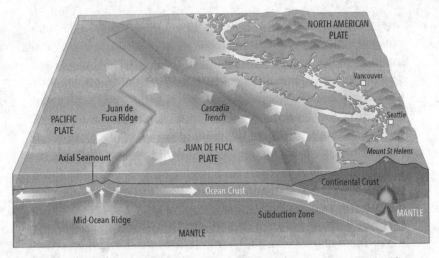

The Juan de Fuca tectonic plate underneath the eastern Pacific Ocean.

Pacific: the Axial Seamount. Deb Kelley is the Director of the Regional Cabled Array, and it's her job to manage this complex network of scientific tendrils that now cling to the geological behemoth beneath the waves, tracking its every move. The problem with the drama of underwater geology, she tells me, is that you're almost never in the right place at the right time to see it, so the plan for this network is 24/7 monitoring, with data whooshing back down the cables to the shore in real time and a system of alerts for the scientists on watch. Humans have not yet seen a mid-ocean ridge erupt on video, and yet this is the process that has built, and is continuing to build, the entire ocean floor. Next time the Axial Seamount erupts, the eyes of the world will be on it.[16]

[16] The RCA isn't just following the geology. It's also tracking ocean chemistry and biology, which is letting scientists study how deep volcanic activity affects the ecosystems above it.

In Ernest Hemingway's novel *The Sun Also Rises*, a character describes how they went bankrupt: 'Gradually then suddenly.' The same could be said of the constant reshaping of the deep sea. The gradual changes are the slow shunting of the whole plates around the globe, at speeds of a few centimetres per year. But what do the sudden changes look like? The Axial Seamount sits on a straight edge between two tectonic plates that are pulling apart at 6 centimetres every year, and when it erupts it produces the new rock needed to fill the gap. In 2015, just as the Regional Cabled Array was getting up and running, it had its first real test. The instruments lit up with thousands of tiny earthquakes, too small to be detected anywhere but at the seabed, and then the solid sea floor on which the sensors sat dropped by 2.4 metres. This was clearly an eruption, but most of the activity was on the opposite side of the ridge to the sensor packages. When things had calmed down, the geologists took a remotely operated vehicle or ROV down to look, and found a huge new black glassy lava flow 127 metres thick. It's a piece of this that is sitting on Deb's desk. The new rock was covered with microbial mats that had popped up to take advantage of the new situation in less than a month, a complete surprise to the scientists. Deb emphasizes the speed of change: the lava lakes, flow channels and steep rocky walls created and resculpted in the blink of a geological eye, gradually then suddenly. Taking an ROV down for a first look is still like opening long-awaited birthday presents . . . every time, beautiful surprises turn out to be hiding underneath the protective cover of the ocean.

The variety of these plate boundaries is part of their appeal. Further down the flanks of the Axial Seamount, an ocean plumbing system is creating a different kind of drama. The new rocks created as the tectonic plates move apart don't form a smooth and solid foundation. They're full of cracks and fissures which function like messy pipework: drawing seawater deep down into the hot raw rock, heating it, leaching minerals into it and then spitting it back out into the ocean. This incredibly hot water, laden with compounds of iron, sulphur and a wide range of other elements, pours back out into the cold and calm ocean to form hydrothermal vents. These are the famous black smokers and white smokers, which sometimes build towering chimneys tens of metres high around themselves as the metal compounds carried by the water are dumped in the cooler surroundings. The extensive swarms of weird life that cover these vents get almost all the attention of human observers, but Deb is keen to show me something else. She goes over to her computer to find a video from one of the vents she has dived on. 'We see upside-down waterfalls,' she says. I don't understand what she means at first, and it takes me a few seconds to process the video as Deb keeps talking. 'In those vertical chimneys, the walls crack and then hydrothermal fluids come leaking out and you get something that looks like half a toadstool growing out of a tree in an old-growth forest.' And suddenly I see it. This is a gigantic hydrothermal chimney looming out of the darkness, and hot water is indeed leaking out of its side. But because hot water is less dense than cold water, the hot water keeps flowing rapidly upwards. Where it first hits the cold water, it's clearly dumped some minerals and made a ledge that sticks out – that's the toadstool shape that Deb is referring to. The water following afterwards has had to flow outwards underneath the ledge before it can carry on upwards. But the ledge has developed a hollow on its underside like an upside-down bowl, so there is a pool of hot water there, held in the hollow as if it were filling up the inside of an umbrella. The boundary between hot and cold water shimmers like a mirror. And then the hot water is spilling out of its hollow and continuing upwards into the gloom. It really is an upside-down waterfall. 'I think they're beautiful,' says Deb. 'Who would have thought that you'd get something like that?' My first reaction is that this is why you need experimentalists and field scientists: to bring back evidence that the natural world has far more surprises up its sleeve than any of us could imagine.

These long, jagged mountainous scars zigzag around the global ocean, marking the underwater divisions between the seven major plates that cover the Earth, along with a swarm of smaller plates filling the remaining gaps.[17] Each one rumbles along with its own dramas and growth spurts, for the most part completely hidden by the ocean. There are places where these ragged seams intrude into the non-ocean world; Iceland is a part of the Mid-Atlantic Ridge, where the North American and Eurasian plates are pulling apart. Iceland's earthquakes and volcanoes give us a hint of what's hiding in the deep, but the outcomes are different without a blanketing layer of water to absorb heat and provide inertia. Either way, these geological construction and destruction sites are raw and harsh, fickle and shape-shifting. If you look away and then look back, something different will be waiting for you. But not all of the deep sea is so busy.

A field of sea potatoes

We humans have a deep and visceral respect for scale, both the colossal spatial scales of the Milky Way or huge mountain ranges, and also the giant timescales of ancient trees or Stone Age monuments. There is calm and reassurance in permanence, in slowness, and the liberating relief of a refuge from the complex and fleeting details of our messy lives. Putting ourselves in perspective is almost like a drug, but instead of inserting chemicals into our body, it involves mentally inserting our body into

[17] The seven major plates are the African, Antarctic, Eurasian, Indo-Australian, North American, Pacific, and South American.

the universe and then standing far enough back to be amazed at the comparison. Old trees, stars and monuments broadcast their status in rays of light so we can see their scale and feel their age, but the largest and oldest wilderness on Earth is hidden inside thousands of miles of darkness. If you crave stability, this is the place for you. Across the immense extents of the ocean abyssal plains, the outlines of the scene have hardly changed for millions of years. But that doesn't mean that nothing is happening here.

In the eastern Pacific, across the thousands of miles between Hawai'i and Mexico, the abyssal plains are not perfectly flat, but roll very gently up and down shallow ridges. I've referred to the abyssal plains as 'flattish', but many of them carry giant pimples many hundreds of metres high known as seamounts. The Axial Seamount is an active volcano, but after activity has ceased, the mound of lava continues to sit on top of the oceanic crust, poking up into the rest of the ocean. These seamounts are sprinkled across the abyssal plains, and they're surprisingly common.

The sea floor itself is soft and muddy, but its most distinctive feature seems conspicuously out of place: a single layer of hard, dark nodules, each one about the size of a potato and separated from its neighbours by only a few centimetres. This layer of nodules stretches out into the distance, for tens, hundreds, thousands of miles. The mud is an accumulation of the meagre leftovers from the ocean above, the dregs of surface life that somehow avoided the clutches of microbes, jellyfish and other hungry organisms as it fell more than 4 kilometres to the sea floor. Out here, in a region known as the Clarion-Clipperton Zone, it takes more than a thousand years to add one extra centimetre of mud. The age of the nodules makes that seem like the blink of an eye – each one of these metallic potatoes is well over a million years old. And clinging to or crawling through this ancient oceanscape, there is life. Some of it is very odd indeed.

'It's such a privilege, to see things that are totally unique and no one's ever seen before.' Dr Adrian Glover is a deep-sea biologist at the Natural History Museum in London, and we're sitting in his office several floors above the museum's extensive public galleries. Bright winter sunshine is streaming into a space full of labelled boxes and jars and vials, illuminating creatures that lived out their lives in the pitch black of the abyssal ocean and are now preserved for posterity in the museum's collection. Some are the size of a banana, but most are so small I have to ask to check I'm looking at the right thing. They're white or brown or both, and I wouldn't have been able to put a name to a single one of their alien bodies. In life, the nodule-covered plains were their home, 4,500 metres beneath the ocean surface.

Adrian fishes out a plastic box and carefully unwraps a grey knobbly lump that's about the size of my fist but slightly flattened. It's smoother on the bottom than the top, and parts of its surface are covered in a light brown crust. This is a polymetallic nodule, one of the strangest things in the deep sea and also one of the most mysterious. They are found to cover areas of the sea floor which meet a critical pair of conditions: a depth between 3,500 and 6,500 metres and sediment that accumulates at a rate of less than 1 centimetre every thousand years. These sound like strict constraints, but they are met over a significant fraction of the abyssal plains. This nodule started out as a hard fragment of detritus, possibly a shark's tooth or a fish earbone which settled to the seabed, and then over the long quiet aeons, an incredibly slow coating process began. Different chemical processes got to work on both the exposed surface of the nodule and the base buried in the sediment, scavenging scarce metal atoms from the surrounding water and depositing them on the proto-nodule, adding a layer perhaps only 100 atoms thick every year. In the dark, as species on land came and went and as continents meandered across the planet, more atoms were added in complex arrangements: mostly manganese, but also iron, silicon, aluminium, nickel, copper and cobalt. It may have been slow, but it wasn't tidy – sometimes mud layers are incorporated too, making the whole nodule more fragile.

On the basis of its size, it's likely that the nodule sitting on Adrian's desk is several million years old, with an inside layered like an onion. That puts its origin long before the emergence of our species *Homo sapiens* (around 300,000 years ago) and probably also before the appearance of *Homo erectus*, the first species in the *Homo* genus.[18] I ask whether it's been the same way up for all of those years, and Adrian tells me that almost all nodules appear to have been turned over at some point, although it's not clear how.[19] It's possible that grazing sea cucumbers (accurately named, as these animals are pretty much the size and shape of a domestic cucumber) occasionally nudge one over while grazing on them, but it's hard to be sure. What is clear is that these nodule fields support a very distinctive ecosystem, and it's Adrian's job to study the life that lives on and next to them.

And that's not a straightforward task. No one has managed to grow any of these animals in lab tanks, so it's hard to find out what they eat or need or like, and that's even if you successfully bring one back from the sea floor. The team here has just returned

[18] Adrian noted that in the nodule exhibition in the museum, there is a nodule with a Megalodon tooth at its core. The Megalodon was a huge species of shark, now extinct, that lived 23–3.6 million years ago, making this nodule at least 3.6 million years old.

[19] One of the other unanswered questions is how the nodules stay on the surface of the sea floor. If the sediment falls from above at about 1 cm every thousand years and the nodules grow at a few millimetres every million years, why don't they just get covered over with sediment? It could be something like the Brazil nut effect (which is responsible for the largest nuts in a bowl rising to the top when you shake it) or the stirring of biology providing just enough movement to keep lifting them up, but no one knows.

from a research expedition to the Pacific, spending time on board a ship floating in the open ocean, a long way from the subject of study. The best tool available to a deep-sea biologist is an ROV which can scoot around the sea floor, controlled by engineers and scientists on the ship above. Tucked in among the samples and papers on Adrian's desk are substantial data drives storing hours of video taken by ROVs as they explore the sea-floor wilderness. The view is of nodules and more nodules, with occasional gaps where there's a small muddy bump, and then even more occasional sea stars,[20] sponges, sea cucumbers, shrimp and fish. When one of those appears on the video, sometimes a metal arm reaches out from behind the camera, picks it up and deposits it in a sample container to be brought back to the Natural History Museum. This is the closest that a deep-sea biologist can get to the domestic arrangements of the life they study: sitting on the ship 5 kilometres overhead, directing the ROV operator to pluck out a sample for later inspection. It's taken years of patient study to even be sure about the basics, so what have we learned? Adrian gestures at the clear space on his desk, a square about 50 centimetres on each side.

'In an area like that you'll probably get 20–30 animals, which is a really low abundance compared with the floor of the Antarctic or the North Sea. And 5 per cent of those animals are relatively common and we see them everywhere. The other 95 per cent we've mostly only got one individual of. You have a system where you have lots and lots of unique singletons. One of the

[20] You have probably been told that they're called starfish but this name really winds biologists up, largely because they're not fish. They're not even closely related to fish. So a 'sea star' is the proper name. Jellyfish, crayfish, cuttlefish, and silverfish also aren't fish, but no one is arguing that their names should be changed. Not yet.

things that drives us mad is that the abundance is so low, but the diversity is so high – it's a critical question.'

It's clear that there are a lot of things down there that both frustrate and fascinate the biologists. There are also microbes inside the nodules, but no one knows how they get enough food, tucked away from the water outside. There are animals growing on top of the nodules, possibly using them as a stepladder so that they can scavenge food away from the sea floor. No one knows how old any of these animals are or how long they live, but they are all living in a very slow system where there is plenty of space and a pitiful amount of food. It appears to be completely dark, but Adrian shows me a photograph of a freshly recovered nodule with a small worm tucked in a crevice. The worm is (or was) peering out at the world – it has eyes, which implies that there must be something for them to see. But what? We go back to the video, and then into the field of view comes the weirdest of them all: a xenophyophore. It sticks up from the seabed like a hand, and it looks relatively rigid. But this relatively large organism is a *single cell*: a type of foraminiferan that sucks up minerals from its surroundings and uses them to build an exoskeleton to contain itself. It's only found down here, on the abyssal plains, and like all of its neighbours it's the source of far more questions than answers. The shape of these parts of the ocean floor – reasonably flat, very deep and well away from the churning ecosystems that might cover them up with detritus – is critical to the type of life that exists here.[21]

The majority of the water in the ocean engine sits over oceanic crust, making these deep basins essential for understanding the engine's workings.

---

[21] For more about the biology of the deep ocean, I recommend *The Deep* by Alex Rogers and *The Brilliant Abyss* by Helen Scales.

THE SHAPE OF SEAWATER

Why no one should ever compare the deep ocean to the moon

Before we move on from the shape of the deep sea floor, this is probably a good point at which to air my heartfelt frustration at one of the phrases we consistently hear about the deep ocean, one that is both categorically wrong and also misleading in a really dangerous way. Here it is: 'We know more about the Moon/Mars than we do about the deep sea.' Hearing it spoken is like fingernails down a blackboard to me. Versions of this sentiment have been knocking around since at least 1948,[22] and some of the current confusion comes from discussions of seafloor mapping. We've mapped the entire surface of the moon to a resolution (the distance between mapped points) of 100 metres,[23] but much of the deep ocean has only been mapped with a resolution of about 1 kilometre (although plenty of it is known far better than that). I love maps, but they're far from the only thing that matters. My basic complaint about this comparison between our knowledge of the moon and the ocean is that it's absolutely not true, for one very simple reason: the moon and the deep ocean aren't comparable. The ocean is a complex and dynamic system which is constantly changing; it's full of water which is also moving around and doing things, and it's full of life. The moon is important, but it is a dead rock that has barely changed for a couple of billion years. We absolutely know more about the deep ocean because there is *more to know*.

---

[22] 'Man's perpetual curiosity regarding the unknown has opened many frontiers. Among the last to yield to the advance of scientific exploration has been the ocean floor. Until recent years much more was known about the surface of the moon than about the vast areas that lie beneath three-fourths of the surface of our own planet' (*Submarine Geology*, 1948, by F. P. Shepard).
[23] The moon can be mapped from some distance away relatively easily, because you can bounce light off it.

Hundreds of scientists like Adrian and Deb have been sampling the deep sea, visiting it and constructing knowledge of it for decades, and it's insulting both to them and to the ocean to imply that we know more about a dead rock that has only (at the time of writing) been visited by twelve people for less than eighty-five hours in total, all during a three-year period half a century ago. The comparison is dangerous because we have all seen the photos from the Apollo missions and it's very clear that the moon is . . . quite empty. The framing of those photos has usually been chosen to emphasize the humans, representing human achievement, or the blue Earth in the background, highlighting just how special our planetary home is in the vastness of space. The moon itself is only ever the backdrop, never the focus. Continuing to make the comparison between the moon and the deep ocean implies that the deep ocean is like the moon: empty, unchanging, dead, just the wallpaper in the background of the Earth. It isn't. But if we continue to make that comparison, we damage the emotional relationship that we could have with the deep sea: the knowledge that it's a vast, changing and fascinating wilderness which is a special part of our planet. So instead, I think we should be saying something like this: 'Our deep ocean is so rich and dynamic that we have only scratched the surface of what there is to know.' We urgently need more exploration of the deep sea, because there is so much more to find out about this fascinating and special part of our world, and of course we need to explore the moon as well.[24] So this is my plea to give the deep ocean the respect it deserves, and stop, for good, comparing our knowledge of it to what we know about the moon.

---

[24] I'm not having a go at lunar scientists. It really is important to understand our only natural satellite, both for pure curiosity and also for what it can tell us about ourselves and our own planet. It's just that there's more to know about the ocean. Much, much more.

The character of the ocean in those deep basins is far more typical of what the ocean *is* than the coastal areas we're more familiar with. That's not to say that shallow coastal seas aren't important – they are. But the coasts themselves have a more specific and critical role, not just because they're where the top of the ocean meets the bottom, but because they're the places where the ocean connects with the land. Before we can look at what crosses that border and why, we need to ask ourselves what the nature of this edge is and why its influence on the ocean is so strong.

Edges

The first known owner of a globe was a teacher and philosopher called Crates of Mallus, who lived in Greece in the second century BC. In order to fit all of the detail on (even the detail known two thousand years ago), he recommended that a globe should be at least 3 metres in diameter – and with all the additional detail we now know, that can only be truer today than it was then. And the most important detail – the most important function of a globe – is the mapping of edges, displaying how the world is divided between land and sea. As civilizations grew and explored the Earth, kings and admirals paid huge sums for these ever more intricate and accurate spherical maps, partly because they were so obviously beautiful and partly because knowledge is power. If you go to a stately home or museum today, you'll see the globes that were a regular feature of school-rooms and drawing rooms in the nineteenth and early twentieth centuries, their wiggly coastlines getting closer and closer to those we recognize today. But until 1946, every single one of them was nothing more than a very educated guess. The position of every coastline was deduced using logic from thousands of measurements of the time and the sun's position, allowing accurate and useful maps to be drawn and shared. But it was only in 1946 that a V-2 rocket was launched from New Mexico in the United States carrying a motion-picture recorder packaged in enough armour to survive a crash landing. The photographs that came back marked the first time that anyone had ever actually seen the Earth from far enough away to see the shape of a whole coastline. By then, the existing globes were so well accepted that no one thought much about it, but the result was a huge endorsement of the scientific method. The coastlines that the first satellites saw matched the coastlines drawn collectively by generations of human navigators, verifying in an instant the product of centuries of painstaking work.

The confirmed shape of those wiggly coastlines is now displayed everywhere: on wall maps, place mats, mugs, beach balls, tea towels and a large marble that I have in my desk drawer. It's become so familiar that most of us don't really look at it any more. But some of the things that we take completely for granted are critical for the way the ocean engine behaves today. First of all, most land – two-thirds of the total – is in the northern hemisphere. That means that the southern hemisphere is 80 per cent water and 20 per cent land. Tilting the world over, we can see that the Southern Ocean (the ring of water around Antarctica) is the junction box of the world ocean, an uninterrupted connection linking the Pacific, Indian and Atlantic Oceans. This brings us around into the Pacific, an underappreciated ocean in much of the western world because it is split in half by most modern rectangular maps (to the immense frustration of Pacific peoples). The Pacific is *huge*: the part of the equator that crosses it is a full third of the entire circumference of the Earth. If you look at a globe from above the middle of the Pacific, you see almost no land at all. And then if we tilt the globe again to look down on the northern hemisphere, we can see that the Arctic Ocean is almost entirely enclosed by land, with small outlets into the Pacific (at the Bering Strait) and the Atlantic (between Canada and Norway). The general pattern is clear: the continents are big lumps, between which large spaces are left for the ocean to fill. The floor of these large spaces is mostly oceanic crust, and that makes the deep ocean basins deep as well as wide. But the ocean also has bottlenecks: particularly at the southern tip of South America, the islands between Asia and Australia, and at the entrances to the Arctic Ocean. These big ocean basins and the connections between them dictate how water can move around the world, forcing the hand of the ocean engine. But what of the edges themselves? Our culture still thinks of them as sharp lines on a map, but the reality can be much messier, and the boundary between the land and the sea is often fuzzy, with its own complex and distinctive character.

The in-between places

As we pick our way down a narrow path which crosses a grassy hillside while winding towards the sea, it feels as though we have stepped through a portal into an alternative version of the world. On the other side of Cornwall, a huge gale is bashing the coastline and the news is full of warnings of fallen trees and flying debris. But down here the wind is quiet, the sky is blue, and the sea is barely ruffled by waves. We pass the last hedge, and a band of grey appears: a 50-metre-wide ribbon of knee-high boulders that splits lush green land from charming turquoise sea. It does not look hospitable. It looks harsh and empty, unwanted, as though neither the land nor the sea has any time for it. But this is the home of a robust and beautiful ecosystem, one that plays by its own rules in the twilight zone between wet and dry.

The places where land touches ocean are often messy and ambiguous, defying attempts to draw a nice tidy line on a map dividing land and sea.[25] They're also changeable. In 2015,

[25] The most obvious ambiguity was made famous (although it wasn't a new idea) by the mathematician Benoît Mandelbrot. In 1967, he published a paper

Porthleven beach in Cornwall was completely emptied of its sand by a big winter storm, effectively vanishing overnight. The sand was soon redeposited, having reminded the locals not to take it for granted. The edges have everything thrown at them: they're thumped by heavy waves from the ocean and sometimes by rocks and soil falling off the cliffs on land. They can bask in hot summer sun at low tide, parched and scorched by damaging ultra-violet light, or they can sit underneath metres of water at high tide, weighty pebbles rolling around the bottom as they're shoved in and out by the surge of the waves overhead. As sea and rain alternate, they face salt water and then fresh water and then salt water again. Survival here demands extreme resilience. There is one group of organisms that stand out as the absolute masters of this environment: the seaweeds. And Tim van Berkel believes that the same features that help them survive out in this border zone could also help us solve many of the problems we are creating for ourselves on land.

We clamber over the dry boulders and pass the high-tide line, and then I start to see the dark clumps of seaweed adorning the still-damp rocks. We are here at the lowest of low tides, so six hours ago these rocks were covered with a few metres of water. Tim hops over the rocks and starts pointing out seaweeds, and each dark clump turns into a medley of colour and texture. There are large smooth fronds of mahogany-coloured kelp, each the size of my forearm. Deep red jumbles of delicate threads

---

entitled 'How Long Is the Coast of Britain?', using the example of the British coastline to develop his ideas on what later became known as fractals. The problem, as he pointed out, was that the coast of Britain (or anywhere else) doesn't really have a definitive length, because the measurement depends on the length of the ruler you use to measure it. If you use a big one, it strides across all the wiggles to produce the large-scale picture only, giving you a relatively short total length. But if you use a smaller ruler, you have to detour in and out of all the wiggles, which is inevitably a longer path. The smaller the ruler, the longer the path; so there is no single answer.

cling in the hollows next to shiny purple ribbons. Bright glossy green folds poke out from underneath, jostling for space with flat olive lobes, and all of this is anchored directly to bare rock. Some of the rocks are painted with rough pinkish splotches.[26] As we get closer to the sea, pools of seawater appear and the seaweed colours intensify as the seaweed is buoyed up by the water, showing the full glory of the underwater garden. 'It's a wilderness out here,' Tim says; 'we see birds overhead, and there are seals just behind those big rocks. It's an amazing place to spend time.'

He bends down to pick up a broad kelp frond, one that ends at an oddly straight edge halfway up its length. 'This one has been harvested recently, and it's already starting to grow back,' he says. This is what he has spent the past ten years thinking about. In 2012, Tim co-founded the Cornish Seaweed Company, whose purpose is the sustainable harvesting of seaweed. One of the striking features of seaweed is how quickly it grows – sugar kelp can grow 3 centimetres in a day, in the right conditions. They've been harvesting here for years, moving back and forth along a short stretch of coastline, and there has been no decline in seaweed populations or health. There's no need for fertilizers or pesticides, because the ocean looks after the seaweed. Harvesting is done by hand with a pair of scissors, only ever taking part of the seaweed and leaving the rest to flourish. Then it's dried out, flaked and sold as food. But the uses of seaweed don't stop there. Evolving to cope with their harsh environment has given them a range of habits that are very useful for us.

Seaweeds are 'macroalgae': algae that are big enough for us to see and pick up. Flowering plants can't cope with the salinity of ocean water, and so are almost all restricted to land. Along the

[26] The bright green folds were sea lettuce, the flat olive lobes serrated wrack, the delicate threads 'bunny's ears', the shiny purple ribbons dulse, and the pinkish splotches common coral weed.

coast their ecological niche is taken by the seaweeds, which are actually three barely related groups: red, green and brown seaweeds. All of them use the energy from sunlight to photosynthesize, building themselves from the raw materials in the water and providing food and a habitat for lots of marine creatures. They don't need roots because they take what they need directly from the water around them. This boulder-covered beach is perfect because seaweeds need rocks to cling on to – sand won't do. They must withstand the pounding of the ocean when the tide is high and then salt-encrusted desiccation when the tide is low, and everything changes four times each day. Tim points out the gradual changes in seaweed species as we get closer to the water: bladderwrack near the high-tide line, serrated wrack further down, mixed in with dulse and nori, and then a little way out to sea, never fully uncovered by the tide, is the sea spaghetti. The occupants of this narrow intertidal zone – the ribbon of coast between high and low tide marks – inhabit even narrower bands determined by how much air and ultra-violet light the seaweeds can tolerate.

I put one hand at either end of a big kelp frond and cautiously tug on it. It stretches. I pull harder. It stretches further. It's tough and rubbery, able to hold itself together while the top is yanked by the surge of the water. Healthy stretches of seaweed are important for many marine creatures, not just as food but because those stretchy tough fronds absorb wave energy, sheltering the areas underneath and behind them. The internal structure giving the kelp cells their strength contains alginate, a substance that we extract and use as a thickening agent in food. Agar and carrageenan are similar substances found in the red seaweeds,[27] and you'll find them in all sorts of modern food products. In this form, almost all of us eat seaweed every day.

---

[27] This is the same agar that is used on petri dishes in labs as a growth medium for microbiologists.

But in Britain we don't have the habit of eating the whole thing (although laverbread is still popular in Wales), something that Cornish Seaweed want to change. Tim hands me a small branching red frond, a fragment of pepper dulse. It tastes like tangy lettuce, not 'seaweedy' at all, and it's something I would love in a salad. Many cultures regularly eat seaweed and it's farmed all over the world.

Seaweed can't always resist the tug of the waves, so being able to regrow incredibly quickly is critical for survival. It's also why harvesting is easy to sustain – the harvesters only take part of the seaweed and leave the rest to start again from the cut surface. Its speed of growth means that seaweeds take in nutrients and carbon very rapidly, concentrating nutrients and minerals from their surroundings. Seaweeds are particularly famous for being a very good source of iodine, something which is abundant in the ocean but rare on land.[28] The only downside of us all eating more seaweed that Tim will admit to is that if you ingested huge quantities, the consequent iodine excess would cause problems. But for most of us, a bit of extra iodine would just act as a useful supplement – we need it to stay healthy, and it has to come from our diet. The seaweed Cornish Seaweed sells goes in soups and salads, quiche and curry, an addition that melds well with our more familiar ingredients. Seaweed has also been used as a natural fertilizer, to make edible packaging, for skin care products, cattle food and in many other ways besides.[29] As we poke around in the tide pools, uncovering more seaweed species as well as the marine life that feeds on them, Tim talks enthusiastically about the potential for seaweed to provide food, raw material, nutrients and support for marine ecosystems. We

[28] This is particularly important for vegans, since most people get their iodine from dairy products and seafood.

[29] It's reported to reduce the amount of methane that cattle produce, which would help our efforts to reduce greenhouse gases in the atmosphere.

agree that it's probably just as well that seaweed looks a bit brown and slimy to most people, because it means that we humans haven't ruined it yet. But if we understand enough about how it connects to both land and sea, maybe we can find ways to scale up seaweed harvesting and farming in a way that works with its natural environment, rather than in spite of it. The harsh edge between land and sea is a very specific habitat, and our underappreciation of seaweed – this super-survivor – reflects our underappreciation of this whole transition zone.

The fuzzy area between land and sea can take many guises. We in the UK tend to think mostly of sandy beaches protected by cliffs or seaweed-covered rocks, but the world offers huge coastal riches in many other forms, such as salt marshes, mangrove swamps and seagrass meadows. These all form broad boundaries that are part land and part ocean, inhabited by hardy organisms that belong fully to neither one world nor the other. All of them provide critical habitats for wildlife, and increasingly they are seen as huge assets in the effort to stop further damage to our climate. We may think that the ocean is always somewhere else, but it creeps on to the land, creating unique ecosystems that are beautiful, diverse and valuable in their own right. But they also act as buffers between us and the real nature of the open ocean. The coast is not the ocean, but it is our strongest connection to it. I think that we should see it for the gateway that it is: an opening to another world beyond, but one that's familiar enough not to scare us away.

Following the seaweed

Eating seaweed certainly isn't a modern fad. In 1977, the American anthropologist Tom Dillehay started excavating an archaeological site in southern Chile known as Monte Verde. What he found was astonishing, and astonishingly well preserved: large stone hearths, the remains of wooden huts, plants, a lump of meat, wooden tools, clothing, fruit, berries and animal hides. It was a small human settlement, and the original owners had even left their mark in the form of footprints. These were unimaginable riches to archaeologists used to spending years piecing together tiny nuggets of information from rare stone and bone fragments, which are often all that survives the destruction of the centuries. But a nearby stream had risen to blanket this site with warm, acidic, oxygen-deprived peat bog, a harsh preservative that kept the microbes at bay for millennia. The fragile organic remnants of everyday life were protected until Dillehay dug them up, a tangible bridge to the deep past. The excitement was only marred slightly by one oddity which many people at the time chose to ignore. Radiocarbon dating suggested that the site was 14,800 years old, and this was a problem. The consensus at the time was that no humans had arrived in the Americas until at least one thousand years after that. So who were the people who had built the wooden huts? As investigations continued, one particular ingredient of their domestic lives stood out: nine species of seaweed. But the significance of the seaweed is buried even further back in history.

Around 26,000 years ago, much of the Earth was extremely inhospitable by today's standards. A quarter of all land was covered by permanent ice, and the average global temperature was 6°C lower than today. All of modern-day Canada and a significant chunk of the modern United States were covered by ice that formed a thick white barricade from the Atlantic to the Pacific. That water had come from the ocean, evaporated by the sun's energy and then rained and snowed back down to freeze solid on land. So much water had been removed from the ocean and piled up on the continents that the sea level was around 120 metres lower than it is today. The Pacific Ocean was entirely cut off from the Arctic Ocean by new land which had been uncovered between Russia and Alaska. All the humans on the planet were on the western side of the vast Canadian ice barrier.[30] All of North and South America, with its giant sloths, sabre-toothed cats and woolly mammoths, lay on the other side. But then the world started to warm, and the ice started to melt.

As solid ice turned to liquid and the borrowed water flowed back into the ocean, the shoreline of the northern Pacific crawled back northwards, eating away at the Russia–Alaska land bridge. Ocean species that had been banished southwards by the bitter cold of the ice age drifted back into the new shallow coastal waters, carried clockwise from east to west by a large aquatic carousel called the North Pacific Gyre. This new expanse of cool, shallow, nutrient-rich water covering a rocky sea floor was perfect for one particular organism: kelp. Given the right raw materials, kelp will construct the foundation for some of the richest and most dynamic ecosystems in the ocean.

Imagine yourself drifting in bright greenish-blue water, just a few metres below the still ocean surface. The water seems slightly fuzzy, so you can't see huge distances through it, but that doesn't matter because you are surrounded by thick vertical ropes just a metre or so apart. Each rope is a garland of yellowish-brown fronds about a handspan across and attached to a central stalk. The ropes are soft and pliable if you touch them, but return to their upright stance as soon as you let go. This is giant kelp, the largest algae in the world, and it's impossible to dive into the environment it creates without being reminded of forests on land. If you look upwards, bright sunbeams sneak through the gaps in the canopy to stab into the forest, turning any kelp they

---

[30] For the purists, this was mostly the Laurentide Ice Sheet, which covered the largest chunk on the eastern side. The section to the west of the continental divide and reaching to the Pacific was called the Cordilleran Ice Sheet. The continental divide is marked by the rivers: all rain falling to the west of it drains into the Pacific, and all rain falling on the eastern side drains into the Atlantic.

touch to gold. If you look downwards, vertigo washes over you because the ramrod straight garlands keep going and going until they disappear into the gloom, looking as though they will never stop. The slight fuzziness of the water turns out to be caused by tiny fragments of life and death – larvae, broken kelp and organic particles. Shoals of tiny fish dart among the fronds, munching on the drifting bounty and competing with the larger fish, crabs and molluscs that make the forest their home. Larger fish like sea bass sometimes cruise slowly into view, but disappear with a flick of their tail if you disturb them. An occasional seal twists through the kelp, curious about everything and on the hunt for dinner. Giant kelp is one of nature's most impressive builders, able to grow by 30 centimetres in a single day.[31] Firmly anchored on the rocky sea floor and filling the whole of the water column, it drags on any water pushing past it, dampening waves and currents and so providing calm, food and thousands of hiding places for hundreds of species. This includes other seaweeds which create the equivalent of a forest floor, so there are a rich variety of bottom-dwellers too. A healthy kelp forest is a seemingly endless showcase of nature's riches.

There is one major character difference between kelp and trees. A well-established forest on land is a thousand-year commitment, with individual trees guarding one fixed spot from seedling to ancient stump. But kelp has a more flexible approach to existence. A single individual might last for just ten years, and patches of kelp forest can come and go relatively quickly, perhaps destroyed by a particularly violent storm but bequeathing new generations that start from scratch near by. So a region may have stable areas of kelp forest, but the exact locations will vary from year to year. Both the kelp and the species it supports can travel relatively easily from place to place. As the peak of the ice

---

[31] That's an average, in good conditions. It has been recorded growing up to 60 cm in a day.

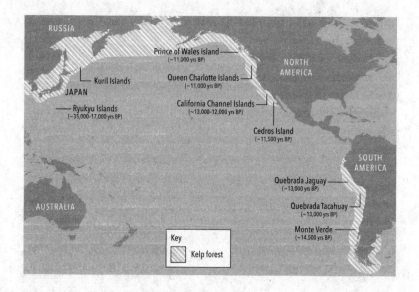

The extent of kelp forest around the Pacific today and the locations of the earliest coastal archaeological sites.

age receded and the ocean recolonized the northern Pacific shore, the kelp and its riches moved in.

So it seems likely that the warmer post ice-age world arrived wearing a giant crown of kelp forest arcing across the northern Pacific Ocean,[32] stretching 5,500 miles from Japan across to Alaska and all the way south to California. Today, a high proportion of the seaweeds found in California are also found in Japan, and the ecosystem along the entire arc would

---

[32] It's hard to find direct evidence because areas covered in kelp forest then are now submerged under many more tens of metres of water, and the ocean is so good at recycling that no one expects to find buried remnants of kelp. But there are indirect lines of evidence, and kelp forests are abundant in these areas today, demonstrating that the conditions are favourable for it.

have been relatively consistent. Any humans on the coast who knew how to live off a kelp forest by gathering seaweeds, hunting fish and seals, and foraging for snails, abalone and crabs, could have moved quickly along that long coastline. Once the great ice barrier covering what is now Canada released a thin ribbon of ice-free land along the coast, there was no need to wait for a central ice corridor to melt. It's now thought likely that the first Americans passed the ice at least 16,000 years ago and took a coastal route along a familiar ecosystem,[33] possibly following the kelp forests. They might have been creatures of the land, but their route to survival was to stay next to the ocean.

And yet the settlement at Monte Verde was 56 miles from the coast when its last inhabitants left. They must have traded with coastal foragers, suggesting a sophisticated network for gathering and sharing resources. It seems that they were confident in the use of seaweed for both food and medicine, and the seaweeds were found all over the site: trodden into floors, scattered near hearths and stuck to stone tools. There were plenty of land-based plants as well, and there isn't enough evidence to be certain of their balance between land and ocean foraging. But they were certainly very familiar with the ocean, and the seaweeds they kept are still used for medicinal purposes by local indigenous people today. We may not be an ocean species, but the coastal seas have nourished us for many hundreds of generations.

For much of the twentieth century, it was accepted that the first humans to arrive in the Americas were the Clovis people, making their way through an ice-free corridor separated from the ocean by the huge Rocky Mountains. Their passage through

---

[33] There isn't enough evidence to be certain about how strong the connection to the kelp forests was, but this idea has a very catchy name: the Kelp Highway Hypothesis.

the ice has been dated to around 13,000 years ago, and DNA evidence suggests that the Clovis people are direct ancestors of 80 per cent of Native American people in both North and South America. But the evidence from Monte Verde and other sites now suggests that they were not the first humans to reach the Americas. Others found their way past the ice at least a thousand years earlier, and they may well have taken a coastal route, being dependent on the ocean. It will be hard to pick this puzzle apart because sea levels continued to rise after the Monte Verde community lived, hiding potential archaeological evidence beneath the ocean. The edges of the oceans change as sea levels rise and fall, so the boundary shifts over time, with the benefits and the dangers of both land and sea constantly washing over it. But there are even more direct connections between these two worlds. The fuzzy edges separating wet and dry are frequently punctured by passageways across another divide that's just as significant: one between water with and without salt.

Entrances and exits

The ocean is the realm of salt water, which makes up 97.5 per cent of all the water on Earth. The other 2.5 per cent is fresh water, and land is the temporary home of almost all of that. This fugitive is always on a journey; distilled from the ocean, carried through the sky, dumped as rain on to grass and trees and rocks. The process of salt water becoming fresh water is incredibly quick: a molecule just needs enough energy at the ocean surface for an instant and it will escape into the atmosphere, leaving all the salt behind. But the journey back can be complicated. If it seeps deep underground to fill the tiny voids left behind by geology and life, or if it's deposited frozen and trapped in solid form without enough energy to remobilize, the journey might pause for a few decades or even a few centuries. But energy and gravity will inevitably kick back in, and the trickling, evaporation, condensation and flow will continue until every water molecule rejoins the ocean. All of our fresh water is borrowed from the ocean – every cup of tea, every waterfall, 60 per cent of you and me, the most expensive champagne, your dog's territorial liquid markers, and the snow covering the top of Everest. And this distinction between fresh and salt water isn't trivial, as we saw with the tearful turtle. More than anything else, the feature that divides life on land from life at sea is the boundary between too much salt and not enough. But this is also a complicated and messy transition region. Our guide to the mysterious hinterlands that separate the salty blue ocean machine from our own familiar freshwater existence is a strange creature of many guises, capable of navigating both worlds.

An epic voyage across the divide

Towards the western side of the North Atlantic Ocean, the smooth surface of the royal blue water is broken by flecks of gold: floating clumps of orange sargassum seaweed. This distinctive colouration extends across an oval that's 2,000 miles wide from east to west and 700 miles north to south. It's known as the Sargasso Sea,[34] the only sea on Earth with no land boundaries. Sunlight penetrates the lukewarm water easily here, because there's so little in it except for salt, and so the only hiding places are in the floating sargassum or in the gloom far beneath the surface. About 100 metres down, a transparent gelatinous leaf the size of an apple pip flexes as it searches for food. This is the great voyager, the expert navigator, the master of multiple disguises and, for centuries, one of the most

---

[34] Named after the sargassum seaweed. Early sailors complained about this patch of ocean because the lack of wind meant that they often got stuck there.

puzzling creatures of both land and ocean.[35] But for the moment, here in the calm Sargasso Sea, it climbs perhaps 50 metres towards the surface at night and sinks back down when the sun rises, relying on almost complete invisibility to avoid being eaten. As it grows, it finds its way northwards and eventually leaves the sargassum behind as it enters slightly warmer water. The larva only has to float and feed and grow; it has hitched a ride on the great Gulf Stream, the mighty river of warm ocean water that curls away from the North American coast and then snakes across the Atlantic, and so at a luxurious speed of around 6 kilometres per hour it is carried 5,000 kilometres across a great ocean basin 5 kilometres deep, over the jagged mountain range of the Mid-Atlantic Ridge and on towards Europe. As it goes, the Gulf Stream mixes with the water on either side of it, cooling as it goes, until the sea floor beneath it suddenly rises, squeezing the ocean engine from a depth of 3 kilometres deep to one of just a few hundred metres. This is the continental shelf, and just as the larva drifts across this geological border, it completes its first metamorphosis, from a thin transparent ribbon around 5 centimetres long into a compact cylinder with bone instead of cartilage and a sharp awareness of electric and magnetic fields, light, temperature and taste. Finally recognizable as a glass eel,[36] one of the juvenile stages of the European eel, it's now equipped to leave behind the calm, quiet world of the

[35] Quite a lot of the details of its life-cycle are still unknown, especially how long each stage takes. It turns out that it's easy to hide in the ocean, just by being small and hard to tag.

[36] For centuries, the lack of baby eels puzzled everyone. The western Atlantic was eventually suspected as a potential spawning ground, but it took a persistent Danish biologist, Johannes Schmidt, a decade to piece the picture together in the early 1900s. He assembled and analysed trawls at different depths and at different times of year all over the North Atlantic and from many different ships, and was able to show that the smallest eel larvae were in the west and that they got bigger as they moved east later in the year. His 1921 paper tells an

open ocean and propel itself across the border that separates it from a safe adolescence as a fresh-water fish. At this point, it could be anywhere from Norway to North Africa.

The first task is to find the fresh water. Using the Earth's magnetic field and the phases of the moon to navigate, the glass eel crosses the continental shelf, passing through murkier water, pushing inwards over sea-floor channels and hills that were dry land only twenty thousand years ago. Now it's time for its spectacular sense of smell to take the lead. All the water here is tainted with dilute chemical signatures of land, providing the eel with the equivalent of teasing shop window displays as it chooses a home. The eel can't see the meadows, forests, agriculture or marshes that sit atop the large land masses to either side of it. But the land is regularly rinsed by the weather, and the run-off – the land's dirty shower water, if you like – will eventually find its way into the open plumbing system formed by streams and rivers. The scents of the land – flowers, plants, decomposition and soil – are carried down into the ocean, providing tiny clues which get more and more concentrated as the eel swims towards their source. Closer to the edge, the eel senses the dilution of the ocean, a decrease in salinity due to the fresh water pouring off the land. There might even be a whiff of other glass eels who are ahead on the same journey. And as the seabed gets shallower, the water gets murkier and the earthy smell gets stronger. The eel finds itself in a new environment: the wide mouth of a funnel where a substantial fresh-water drainage channel meets the ocean, a place constantly shaped and reshaped by the tussle between the two. Twice daily tides shove the sea inland, overwhelming the fresh water coming the other way; but then, as the tide ebbs, the channel is left half empty and the

extraordinary story of deduction and deep commitment to the art of discovery.

land drains freely and rapidly. This constantly shifting transition zone of daily and hourly switches between extremes is an estuary. This is the great border between the worlds of salt and fresh water, a dynamic, complicated and constantly changing environment. To reach the safety of fresh water, the eel must navigate 60 miles of this obstacle course.[37]

There is a price for the crossing: a fundamental transformation of the eel's inner cellular machinery. Every living cell must maintain inside it a critical balance between salts and water: less salty than the ocean but more salty than fresh water. So as the glass eel swims towards the fresh water, its gills, kidneys and intestines switch function. Before, the priority was to eliminate excess salt and conserve water. In fresh water, it must eliminate excess water and hold on to salt.[38] The cellular machinery must always match the surroundings as the eel wriggles upstream, because any miscalculation will cause death by either dehydration or water poisoning. By now the glass eel is about 6 centimetres long, and it is swimming into a torrent of nearly five million tonnes of fresh water per day. This is the run-off from 6,000 square miles of land, about 12 per cent of all of England, because this particular eel's chosen home is the mighty River Thames. The water is opaque, reeking of life, full of natural sand and sediment as tiny broken bits of land from far upstream are washed away. And so the journey to the other side of the great salt/fresh divide begins.

The eel will ride inwards on the tide and then swim down to the bottom to anchor itself and wait as the tide switches, progressing upstream one tide at a time. It follows the salinity gradient and the smells, taking perhaps ten days to get through

---

[37] There are many continuing mysteries about eels, one of which is that only some eels choose to cross this boundary into fresh water.

[38] One of the consequences of this is that eels urinate much more in fresh water than they do in the ocean.

the estuary to a promising tributary where it can leave the ocean behind,[39] adapting a brownish colour and then carrying on many miles upstream to a quiet life as an adult fresh-water fish. After its phenomenal journey across a huge ocean, a coastal sea and a river estuary, the eel settles into a slow, nocturnal and secretive life, spending up to twenty years hiding under stones and roots, rarely venturing more than a couple of hundred metres from its chosen home as it feeds on small living things: insects, crustaceans and small fish.

The transition zone navigated by our eel is a vital connection between the land and the ocean. An undisturbed estuary is a broad shallow fan of channels that are constantly re-sculpted as sediment is deposited and then washed away again. The vast expanses of shallow mud are perfect homes for worms, shellfish, and other small crawling and wriggling fragments of life, which makes them critical feeding grounds for wading birds. The braided channels are surrounded by regularly flooded salt marshes, huge flat half-and-half areas that are home to salt-tolerant plants and act as nurseries for hundreds of species. It's not land and it's not ocean, but it matters. And not just to eels.

It was marshland and thick forest that the Romans found when they sailed up the Thames in AD 43. By building the first bridge to cross the Thames in the area that became London, they didn't just connect the two sides of the estuary.[40] They also made a human connection between land and sea by establishing a trading post that grew to become one of the world's largest

---

[39] The Thames today is tidal all the way up to Teddington, far past central London. Technically, if you live in Battersea, Putney or Richmond, you're living on the coast of Britain. The Ordnance Survey defines the coast by the tides, and even in Putney, 4 miles upstream from central London, the difference between high and low tides is 7 metres.

[40] That bridge was near the site of today's London Bridge.

ports. As the centuries passed and successive generations of eels swept past on the tide, the human settlement grew – in spite of the Viking and Norman attackers who also came in with the tides. At first, access to the European seas and port cities was enough for these settlers, but the ocean was calling. The transformation between fresh-water land-based creatures and seagoing adventurers occurred in the estuary for humans just as for eels, as Thames shipwrights constructed hundreds of the wooden exoskeletons that gave humans a self-contained existence on the open ocean. Instead of navigating by smell and the moon, the humans built accurate clocks (John Harrison's groundbreaking clock H4 was completed in 1761) and sent out explorers like Captain Cook to map the continents and islands of the world in unprecedented detail with sextants and compasses. By the eighteenth century, English ships were leaving from London to claim colonies and to trade within the new British Empire, even as abolitionists like Olaudah Equiano were protesting against the currency of this enterprise: human lives. The Port of London was the link with the rest of the world, a dynamic and messy melting pot of resources and competition and life that came in and went out, a junction box that created possibilities, redistributed wealth and crushed the lowest in the food chain. Just like the estuary.[41]

Far upstream of all the fuss, after a decade or more of living off the land and being gently rinsed by rainwater and meltwater and the soup of the land, the eel (now a metre long) will start to change again.[42] Its long thin body turns silver as it prepares to

---

[41] I find it very distressing that many of the citizens of London today see the Thames as an *absence* (of roads, shops and access) rather than the wonderful natural *presence* that it is. This is the feature around which London is built, and we should celebrate it accordingly.

[42] There seems to be no clear consensus on how long eels can live. There are reports of individuals living in captivity or semi-captivity lasting 55 and

fulfil one last aim: to breed. But before that, it must once more navigate the boundary between the worlds. In autumn, the eel slips out of its protective hollow and weaves its way downstream. The stream is joined by other streams, and the flow gets faster as the channel deepens, collecting and dragging all the detritus of the land – twigs, sand, bones – with it. The eel gets its first taste of salt for twenty years as it approaches the city of London again. With the eel comes far more than the river's natural cargo of woodland detritus and rocky sediment. Rivers provide a critical input to the ocean, of nutrients, fresh water (which balances evaporation) and sediment. The edge is a porous one and our lives on land leak into the ocean.

Once poured out with the murky fresh water to rejoin the ocean, and with its cellular machinery reset to cope with salt water, the eel makes its final journey. It swims back across the continental shelf and out into the Atlantic, sinking deeper into the gloom than ever before – down to the cold, black water up to 1,000 metres below the surface during the day, rising to 200 metres below at night. The path is westwards, and it will spend many months swimming in the dark without feeding before it finds the Sargasso Sea once again. It has finally reached sexual maturity, and after a round trip of at least six thousand miles it will spawn and die. Some time later, a tiny gelatinous leaf will hatch and wriggle, and the process will start all over again.

In the western world, humans haven't traditionally been huge fans of estuaries, seeing them as inconvenient, fickle and untidy places, with wet bits that are too dry and dry bits that are too wet. But other civilizations have been far more appreciative, celebrating and strengthening these transition regions, rather than trying to dredge, blockade and reshape them.

---

perhaps even 155 years. But it's generally agreed that eels spend 5–20 years in fresh water before starting the journey back to the Sargasso Sea.

Sculpting the edges in their own image

The ocean is beautiful at sunrise. Whenever Kimokeo announces that we are going to meet at the canoes at 5 a.m., part of me always mourns the loss of a precious extra hour's sleep. But at 5.45 a.m., as the sky turns pink over a calm ocean and we roll the canoe into the water to meet the long smooth swell gliding across the glassy surface, this is the only place to be. On this particular morning, he's decided that we're taking the big six-person canoe a little way down the coast to play in the surf – because if you've got a canoe and an ocean and time, why *wouldn't* you frolic? We paddle parallel to the beach for half an hour, past condo blocks and houses towards the surf break. The destination is a shallow sand bar about 300 metres offshore. This subsurface bump forces the long rolling swell from the open ocean to steepen into sharp ridges 2–3 metres high that come barrelling into the shore, eventually curling and breaking as they crash into the beach. Kimokeo steers the canoe around to the seaward side of the break, and then the calm morning shifts gear into serious play mode as we line up with the waves behind us. 'Imua, imua, imua!' is the urgent command for our paddles to connect with the water and accelerate the canoe forward, racing to match the speed of the ridge of water looming behind us. As the wave lifts the back of the canoe and the nose tilts downwards, we paddle harder to match, to catch up with nature, and then we're flying and Kimokeo shouts 'Lava!' – stop – and we rest our paddles as the canoe soars down the front of the wave at astonishing speed, riding almost all the way to the beach. Then we paddle again to turn the canoe as more waves chase us in, somehow getting out of it all without capsizing and then going right back around and doing it all over again. It's the adult version of being a child who has just discovered playground slides and is running back round to the ladder as soon as each thrilling whoosh is over. But this canoe of adults has skill and experience and strength, and is reaping the benefits of years of working out the best ways to play with the waves. So it's even better.

After half an hour of surfing, we turn towards the most distinctive feature of this part of the beach, something even more noticeable than the white foam from the waves. Stretching out from the shore and around in an oval approximately 200 metres across, there's a wall built of stacked lava boulders. It's got broad sloping sides, and pokes up about a metre above the current water level. We paddle through the only gap in the wall and pull the canoe up on the beach. Kimokeo has decided that it's time for a lesson about the Kō'ie'ie fishpond, because that is what we've just paddled into. Like all Hawaiian stories, it doesn't start with where we are now. It starts with where it all came from.

First, he draws a compass on the beach with his paddle, and talks about the local winds. The dominant winds on Maui are the trade winds, which blow from the north-east. Then he draws a circle that represents the volcano just behind us, and splits it up. The side of the mountain that faces the trade winds gets lots of rain, and so it's lush fertile land. This is where the farmers live. On the side where we are, facing away from the trade winds, it's arid, and so the people living here make their living from fishing. Kimokeo splits the circle up into sectors and explains how traditional cooperative living worked in Hawai'i. The farmers and the fishermen were dependent on each other, needing to trade so that each had everything their community required.

At this point, we're interrupted by chanting coming from a little further down the beach. Another Hawaiian is there with a small group of visitors to Maui, his deep, melodious voice addressing the ocean, and Kimokeo pauses to reply. After a sung exchange that lasts a couple of minutes, the second Hawaiian comes bounding over to us, enthusiastically thanking us for coming to the fishpond and for sharing his appreciation of it. He gestures out to the stone wall and cheerfully tells us that every stone was touched by ten thousand hands as it was carried down from the mountain, and that this fishpond is the proudest symbol of their community. And then we get into what Kō'ie'ie actually is.

When Captain Cook came to these islands in 1778, he found hundreds of these fishponds spread out along the coasts.[43] These were the basis of a sophisticated system of aquaculture that formed a vital part of Hawaiian culture, designed as it was to produce huge harvests of fish conveniently close to the shore. But instead of building barriers between nature and their fish farms, the Hawaiian way was just to extend what nature had already put in place – a beautifully elegant, effective and sensible arrangement.

The critical observation was that some of their most prized fish, 'ama'ama (mullet), awa (milkfish) and āholehole (Hawaiian flagtail), spent significant amounts of time living in estuaries as juveniles, thriving in the dynamic brackish water that you get where a river or a stream joins the ocean. These fish spawn at sea, come into estuaries to feed on algae and small shrimp-like creatures as they grow, and then go back to the open ocean to reproduce. The Hawaiians discovered that by building a wall around the mouth of a stream, they could effectively extend the estuary, keeping the fish in and the predators out. The nutrients carried by the stream water acted as fertilizer, so the enclosed, protected water became a farm for algae – perfect fish food – and any juvenile fish that chose to swim in had an easy life. Each fishpond had at least one sluice gate which the kia'loko, the caretaker, could open and close to control water temperature and salinity, and to oxygenate the pond. But the sophistication

---

[43] The Hawaiian term for them is loko i'a.

wasn't limited to the fishpond itself. By growing taro upstream,[44] the Hawaiians could make sure that sediment was kept out of the stream and that even more nutrients were carried into the pond. The *kia'loko* needed a wealth of expertise to manage the many variables successfully, but when the fish grew large enough to start to congregate at the sluice gates in search of a route out to the open ocean, the harvests could be enormous. Conservative estimates suggest that at the height of this system, these ponds produced at least a thousand tonnes of fish per year overall.

Building these fishponds was an enormous undertaking, and a stunning engineering achievement. Some of the biggest had walls over 4 metres wide and 2.5 kilometres long, and constant pounding by ocean waves meant that maintenance never stopped. The only source of rock was higher up the mountain, and so chains of thousands of humans would be assembled to bring the rocks down to the sea. This was truly a community effort, although for much of Hawai'i's history the major beneficiaries of this extension of nature were the Hawaiian royal families.

In the late nineteenth century, as the Hawaiian monarchy and the traditional systems of subsistence disappeared, the fishponds fell into disrepair. But in the past thirty years, interest in them has returned, and so Hawaiian communities are rebuilding them. Kimokeo is keen to emphasize the distinction between revitalization (human hands only) and renovation (which includes the use of machines). The fishpond we paddled into, Kō'ie'ie, has been in existence for more than five hundred years and was constantly maintained until around 1860. Since then,

---

[44] Taro is a tropical root vegetable that's a staple local food in Polynesia. All parts of the plant are eaten and it's astonishingly adaptable. I never thought I'd break open a bread roll and then have to ask why it was a bright purple colour on the inside. But taro is purple, and so is bread made from it.

waves, tides, storms and swells have taken their toll, but Kō'ie'ie never disappeared completely. Now it's slowly being rebuilt by volunteers using manual labour only. The point is as much to rebuild the community as to reconstruct the fishpond itself, and it stands as a symbol of that shared effort and shared appreciation of their natural environment. People fish here now for food, even though this pond isn't yet equipped with the most critical tool for management, the sluice gates. And the locals are genuinely proud of this ancient feature. There is a clear lesson here: observe first and learn from nature, and then, if a community needs resources, the way to gain them is to extend the natural mechanisms that already exist, strengthening the human ties within the community at the same time as strengthening human ties with nature, while always maintaining nature's own internal connections. The changeable, flexible, porous boundary between land and sea is an ideal place to navigate these constantly shifting balances. Coasts are where our human connection to the ocean is most obvious, and you could see them as the practice grounds where we can literally dip our toe in the water before we venture out into our relationship with the whole global ocean.

We have already seen that although maps of Earth make the ocean look like a tidy filler with hard edges, the truth is quite the opposite. The great ocean basins, shaped by the slow shifts of plate tectonics and moulded around our spherical planet, form vast flat-bottomed containers pockmarked by mountains and trenches, but they cannot confine the ocean. Our planet's seawater spills over on to the continents and creates a shallow ledge of coastal ocean around the great land masses. The edges, where land meets ocean, are fickle and blurred boundaries which grow and shrink and change and are home to a broad transition zone that isn't represented by clear crisp lines on a map.

The shape of the ocean matters because the blue machine

must operate within these constraints, interrupted by great continents and funnelled by narrow straits. These irregularities are responsible for the rich variety of ways in which ocean processes express themselves, giving the Earth a global ocean with distinctive local characteristics. And so following these mechanisms is the way to understand the beautiful internal structure of the ocean: its anatomy.

3

The Anatomy of the Ocean

THE UNIVERSAL HUMAN SYMBOL OF the sea is an anchor, representing our best (and frequently futile) attempts to remain stationary and in control while riding a capricious body that is never at rest. We know that ocean water moves, and we frequently use it as an analogy for things that constantly shift and change in ways too powerful to stop. On the face of it, it seems that all this movement would ensure that the water in the ocean mixes up very quickly, just as swirling the water in your bath rapidly gives you a pool of uniform temperature. But this is emphatically not what we see: as we are starting to discover, our ocean is an intricate mosaic of flowing features that are defined by their differences and weave together to create the blue machine. The huge ocean basins have their distinctive characteristics, and the polar waters are different beasts again; together, they make one interconnected engine that surrounds the globe. It's time to look at those connections and examine *how* the engine turns. But we need to start by asking *why* the ocean has anatomy at all – and we will begin with a turtle.

How different is different?

A loggerhead turtle in the starlight, hauling herself out of the warm ocean and up on to the beach to lay her eggs, is a special and intimate sight. Turtles have been doing this for at least 100 million years, and can reliably navigate back to the beach on which they hatched, and therefore the beach on which their mother, and their mother's mother, hatched. Apart from their rare forays on to land to lay eggs, the turtle's long life is still largely mysterious. Loggerheads can live to be around 50–60 years old, plenty of time to amass stories to tell. As the starlight glints off this turtle's shell, confirming her presence in the here and now, the rest of her story is a blank space that is full of possibilities. It seems as though she could have spent her years anywhere out there in the big wide ocean, and we will never know. Her story will never be told. But she carries a hitchhiker; and this passenger, snugly ensconced near the edge of her carapace on the left side, has been keeping records.

It's peaceful, being a barnacle, at least once you've made it to adulthood. The early dramas of drifting, moulting, avoiding becoming someone else's lunch and choosing a solid surface to settle down on are over. You grow a protective conical tent of six carbonate plates arranged in a ring, and forage by reaching out into the water through a narrow gap at the top, feeding off whatever the ocean brings to you. As you grow, the bottom edge of each plate also grows, pushing the top further from your foundation and making your cone bigger and wider without the need to rebuild anything. It's a sedentary life, spent cemented to the underlying surface, so a barnacle must choose their spot with care, because once stuck in place they're stuck there for ever. But nature encourages the glorious exception, and the barnacle species *Chelonibia testudinaria* must rank as one of the most intrepid armchair voyagers out there. This is the loggerhead's record-keeper, fat and happy while locked to her shell because it can pick at everything that the turtle swims through instead of waiting for the ocean lottery to bring edible gifts to a stationary rock. Loggerheads can cover thousands of miles as they migrate from feeding to breeding grounds, and along with them go their barnacle hitchhikers. The record-keeping is a consequence of the way the barnacle grows; the new carbonate which is continuously added to the base of each plate carries a chemical signature that represents the water it grew in. The turtle herself is continuously rebuilding her body, so she mixes up her atoms with time, muddying any messages they might carry. But the carbon and oxygen that the barnacle takes from its surroundings are fixed in place at the moment of construction,[1] providing a continuous history that runs from the tip of the barnacle's plates down to its base. These barnacles can live for about two years, and the oxygen and carbon carry information about the temperature and salinity of the water through which the turtle has passed. An Australian study from 2019 showed that these signatures can be used to identify the foraging grounds in which a turtle spends most of its time, although the accuracy of the result depends a lot on how much variation there is between the water in different places.[2] The barnacle passively records the water, and there is enough variation in that water that the signatures can be very distinctive. In the last two years, this particular turtle has spent time in a coastal foraging spot close to a river outflow and then out in the open ocean before coming here to lay her eggs.

Turtles don't swim that quickly (perhaps 1–2 miles per hour), but even at these speeds they can easily pass between huge ocean features, from one distinctive water mass to the next. A turtle definitely knows that ocean water is different from place to place. The ocean water itself is full of character, enough that even a barnacle

[1] Both oxygen and carbon come in different flavours, known as isotopes: carbon-13 is the same type of atom as carbon-12 but it has one extra neutron in its nucleus. Oxygen-18 and oxygen-16 are both oxygen, but with different numbers of neutrons. The ratio between these two types of carbon, or these two types of oxygen, is affected by the temperature and salinity of the water.

[2] If site A has identical temperature and salinity to site B, this method won't be able to tell them apart. But generally, there is enough variation to extract quite a lot of subtlety.

can carry a distinctive signature which records enough of those differences to track where a turtle has been. But the ocean is all contained in a single connected global pool, without any physical barriers, so *why* does the ocean vary so much? Why isn't it mostly the same, with everything in the pot all mixed in together?

One way of looking at this conundrum is to recognize that energy is constantly pouring in and out of the blue machine all over the globe, unsettling the system, with the consequence that the ocean just can't adjust quickly enough to ever catch up with itself. But a more pragmatic view is to consider two sets of ocean processes: the ones that cause separation and difference, and the ones that mix everything together and try to even it all out. The balance between the two dictates what happens and where, sculpting the large- and small-scale anatomy of the ocean. We'll start with the mechanisms causing separation, and their possible impact on one of the most significant battles of antiquity, one sometimes considered to mark the transition between the Roman Republic and the Roman Empire.

The unseen hand of Neptune

Two millennia ago, the warm, salty waters of the Mediterranean were full of life: dolphins playing along the coastlines, huge tuna seeking refuge from the open ocean while spawning, and dappled grey dogfish rasping molluscs off submerged rocks. The Mediterranean barely touches the Atlantic, being connected to the rest of the world ocean only by the narrow Strait of Gibraltar. Water is evaporating from the sunny surface of this enclosed sea faster than it's pouring in from the rivers around its coastline, making the salty seawater more concentrated.[3] And this

---

[3] The reason that the Mediterranean doesn't dry up completely is that water flows in from the Atlantic to keep the water level the same as the other side of the Strait of Gibraltar.

warm, sheltered layby of the global ocean has been the stage for some of the greatest dramas of western history, as the Phoenician, Greek and Roman civilizations rose and fell around its coastline.[4] Those societies saw the hand of the sea gods Yam, Poseidon and Neptune in their daily affairs, and in appreciating the role of the water they were not far wrong. But no amount of human worship can change the fundamental rules of the ocean, and so the blue machine just kept turning as the humans yelled their puny battle cries at its surface, oblivious to the inner workings of the waters that carried them.

The critical day is 2 September 31 BC in the brand-new Julian calendar, and as the sun rises, the sea surface at the entrance to the Ambracian Gulf is full of obstacles. Fish are scooting around hundreds of dark wooden ovals that are 50–70 metres long and bulge 2–3 metres down into the water. At the front of each one a huge bronze mass thrusts forward into the water, curved, sharp, and signalling menace. These are the warships of Mark Antony, the great Roman politician and general, and Cleopatra, queen of the Ptolemaic Kingdom of Egypt. The ships are lined up in an arc facing out of the Ambracian Gulf, each bearing a huge bronze battering ram designed to destroy any ship in its way. A few miles further out, the enemy is waiting: Octavian, the named successor to the assassinated Julius Caesar, and his fleet of lighter, more numerous ships. This was the culmination of a power struggle that had lasted more than a decade, and which would ultimately decide the fate of the Roman Republic.

The year had not started well for Antony and Cleopatra. Antony had been camped near by at Actium for months, waiting for support to arrive while disease and desertion stripped him of manpower. Octavian had just been waiting, and when Antony finally decided to take his fleet back to Alexandria,

---

[4] Along with plenty of others.

Octavian was not going to let him go without a fight. With the fleets lined up and ready, it was clear that this was the day. Battle would normally commence at dawn. But the sun rose and the shadows shortened, and Antony's fleet did not move. The ships remained stationary, as if rooted to the sea floor. And when they did eventually move, they engaged with the enemy by getting up close and launching projectiles rather than manning the oars with vigour to drive the great battering rams through the water and into the enemy fleet. An afternoon of sluggish fighting dragged on before Cleopatra's fleet escaped from the rear, and Antony was eventually able to follow, abandoning most of his ships. Octavian, triumphant, ordered the bronze battering rams of the defeated ships to be incorporated into a grand monument overlooking a new town near by, broadcasting his dominance over the Mediterranean Sea. The fallout over the next few years saw Antony and Cleopatra both commit suicide, while Octavian went on to lead what was now the Roman Empire, changing his name to Augustus as the noble ideals of the democratic Roman Republic were finally crushed beneath the new dictator's heel.

The official historical accounts vary in their details, and two thousand years later there is very little hope of definitive new evidence to unravel what actually happened. But it seems that Antony's ships were stopped as they emerged from the Ambracian Gulf, that something prevented him from pursuing the most advantageous battle plan. The historian and general Pliny the Elder suggested that the responsibility lay with remora,[5] long thin fish that attach themselves by suckers and might act as brakes on a ship, although modern studies rule that out. But

---

[5] *Remora* is a Latin word that literally means 'delay', and an alternative Greek name, *echeneis*, means 'ship-holder', although it's not clear whether the origin was intended to denote a fish that holds on to a ship, or a fish that holds a ship back.

recent investigations have suggested that there might have been another contributing factor that day, one arising solely from the anatomy of the sea itself: the phenomenon of 'dead water'.

How water can stop a ship

In 1893, Fridtjof Nansen, the polar explorer, gave the first scientific description of a weird braking effect experienced by his ship *Fram* while in an Arctic fjord:

> When caught in dead water *Fram* appeared to be held back, as if by some mysterious force, and she did not always answer the helm. In calm weather, with a light cargo, *Fram* was capable of 6 to 7 knots. When in dead water she was unable to make 1.5 knots.

What Nansen is describing is a situation where, although the engines were pushing the ship as normal, giving it energy to move, that energy was leaking away so that the ship barely crawled along. It had been thought that dead water was a seafarer's myth, but here was one of the best oceanographic observers of his day experiencing it for himself. The first clue was quickly identified: layers in the water. Arctic fjords are often surrounded by snow-covered mountains, and when the snow melts it forms a fresh-water layer just a few metres thick on top of the ocean water. It's less dense than the water beneath and so it floats on top. Oceanographers have a phrase to describe a part of the ocean which is separated into density layers: they say it's stratified. The more pronounced the layering, the stronger the stratification. This principle is familiar to cocktail aficionados the world over: if you use a spoon to control the flow of a creamy liqueur on to the top of a drink, so that the liquid arrives slowly, sideways and without any downward momentum, you'll get a separate layer which will remain perched on top. If you mess up the pouring, or stir or shake the drink, everything will just mix together, showing that in principle there's no reason why the layers *can't* mix. But they don't, because mixing requires extra energy, and if there's no source of energy they'll stay separate. So in the right conditions, ocean water masses with different characteristics will form layers. And if the top layer is extremely shallow – just a few metres – it can stop a ship in its tracks.

If you watch a boat or a duck progress along a water surface, you'll see a wake behind it: an outward-spreading pattern of waves in a V-shape. As we have seen, waves represent moving energy, so the duck must donate energy to those waves as a condition of passage. Waves like this form at a boundary between different layers of fluid – in the duck's case, the boundary between the liquid ocean and the gaseous atmosphere. Nansen's collaborator Ekman quickly realized that the key to the dead water problem was that the waves on the surface weren't the only ones. Waves could also form between the layers *inside* the water. They travelled much more slowly, and they could be much higher than the surface wake, but if you had layers, you could make what ocean scientists call internal waves. You can see something like this if you create a stratified ocean in a jar: pour in some water and an equal amount of oil and slosh the layers around. You'll see that the surface where the oil touches the air swirls around much as you might expect. But if you look at the boundary between the water and the oil, you'll also see another set of waves swirling around on that boundary, behaving differently from the ones on the surface. These are the internal waves.

The phenomenon of dead water occurs when the depth of the top layer, and the speed and size of the ship, hit a critical combination. When everything is just right (or just wrong, if you're the

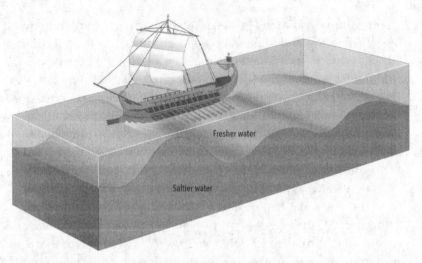

Fresher water

Saltier water

'Dead water': internal waves in water layers that can slow down a ship.

one trying to go somewhere), the submerged wake generated in the water layers beneath the ship – the internal wave – becomes a fixed dip that sits right beneath the ship, and the ship must keep putting in the energy to drag that wave along. There isn't much to see at the surface, but it takes a huge amount of energy to maintain the internal wave, which the ship must carry along like an unwelcome extra anchor. It's only noticeable when the right combination of ship, layer depth and speed is reached, so most of the time, these surface layers in fjords don't matter much. But if you're unlucky, your ship becomes sluggish and there's nothing you can do about it. This was what had happened to Nansen in the fjord.

But what has this got to do with an ancient sea battle in the Mediterranean? The Ambracian Gulf is on the western side of Greece and at its entrance, fresh water from the land meets the sea. This Gulf is an enclosed bowl 25 miles long that barely touches the Mediterranean, like the smallest bowl on an ornamental fountain that pours into a bigger bowl, which pours into an even bigger one. It's filled up by two rivers, and the warm fresh water from the land spills out across the top of the cooler saltier water below. There is no reason for them to mix: the less dense fresh layer floats on top of the denser salty water, and unless something contributes the energy to push it downwards, on top is where it will stay. So inside the Ambracian Gulf and spilling out at the entrance, there's a fresher, warmer layer a few metres thick that's parked at the top of the sea. Fish and dolphins dance in and out of it, feeding and travelling and breeding and living. To the humans on the surface, the layers are invisible. But a ship of the right size will feel their influence.

Could these layers explain what happened to Mark Antony's ships as they came out of the Ambracian Gulf? Did they hit a patch of dead water that completely stopped them reaching ramming speed? It's reasonable to suppose that Octavian's ships would not have suffered the same problem – they were all smaller and sat higher in the water, so would have avoided the critical combination that would cause extra resistance. This region is known to have exactly the sort of layers that could cause dead water, and so it might be that a layer of internal waves, invisible to even the sharpest-eyed generals, could have changed the course of this battle. We'll never know for sure, although historical and fluid dynamics research is under way exploring how likely it was at that location. But it's a clear example of the innards of the ocean affecting the humans scooting around on the surface while those humans thought that they were in control of their movements.

The whole ocean is stratified to some degree, but it's rarely as extreme as the situation required to form dead water. The most obvious layer is the mixed layer: the warm surface layer, often 50–500 metres thick, which acts as a lid on the ocean. The density of ocean water is critically important, because it determines whether a packet of water will stay where it is, or whether it will shuffle up or down in the water column so that the densest water is always at the bottom. So the first reason why the ocean has different components is that water of different densities will tend to separate out so that the layers are stable, just like the cocktail: most dense on the bottom and least dense at the top. Once the ocean reaches this equilibrium, everything is stable and there's no reason for anything to move. Stratification keeps different parts of the ocean apart from each other, so they can maintain their own characters without disturbance. But of course, a wave can move in more than one direction, and although the layers pretty much stay in the right order from bottom to top, sideways movement is much less constrained.

Sideways

Once you go below the upper ocean layer, sideways movement tends to be very slow – the deeper parts of the ocean engine aren't in a hurry – but it's definitely happening. We know it is, because we inadvertently invented a way to track it. This tracking method is a slim silver lining peeping out from behind the colossal black cloud that makes up the legacy of the inventor Thomas Midgley. In his day, he was just a man doing his job rather well, acclaimed and encouraged by his peers and his employers. But with the benefit of hindsight, his personal scientific achievements must surely rank as some of the most disastrous in history. He was never trying to destroy the world, only to make it a better place. And to be fair to him, it was what others did with his inventions that mattered. Thomas Midgley just provided the raw material and then told everyone, loudly and convincingly, that it was safe.

Born in 1889 as the son and grandson of inventors, Midgley was ready and willing from an early age to carry on the family tradition. He graduated with a degree in mechanical engineering from Cornell University in the United States, and by the age of twenty-seven he had found himself a job at General Motors. Around this time, the internal combustion engine was roaring into the world with enthusiasm. But the roar was accompanied by an increasing number of weird pinging noises, and engineers were looking for a way to eliminate this 'knocking' inside the cylinders. The cause turned out to be small periodic detonations that could cause considerable damage if they got out of control. It was Thomas Midgley who discovered that adding tetraethyl lead to gasoline reduced the knocking, and General Motors was delighted. There were questions about whether this 'leaded gasoline' was safe, but Midgley assured anyone who would listen, confidently and repeatedly, that it was fine. There are reports of him taking out a container of tetraethyl lead at a press conference, washing his hands in it and then inhaling the vapour, to prove the point.[6] He and his employers made sure that their proclamations of safety were seen as convincing, in spite of the known dangers of lead,[7] and so leaded gasoline roared into the world along with the cars. But the lead belching out of exhaust pipes was a catastrophic addition to the atmosphere and it now had a direct route into people's lungs and therefore into their bloodstreams. The consequences were a vast swathe of neurological problems and millions of premature deaths. It wasn't until the 1970s and 1980s that the problem began to be taken seriously, and leaded gasoline was (mostly) phased out. Kickstarting this story of commercial pollution, corporate pressure on regulatory agencies and horrendous health effects is more than bad enough, but Thomas Midgley hadn't finished yet.

[6] Do NOT try this at home.
[7] And in spite of the number of cases of serious lead poisoning in the GM factories and labs. Midgley denied that any of this was related to lead, instead blaming the workers themselves. But he himself spent months off work, recovering from lead poisoning.

One of the other mechanical foundations of the modern world, the refrigerator, was growing in popularity at the same time, even though a domestic fridge then cost about the same as a car. If you can reliably keep food cool, you can store it for longer, have access to a wider range of fresh food and reduce food waste, all while reducing costs and improving general nutrition. It's an enormously beneficial technology, if it's accessible, affordable and safe. But the refrigerators of the 1920s mostly relied on ammonia or sulphur dioxide to make them work, both toxic substances that posed considerable risks if they leaked. The hunt was on for an alternative. And right on cue, along came Thomas Midgley, who had moved on from cars but was still employed by General Motors. Working with Albert Leon Henne in 1928, he identified chlorofluorocarbons (CFCs) as the solution everyone was seeking, and developed a synthesis method for the first modern commercial refrigerant: CFC-12, marketed as Freon. Freon was an immediate hit. It was non-toxic, non-flammable, had very low reactivity, and didn't seem to do anything very much other than what a refrigerator needed it to do. Compared to the known dangers of what was in existing refrigerators, it seemed like a slam-dunk win. Having learned the value of a memorable public stunt when it came to safety discussions, Midgley dramatically inhaled the gas deeply at another press conference and then blew out a candle. He didn't suffer any obvious ill effects, the gas didn't explode in the candle flame, and the home refrigerator was off to the races.

The term 'CFC' covers a family of very similar molecules, and over the next few years almost all of them were tested and commercialized. It turned out that the chemical structure of CFCs made them useful for all sorts of things – not just refrigeration, but also in aerosol cans to push out shaving foam or deodorant, in fire extinguishers, as an industrial solvent, and in a whole range of other more specialist applications. It wasn't long before CFCs were everywhere. Thomas Midgley basked in the approval of society, receiving almost every medal or award possible. Before the age of forty, he had invented two of the most consequential chemicals of the twentieth century. He was elected president and chairman of the American Chemical Society, and as far as he was concerned, his place in history was assured.

One of the major advantages of CFCs as an addition to the world's chemical toolbox was their stability. They didn't react with anything else and they didn't break down; you could rely on them to be long-lasting. But this immediate advantage turned out to be their greatest long-term disadvantage. CFCs did their job for a short period, and then they all inevitably leaked away into the atmosphere. Once there, there was almost nowhere for them to go, and no mechanism to eliminate them. So they accumulated. As the 1960s and 1970s progressed, the concentration of CFCs in the atmosphere rocketed, although for a long time no one noticed.[8] The story of what happened next is familiar: a hole in the ozone layer over Antarctica was discovered, and CFCs were found to be responsible. They lasted so long that they found their way up into the stratosphere and helped to destroy part of the Earth's shield against ultra-violet radiation. As awareness grew and the seriousness of the consequences became inescapable, civil society kicked over the objections of industry and the 1987 Montreal Protocol was signed, committing the world to a complete phase-out of CFCs. The amount of CFCs in

[8] It only really became possible to notice them after James Lovelock – later the creator of the Gaia theory – invented a technique called electron capture gas chromatography in 1957. This is a spectacularly sensitive way to detect CFCs and other similar molecules, even when they're only present in very tiny quantities. Although Lovelock is most famous for his invention and popularization of the Gaia theory, this detection device is considered by scientists to be far and away his most significant achievement.

the atmosphere is finally on its way down, and the hole in the ozone layer seems to be recovering, slowly. The scale and success of this global agreement to regulate a dangerous chemical is held up as a glorious example of what international cooperation can achieve. But the CFCs haven't gone away. Most of them are still in the atmosphere. And a decent chunk is now also in the ocean.

As we have seen, the ocean has two major layers: a thin, warm one on top that's constantly in contact with the atmosphere, and colder, thicker layers below which are detached from the air. Any gas in the atmosphere can pass into the top layer of the ocean, and the CFCs certainly did that. Once in the water, they'll generally just stay in that top layer, parked at the top of the rest of the ocean.[9] But we have already visited one of the rare places where water does penetrate below the surface layer and down into the deep ocean basin beneath: the Denmark Strait Overflow, the biggest waterfall in the world. There are also other smaller 'sinking sites' near by, which act like plugholes into the deep ocean, slowly draining surface water into the depths. Whatever the water is carrying when it flows down those plugholes has to go with it. So the sinking water carries the fingerprint of the atmosphere it has left behind, including the CFCs. As the water slides down the waterfall and then southward into the depths of the Atlantic Ocean, it takes the atmosphere's last fingerprint with it. And so, because the atmospheric concentration of CFCs was increasing as the years went by, the year in which each water packet left the surface is imprinted in its CFC markers.

All this has allowed scientists to make maps of the *age* of the water that has washed down the North Atlantic plugholes.

---

[9] Gas transfer is a two-way process, so they will actually travel in both directions across the surface, keeping the ocean surface in equilibrium with the atmosphere.

The younger water is just following the path of the older water, and so we can follow its progress and see how long it took to get there. After cascading down the Denmark Strait Overflow in the northern North Atlantic, a tongue of CFC-enriched water is slinking southwards. There are no natural sources of CFCs, so it's all from humans: from our fridges and aerosol cans and fire extinguishers. The CFCs are inert, so they don't do anything except sit in the water at very low concentrations. The CFC-labelled water moves fastest on the western side of the Atlantic, sliding southwards at considerable depth while the Gulf Stream flows northward overhead. This southward-bound flow is travelling at around 1 centimetre every second, so just over half a mile every day. So it's not moving very fast; the blue machine turns slowly in the depths. Water that left the surface near Greenland forty years ago is now somewhere just off the coast of Brazil, 6,000 miles south of where it started. And it's still going, the CFCs it carries tracking the water's sideways movement on its journey around the globe.

The existence of this CFC marker is temporary, because we humans have finally started cleaning up our act, and now that CFC levels in the atmosphere are dropping, it's losing its usefulness as a way to track the sinking water. But for a few decades, humanity's great blast of these distinctive molecules tainted a chunk of ocean water that is now trapped inside the great ocean engine, and we will be able to follow its sideways journey around the planet. The inner, deeper ocean waltzes at a different tempo from the surface layer, and it may be hundreds of years before this water touches the atmosphere again. The water that drained down the North Atlantic plughole is now part of a huge overturning circulation, driven by the wind and by density. Anything that gets into that deep ocean water is disconnected from the surface – and from us – until it finds its way back upwards. This system of deep currents – known as the thermohaline circulation – constantly moves warm water from the tropics towards the poles at the surface, and cold water away from the poles in the depths.[10]

This is the slowest and most majestic foundation of the blue machine: the slow sideways shifting of water around the planet at depth, on timescales of hundreds of years. Once the deep cold water from the North Atlantic reaches the Southern Ocean, it can move around Antarctica and mix with waters in the Indian and Pacific Oceans, looping into those ocean basins and eventually finding its way back to the surface. It is this deep, slow circulation that connects the global ocean together.[11] Once those waters leave the surface, they are separated from the atmosphere and the world of humans as they progress slowly around the depths, until they reach the surface again hundreds of years later, and are once more exposed to us and the world we've created. This slow conveyor belt shunts heat around the Earth, and its speed and the subtleties of how it works are critical influences on our climate. The CFC spike is helping us track this deep circulation and understand it,[12] the tiny sliver of a silver lining on Thomas Midgley's grim contribution to human civilization.

---

[10] The name 'thermohaline' is contentious, because it implies that the system is driven only by heat and salt – 'thermo' and 'haline' – but the wind blowing on the surface of the ocean is critically important to making the whole thing work.

[11] The overall path around the global ocean isn't simple: it branches and mixes, with components at different depths sliding over one another.

[12] This isn't the only example of an accidental marker introduced by humanity. During the intense period of atomic weapons tests in the 1950s, a huge spike of 'bomb carbon' (carbon-14) and tritium also entered the oceans. The evidence of our atomic mess will be drifting around the deep ocean for many, many years to come.

The tyranny of gravity

We have now built up the big picture of ocean anatomy, and the lesson is that the ocean is just too large to be easily mixed up – or, to put it more accurately, there isn't a mechanism that can stir up the whole ocean quickly enough to make it all the same. That's why ocean water has different characteristics in different places, and can be labelled by its temperature and salinity. The greatest changes to ocean water happen at the surface, and when a water packet leaves the surface behind it carries a signature that can only change very slowly. But alterations are possible at depth, even hidden in the ocean's innards. The water itself can't change, but temporary visitors can come and go and sometimes stay, although gravity ensures they only ever pass through in one direction: downwards.

A sinking ship is perhaps the largest and heaviest single item that will ever fall all the way from the ocean surface to its depths, its engineered steel capable of plunging through ocean water far more quickly than any natural object. It will sink quickly, relatively untouched by its surroundings. But even then, the dense ocean water will rarely let it fall along a straight vertical line. The water may part for the sinking object to pass, but it can still exert substantial forces to delay or deflect its downward progress. The smaller the object, the greater the influence of the ocean on its path and speed. And so the combination of dense ocean water and gravity leads to a sorting process, one to which even something as large as a sinking ship is subject. This sounds like a trivial detail in the story of a disaster the size of a lost ship, but it certainly mattered to Robert Ballard, because it was what allowed him to find the wreck of RMS *Titanic*.

In 1912, the brand-new RMS *Titanic* was the largest ship in the world,[13] a luxurious British passenger liner and the pride of the White Star Line fleet. When she set off from Southampton on her maiden voyage to New York there were 2,224 people on board, and the ship was bursting with confidence, not least in her engineering: a majestic symbol of a modern age. The transition from dream to nightmare was sudden and brutal. Four days into the voyage she struck an iceberg in the middle of the night and within three hours had sunk, with the loss of two-thirds of those on board. The shock reverberated around the world, leading to a complete overhaul of maritime safety procedures, endless speculation about what had really gone wrong, and a steady flood of books, articles and films chronicling the drama of that night. But a crucial piece of evidence was missing: the ship itself. It had disappeared beneath the waves in a region of the north-west Atlantic around 450 miles off the coast of Newfoundland, in an area where the ocean was around 4 kilometres deep. The thought that the wreck might never be found was almost unbearable to many, and a wide range of search and salvage proposals came and went over the following decades. They varied in their degree of technical soundness, but shared the universal trait of being eye-wateringly expensive.

In the 1980s, after the *Titanic* had lain undisturbed for seventy years, a prime opportunity was finally crafted by Dr Robert Ballard of the Woods Hole Oceanographic Institution. It came about courtesy of the US Navy, even though the Navy had zero interest in finding a historic civilian wreck. What they did have was quite a lot of interest in was surveying the wrecks of two military nuclear submarines which had been lost in the 1960s close to the *Titanic* site: the USS *Thresher* and USS *Scorpion*. The Navy wanted to know what had happened to the nuclear reactors that had powered the submarines. So they funded Bob Ballard to

[13] She was 269 metres long, in case you were wondering.

design and build the Argo, a robotic camera that could be towed behind a ship and would send real-time video up to the controllers on board. As long as he found and mapped the submarines, he was told, he could do whatever he wanted with the remaining ship time. It worked for everyone: the Navy got to survey Cold War wrecks under cover of a search for the *Titanic*, and Ballard got a well-equipped ship to search the seabed once his other work was done.

Argo represented a revolution in deep-sea technology. Previous camera systems needed to be lowered down, towed to collect data and then brought back on board so that the information could be retrieved, which was a disruptive and time-consuming way to survey a large area. Argo acted like a mobile robotic eye at the seabed – it could remain at depth for days as it was towed back and forward in a lawnmower pattern, while the humans took shifts at watching live video images from the comfort of the ship and guiding the tow pattern in real time. But Argo could still only scan a relatively small area. By the time the R/V *Knorr*, carrying the Argo system and Bob Ballard, arrived at the search area for the *Titanic*, they had only eleven days of ship time left.[14] Sonar surveys had failed to identify any large features that looked like the hull of a ship, and there was still a huge potential area to survey, around 100 square miles. The task of finding the *Titanic* seemed impossible.

But Bob Ballard had learned a lesson from the wreck of the USS *Thresher*: a sinking ship is not a single object, and objects of different shapes and sizes fall through dense ocean water in different ways. The *Thresher* had imploded at a depth of several hundred metres, and each piece of debris then drifted along its own independent route to the sea floor, which was a

---

[14] They were joined by a French ship, R/V *Le Suroît*, which was using a type of sonar to map the site as part of the same search effort.

further 2 kilometres down. Gravity will pull every sinking object downwards, but in order actually to move, the object has to push the water beneath it out of the way. The faster it can do that, the faster it will fall through the ocean, and there are a few factors that affect the end result. If the sinking object has a larger mass squeezed into its volume and is therefore more dense – if it's solid steel instead of porcelain, for example – then the downward gravitational pull on it is greater. If the object is compact – a cricket ball rather than a tennis racquet – less water needs to be pushed out of the way for it to move downwards, and again the descent will be faster. If it has a shape that will let it glide like a paper aeroplane, that may also slow its descent and push it sideways. Most of all, for smaller objects the drag forces are much larger relative to the object size, so small items will sink much more slowly. And every object will get carried sideways by the currents at whatever depth it's at, so the more slowly it descends, the more sideways movement there is likely to be.

The debris field around the hull of the USS *Thresher* extended outwards for hundreds of metres, representing the wide range of drift paths of objects with different ways of sinking. Some of the survivors from the *Titanic* said that they had seen the ship break in half, and if this really had happened there would be a huge amount of debris scattered across the sea floor. So with just eleven days to search a vast area, Bob Ballard decided to take a new approach. Instead of searching for the *Titanic* itself, they would use the robotic camera to search for the debris field that must stretch outwards from it. The debris would be too small to show up on any sonar images, especially if it was scattered across a lumpy sea floor,[15] but with a robotic camera

---

[15] As we have seen, the deep sea floor isn't often lumpy. But icebergs regularly drift into this region (which is why the *Titanic* had hit one) and as the ocean water heats them up, any boulders trapped inside will drop out on to the sea

perhaps they stood a chance. Argo was lowered, and the humans on deck rotated on four-hour shifts, watching the sea floor as the *Knorr* steamed back and forth doing its lawnmower impression. The days passed, and the stress levels on board rose as the clock ticked down. After a week of looking, metal objects began to appear – and then finally came the unmistakable clue: a large circular plate that exactly matched a photograph of one of the *Titanic*'s distinctive boilers. As more and more debris appeared, the human nature of the tragedy became visible: teacups, wine bottles, serving trays. Ballard and his team followed the debris trail all the way to the hull: a giant mountain of rusting iron rising up from the seabed, a silent and decaying monument to the vulnerability of human life at sea.

Many expeditions have returned to the *Titanic* since 1985, and the entire debris field has now been surveyed. The ship broke in half at the surface and sank in a single location, but the debris is spread over an area approximately 3 miles long and 5 miles wide. It seems likely that as the front half of the ship sank, its shape caused it to glide forwards, sending it away from the stern. The stern section spun around as it sank, and suffered extra damage as the sides imploded into spaces full of trapped air. The bow and stern now lie 630 metres apart with a particularly dense patch of debris surrounding the crumpled stern. The point where the *Titanic* slipped beneath the waves is not thought to be directly above either the bow or the stern, because there is a patch of dense compact objects in between the two that are likely to have fallen straight downwards, and these are seen as markers for the sinking site. Small objects and those with shapes that would cause them to glide have been found further away. And all of this reached the

---

floor. So this region was unusually lumpy, making it even harder to distinguish human debris from natural debris on a sonar system.

seabed over many hours, perhaps several days, as each fragment of this huge tragedy found its own way through the water to its final resting place. A nearby ship had recorded a southward current on the night of the sinking, and the slowest-sinking debris would have been carried by that current during its descent, gravity and ocean physics sorting everything by size and shape.

Most of the artefacts recovered from the wreck site are from the debris field rather than the main sections of the ship. Bob Ballard himself did not retrieve anything, either on that first expedition or on later ones, considering this to be equivalent to grave-robbing. He has been one of the most outspoken advocates for leaving the site untouched, and for letting the wreck degrade naturally as a memorial to those who died.[16] Our society is not generally inclined towards just letting things be, but perhaps the ocean itself, in its slow, steady way, is making the decision for us.

Even large objects are sorted by size and shape as they sink through the depths. But life in the ocean mostly operates at the other end of the size scale, and it's the tiniest scraps that drift downwards most slowly, spending long enough in their host water mass to change its character even once the water is well away from the surface. At these scales, the consequences of sorting are even more dramatic. Sinking is one of the most important processes in the ocean – in fact, it's perhaps the most fundamental reason why life in the ocean is distributed as it is. But even though sinking is critical for life, it's often not actually life itself that is sinking. Often, it's what life has jettisoned.

---

[16] It was expected that the ship would be in a much better state of preservation, given the depth and the cold water in which it resides. However, there is plenty of life at those depths, and almost everything organic is already gone. The steel hull itself is also slowly rusting away, and no one knows how much longer it will last.

A leaky surface

As the slow dusk of early summer creeps over the cool North Atlantic, a transparent speck of life begins to paddle with its jointed legs, propelling itself upwards towards the darkening sky. This is *Calanus finmarchicus*, a minute crustacean 2 millimetres long with a rounded oval body and two long antennae poking out sideways from its head. It's just one species of a group called copepods, pronounced *co-puh-pods*, which means 'oar-feet'. Steady sculling gradually lifts this individual from the darkness 100 metres below the waves and up into the warmer surface waters, and here it can start to feed. All around it there are even tinier organisms, mostly single cells which have spent their day using the sun's energy to construct sugars and proteins from the raw material – nitrates, phosphates and iron compounds – dissolved in the water around them. The copepod can't use the dissolved raw materials, but it can feast on the sun-harvesters themselves. And so it spends the night stuffing itself with any small cellular life that it can find.

Life is generally more dense than ocean water, and the default outcome is that it all must sink. And while gravity is pulling everything downwards, the sorting process happens. The single cells are so tiny that gravity hardly affects them. They have an enormous amount of surface area for their tiny volume, and so the gravitational pull of an entire planet can barely overcome the drag of the dense, viscous water around them. The gentle overturning of the upper ocean layer is usually enough to keep them suspended throughout their short lifetime. But *Calanus* is a messy eater, and chunks of half-eaten cell are left drifting in the water column, meandering ever-so-slowly downwards. Bigger predators will chomp at the copepods, releasing fragments of leg or antenna to the whims of the water and gravity. Any organisms that die will also begin a slow descent, over hours or days. But nature rarely lets a good food source go untouched. Bacteria will munch on the detritus, releasing the raw materials – the nutrients, mostly nitrates, phosphates and iron – back into the water column ready for reuse.

At the first glimmer of dawn, the copepod retreats back to the depths. Many of its predators are visual hunters and the daylight gives them a huge advantage over a small, solitary paddler. So *Calanus* sinks back into the darkness to digest the night's feast in peace. But what goes in must come out. Down in the depths, the copepod releases a faecal pellet. Tumbling gently, the tiny poo meanders downward into the abyss, joining the leftover detritus from the upper layer that escaped the surface bacteria and is sinking away from the sunlight. The bacteria, though, have not yet finished. As the leftovers of life spiral slowly in the dark, microbes get to work, gobbling up the valuable food source and releasing its component nutrients back into the water. At this point, the gravitational tug-of-war loses its grip. The raw materials released into the water are now just small molecules, jostling among billions upon billions of water molecules. They are no longer able to sink and so they become part of the water around them, full of potential for building more life, but trapped in the gloom to go wherever the water takes them.

As the copepod rises and falls each day, it's playing its part in a critical feature of the ocean system: a gigantic slow leak. The surface layer keeps hold of most of its nutrients, to be recycled as cells live and die. But there is a continual slow seepage downwards, known as the biological pump. Once biological material has fallen out of the warm surface layer and is digested in the deep water, those nutrients join the deep water, and there is no way back. This changes the character of the deep water, filling it with nutrients at the same time that the surface layer is gradually emptied out. The sorting process is critically important in determining what ends up where – if things fall very slowly, there may be enough time for them to be eaten by something else on the way down, pushing their nutrients into the water mass around them. But if things are larger and denser, they will quickly reach the sea floor, passing through the deep water without adding anything to it. In both cases, anything that leaks out of the top layers of the ocean is transported well away from the sunlight. And so the nutritional character of the deep water is determined by a gravitational sorting process – and the leakage can only ever be downwards.

The paradox of life in the ocean

And so now the fundamental problem of the ocean as a life-support system is laid bare. Energy is essential for constructing living cells, and the energy is up at the surface, where sunlight reaches into the ocean to glimmer and glitter with potential. But gravity, hand in hand with life itself, gradually strips the warm surface water layer of its nutrients, dispatching them into the depths. The nutrients – critical raw materials for life – reside in the layers of deep water, where they will never be touched by sunlight. This nutrient-rich water stays at the bottom because it is more dense than what's above. The great paradox of a layered ocean is that the ocean engine itself forces the separation of the two essential ingredients: energy and materials. Life is scuppered.

This is not a theoretical situation. Out near the centre of the North Pacific, the water is a startlingly bright royal blue, reminiscent of a cheerful blue crayon or the purest lapis lazuli. If you dip a camera below the water line you can inspect the entire hull of the ship you're on, almost as if the water were not there. Visibility can extend for hundreds of metres down in the perfect blue – but there is nothing to see. The surface waters are almost completely empty, because although they are bathed in bright sunlight they are empty of nutrients. All you can see is the water itself, unsullied by life and organic detritus. The centres of the great ocean basins are generally the ocean equivalent of deserts, with minimal life clinging on in impoverished conditions. Only a few hundred metres beneath there are nutrients aplenty, but the water down there is more dense than the surface waters, and so cannot rise up towards the sunlight. If the layers of the ocean are intact, and the nutrients are all trapped below, their potential cannot be realized. And if there's no way for phytoplankton to do their work of converting the sun's energy into neatly packaged molecular energy, there's no source of food for anything else, either.

But as we have seen, all is not lost. Where the upper layer of the Pacific is pushed away from the coast of Chile by tropical winds, the cold, nutrient-rich water can escape from beneath, pushing upwards into the flood of energy from the sun, and the result is a smorgasbord of life. We have considered the processes forcing different parts of the ocean to separate, but now we can see that the distribution of ocean life must be dictated by the exceptions – by the places where the nutrients *can* connect with the sunlight. And so now it is time to consider one of the most important but unseen processes in the ocean: mixing.

Recombining the critical ingredients

The great paradox of separation doesn't just apply to the requirements for life. The ocean engine itself would not turn if all the cold salty water accumulated at the bottom, and the warmer fresher water floated on top. Fortunately for us, the ocean engine *does* turn, because it receives regular nudges that prevent it ever reaching stasis. But we shouldn't take those nudges for granted, because they come at a high cost.

As anyone who has ever lifted weights or even climbed stairs knows, lifting things up takes energy. The larger and more dense the object, the more energy it takes to lift it up.[17] So mixing water layers that are separated by density isn't just about doing a bit of stirring. Something has to supply enough energy to lift the colder, denser water upwards so that it can mix in with the warm water above. When you stir your tea, this takes a minuscule amount of energy, which comes via your spoon without your ever noticing. But the ocean is much bigger than a cup of tea. Stirring up even a fraction of it takes a huge amount of energy,[18] because it's necessary to lift a huge amount of cold dense water upwards. Two things are needed to avoid a static ocean: a gigantic and continuous flow of energy to mix things up – and a mechanism to get that energy into the ocean. There must be something that supplies at least some energy to do this job, because we can see that the ocean is not just a static layered pond.

Now we have a big and serious question. What is the source of the energy doing all this mixing? This gigantic energy source must exist, because the ocean engine is doing something rather than nothing. It turns out that there are several sources, but one of the biggest conduits for it is also one of the most unexpected. The corals at the bottom of Hanauma Bay in Hawai'i feel its effects on a daily basis.

[17] You get that energy back when you drop the mass again. When the object is up in the air, it has gravitational potential energy, so it's a temporary store of energy. When you move it back down, you convert that energy into another form. This is how hydroelectric power works – lifting water up stores energy, and the turbines at the bottom of the dam can extract most of that energy and convert it into electricity. If you drop the mass on your foot, that energy will pretty much all be turned into heat, so it hasn't been lost but it isn't in a useful form any more.
[18] Oceanographers have estimated the amount of energy that has to be put into the ocean just to mix it up enough to generate the ocean that we see and it's gigantic – about 2 trillion watts.

Time and tide

On the eastern side of the island of Oʻahu, about 10 miles from downtown Honolulu, the last of the volcanic activity that formed the island has left its mark. Most of the Hawaiian volcanoes slope gently into the sea, the result of lava seeping out slowly over millions of years. But 32,000 years ago, the volcanic eruptions on this site were explosive and urgent. They left a jagged crater that sank downwards as it cooled and was eventually breached by the ocean on one side to form a beautiful round bay surrounded by steep hills. Today, Hanauma Bay is a few hundred metres across and contains water that's a maximum of 30 metres deep, a stunning sheltered inlet visited daily by hundreds of tourists. Beneath the waves, pastel-coloured corals cover the majority of the sandy sea floor, and green turtles and parrotfish glide around the bustling reef. The water is warm, and as springtime progresses it gets even warmer, heated by the sun until it's a pleasant 26°C. But on one particular day, just as the sun reaches its peak, the fish at about 15 metres depth notice a change. The temperature suddenly drops. For a few hours, these corals get a temporary respite from the heat as slightly cooler water floods into their surroundings. Six hours later it has gone, and the corals and fish are bathed in the warmer water again. Another six hours later again, the cooler water is back. Throughout the late spring and early summer, the deeper parts of Hanauma Bay live an alternating existence between the water that's as warm as the surface, and this incongruous visiting water that ... came from somewhere else. Somewhere deeper and cooler. The culprit is just offshore, completely invisible to the tourists on land, but playing a vital role in this part of the ocean.

The Hawaiian Islands sit approximately along a line that stretches from south-east to north-west. The active volcanoes are on the Big Island (where OTEC is), and the islands north-west of there are progressively older and mostly past their days of active vulcanism, sinking and shrinking as they cool. But these islands are only the mountaintops that poke up above the water surface. The mountain chain continues to spread outwards underwater for 1,200 miles, forming a huge narrow ridge of submerged ex-islands. This is the Hawaiian Ridge, a geological knife edge poking up from a sea floor that is mostly around 4,500 metres deep. It matters because it gets in the way. And what it gets in the way of is the tides.

The Earth–moon system generates two tides a day in most places,[19] as the ocean responds to the yank of gravity and the physics of rotation. We tend to think that the major consequence of the tides is that they make the water level go up and down near the coastline, and so it's easy to miss the fact that to make the sea level across a big swathe of ocean go up by a metre or so for a few hours, you've got to shift an awful lot of water from somewhere to somewhere else. And then, six hours later, you've got to shift it back again in the opposite direction. The ocean is deep and wide, so a little bit of movement across the whole depth will move a decent chunk of ocean. But when a mountain range that's half the depth of the ocean, like the Hawaiian Ridge, is plonked in the middle of the tide's path, it makes a pretty substantial obstacle.

If you have ever seen a stream running over a completely submerged pebble and making ripples at the surface even though the pebble is some way underwater, you'll have noticed how the moving water responds to such an obstacle. At and above the submerged Hawaiian Ridge, internal waves are created where the water is forced to squeeze through the

[19] The full story of tides is complicated, because the sun, and the shape of ocean basins and coastlines, all make a big difference to where water can get pulled to, and the timescale of its movement. For a more complete story, see *Tide* by Hugh Aldersey-Williams.

narrowed gap. The internal waves that we met at the battle of Actium were on a reasonably sudden boundary between two separate layers. But in the ocean's middle depths, the density varies more gradually, and so the wave doesn't spread out along a single boundary. Instead it exists at lots of depths at once, drifting up and down at the same time as the wave moves slowly outwards. These waves can be huge, pushing layers of ocean up and down by a few hundred metres in some places, moving horizontally perhaps a couple of miles in a day and taking tens of minutes or even hours to complete the movement of one wave.

This is what's happening just outside Hanauma Bay. Part of the Hawaiian Ridge acts as an obstacle to tidal currents, and so it creates giant internal waves that switch direction as the tide switches direction. This is what's generating the periodic influx of cold water to Hanauma Bay. These waves lift and depress

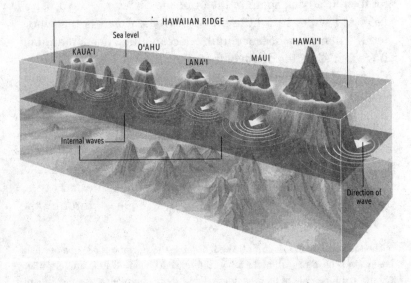

Internal wave generation between the Hawaiian Islands.

ocean layers as they move, and it's estimated that the cooler water flooding the base of Hanauma Bay has been lifted up from 70 metres further down the water column. The waves are created all year round, but it's only in the spring and summer that the surface water gets warm enough for their influence to be obvious in this relatively shallow bay.

Giant internal waves like this aren't tidy things. They may slump and break over the ridge, mixing a mass of turbulent water and so helping heat to move downwards and salt to move upwards. Or they may keep travelling out into the deep ocean for hundreds of miles, taking their energy with them. It's astonishing to follow the path of that energy, because it's almost completely outside the energy flows we generally think about on Earth. The energy started off in the Earth–moon orbit, stored in their rotating dance. As the tides move, they drag on the Earth, leaching energy away from the orbital system so that the Earth's rotation slows by about 17 milliseconds each year, and the moon gets about 4 centimetres further away every year. This reduces the amount of energy in the Earth–moon system and provides the source of energy to move the tides. And then the tidal currents give up some of their energy at the Hawaiian Ridge to form internal waves. If the internal waves break at the ridge, that energy goes into mixing the ocean over the ridge. And if the waves keep moving, that energy is carried outward into the ocean, until those waves break on a faraway continental shelf, donating that energy to the mixing process and eventually to heat. It's hard to visualize the breaking of these waves, because they are so large and travel so slowly through the ocean interior, but they break in a similar way to waves breaking at the shoreline. They may seem sluggish to us, but on the grand scale of the ocean engine, this is a rapid way to move energy around.

Internal waves like this are found throughout the global ocean, frequently formed where currents meet mountains, ridges or the edge of a continental shelf. They glide through the ocean depths mostly unseen, although they do sometimes leave telltale signals at the ocean surface and can even be detected by satellites. The Hawaiian Ridge is particularly effective at generating them, and so the underwater slopes around the Hawaiian Islands often receive pulses of cooler water as the tops of internal waves brush across them, and this water is also likely to be saltier, slightly more acidic and richer in nutrients.

Internal waves are the vehicle for a large fraction of all the energy that mixes the ocean up, and most of the rest is provided by wind and storms mixing and moving the surface waters, near-surface obstacles like islands forcing deeper waters upwards, and friction from currents generating turbulence at the bottom. But ocean mixing is uneven: more of it happens in some places than in others, and this is part of what gives the ocean its character. Most importantly, the total amount of energy available from the tides is nowhere near enough to mix the entire ocean up so that it's completely uniform, and so the ocean we see today reflects a balance between mixing and separation, segregated enough to have a distinctive characteristic pattern, but with enough mixing to keep things interesting. The majority of the ocean has strong layers, keeping nutrients and sunlight separate so that much of the ocean is close to being desert. Mixing from internal waves doesn't often affect the surface layers enough to change that – they make a difference at coastlines and near islands, but most of their mixing effect is in the deeper ocean layers, preventing the stratification from getting too extreme.

Next time you have a view out over the deep ocean, perhaps on a video satellite image or from a plane, think about the giant waves travelling almost unseen in the depths, marching through the ocean in their slow and stately way, and eventually dumping energy that came both from the moon and from the Earth itself.

The ocean's stratification and its layers are a critical part of the structure of the blue machine; but while the equatorial circumference of Earth is just over 40,000 kilometres, the ocean is on average only 4 kilometres deep. We have seen that there is slow sideways movement in the depths, but the most dynamic part of the ocean engine is up near the surface, where the sideways movement is far quicker, and where the ocean is directly connected with the rest of the planet. It's time to look at the nature of the open ocean: the surface of the water that fills the huge deep basins between the continents.

Open ocean

We are used to looking at weather maps that show us what's happening in the atmosphere, and we have direct experience of what it all means, as we often see fluffy clouds being pushed along by the wind, or approaching thunderstorms, or napkins being blown off the tables at an outdoor café. Currents of air swirl around us all the time. Ocean water is far more dense than air – a cubic metre of air has a mass of only 1.2 kilograms, whereas a cubic metre of seawater has a mass of around 1,028 kilograms. So it takes much more force to get water moving, and once it *is* moving it's much harder to get it to stop.[20] But a force is a force, and when wind blows for miles over the vast expanses of the ocean, it's enough to get the ocean surface waters to shift. We have already seen how the wind can push on the surface enough to generate waves, but each gust of wind will also provide a tiny forward push, and days of consistent tiny pushes will create a surface current. The slow accumulation of pushes is an exercise in averaging over long periods of time, so ocean currents are generally far less changeable than the atmospheric ones. Global maps of ocean surface currents are also beautiful but very different from weather maps, with the currents linking

---

[20] It's also much more viscous, so turbulence tends to be slower and larger-scale in the ocean.

into stately patterns that ebb and flow over weeks and months as the blue machine turns. These patterns aren't random; the outcome can vary considerably between different ocean basins even when the same underlying mechanisms are at work.

Round and round

Narragansett Bay is a tiny oceanographic cul-de-sac on the western side of the Atlantic Ocean. It intrudes deep into the state of Rhode Island, taking up a substantial proportion of its area and providing its nickname: the Ocean State. This region is a favourite destination for sailors because of its protected waters, and the geography of Rhode Island's towns and cities makes far more sense when seen from the water than from the land. My professional home for two years after leaving Scripps was the University of Rhode Island's Graduate School of Oceanography, which has a small self-contained campus right next to the water. The biologists spoke the language of the coastal ocean – fisheries, shellfish beds and nutrient run-off – while the physical oceanographers thought on scales of time and size a billion times longer and bigger than the bubble processes in which I was trained. Then there was the physical immersion which accompanied the intellectual one: I had to maintain my scientific scuba qualification with a certain number of dives each year, all in Narragansett Bay.

But the nature of this part of the ocean came as a shock. The beautiful visuals of crinkly coastlines and pretty houses above water aren't matched by the view beneath. The water is darkened by swirling sediment, and it can be painfully cold: a mere 5°C during the winter, matching the cold Atlantic, crawling up to 20°C in midsummer as the sun warms the shallow water. I was used to the cool clear waters of the Californian coast, the huge forests of bull kelp and the expansiveness of the Pacific

Ocean. But here, the rocky shore led into murky shallows full of crevices and boulders, undoubtedly a fabulous home for quahogs and lobsters,[21] but hardly a visual feast for a human. As I was navigating a path through the poor visibility on my first dive, trying to avoid being pushed on to the rocks and wondering why anyone would choose to come down here, a bright flash of yellow scooted out of the gloom and then disappeared. I kicked forward and found a butterfly fish bustling about behind a large rock. Its pointy snout, dazzling colour and dramatic markings screamed tropical reef fish, which is exactly what this cheerful creature is. At this point in the summer the water in Narragansett Bay wasn't that cold, but the harsh brown-black surroundings and the insipid light from above made this perhaps not a fish out of water, but definitely a fish in the wrong water. A bit further along I found another one, apparently oblivious to not belonging here. The only other animals I saw were two crabs and a few marine snails, well adapted to these temperate waters. By the time I surfaced I was not only shivering with cold but thoroughly mystified by what I had seen.

Up at the ocean surface the patterns of movement are very different from what they are down in the depths, and they can be quite confusingly overlain on top of each other. There are the patterns you see if you look over the timescale of a day, and the patterns you see if you look over a timescale of years, and they're not necessarily the same. The beauty of the ocean engine is that it accommodates all of these patterns, so that what you see depends on the scales of size and time that you're focusing on. The biggest and slowest of the surface patterns operate on scales

---

[21] A quahog (pronounced *co-hog*) is a bivalve mollusc that lives on the seabed, a type of hard clam. They can live for hundreds of years and they're the official shellfish of Rhode Island. Rhode Islanders are very proud of quahogs, a pride lovingly expressed by eating huge numbers of them.

of months and years rather than the centuries of the depths, and it was one of these that had delivered the unsuspecting butterfly fish to Narragansett Bay. This component of the ocean engine is formed of giant circular currents which are the closest thing to cogs that the ocean engine has, and they are known as the ocean gyres.

The story of the ocean gyres starts with the shape of the land and the space that this leaves for the ocean to operate in. Take the North Atlantic as an example. It's bounded on the eastern side by Europe and Africa and on the western side by the Americas. Across the southern side is the equator, and since it's rare for currents or winds to cross the equator this acts as another boundary. So there's a nice deep ocean basin which has solid land along its eastern and western sides. The trade winds swoop southwards towards the equator and then bend to the right because the Earth is spinning. Water from Africa is then pushed westwards until it meets the obstacle of the Americas, where it bends to the right. It then flows northwards just off the coast until the spin of the Earth turns it towards the right again, sending it back across the Atlantic. Up here, much further north, the prevailing winds are westerlies, blowing from America across to Europe, and these continue to push the water along until it turns right again and flows back towards the equator. This is a considerable simplification of a complex picture, but the critical point is that the wind pushes on the water surface and can eventually create a current. The shape of the land and the spin of the Earth mean that this current becomes a giant carousel, circulating around the North Atlantic in a clockwise direction. But this is only half the story.

Just as Nansen saw that ice drifted to the right of the wind direction in the northern hemisphere, these surface currents are diverted by the Coriolis Effect. They are moving on a rotating planet, and so, as they trundle along, they are being turned slightly to the right in the northern hemisphere. So, you may say, doesn't the Coriolis Effect mean that all the water in the carousel would turn to the right and inwards towards the centre, making water pile up in the middle? It's not an unreasonable thought – and in fact this really is what happens. There is a hill made of water in the middle of the Atlantic Ocean. It's just under a metre high – we can measure it from satellites. But it doesn't keep growing indefinitely because, given the chance, water will run downhill. All these forces balance, so that while the water progressing around the outside of the hill does experience a rightward push into the centre from the Coriolis Effect, this is exactly cancelled out by gravity pulling that water back downhill.[22] The wind doesn't blow continuously in its average direction, but the water has so much momentum that it effectively smooths that out, and the giant carousel drifts around as the storms come and go – and thus the gyre keeps turning, continually carrying water around the outside of its ocean basin.

There is one last twist in this tale, and this is what explains the presence of the butterfly fish. The Coriolis Effect varies with latitude, so the spin of the Earth matters much more if you are closer to the poles than to the equator. The consequence is that the gyre doesn't turn in a perfectly symmetrical way. The current on the westward side is squished into a fast-flowing narrow band that rapidly shunts water from the equator up towards the north before it turns to cross the ocean.[23] In the North Atlantic, this distinctive warm current is called the Gulf Stream. There's an equivalent in the North Pacific Gyre, an intense western boundary current that flows up the

---

[22] This situation, where the Coriolis force is balanced by gravity pulling water downhill (or the pressure gradient created by the hill of water, to express it in the language of physical oceanography) is known to oceanographers as a 'geostrophic flow'.

[23] This is known as 'western intensification'.

coast past China and Japan called the Kuroshio Current. But on the eastern side of each gyre, something very different happens: the return current going from north to south is slow and very broad, so that it's barely noticeable. The return flow does have a name – the Canary Current – but we don't hear much about it because it's far less dramatic than the Gulf Stream. It slinks lazily around the eastern side of the North Atlantic until it turns near the equator to head back across and join up with the beginning of the Gulf Stream. This completes the circuit dictated by the Coriolis Effect: a rapid south-to-north transit around the western side (the Gulf Stream), followed by a leisurely trundle from north to south around the eastern side (the Canary Current), before starting all over again. Another way of thinking about this asymmetry in the gyre is to imagine the shape of that hill of water. The peak is pushed towards the west, so while the slope on the western side is steep, the slope on the eastern side is very shallow. Water has to rush along to get across that steep slope on the western side, but can then take its time to come back round across the gentle eastern slope.

The butterfly fish will have started its life in Florida, in a warm coral reef environment. But once it was picked up by the mighty and rapid Gulf Stream, it was carried northwards in a great torrent of warm water. The Gulf Stream actually turns across the Atlantic about 350 miles south of Rhode Island, which is why Narragansett Bay normally only connects to the cold water of the north. But as that great torrent rushes along, it sometimes spins off small rotating islands of warm water, known as 'warm-core rings'. These little spinning islands of water travel off northwards and can last for weeks or months. They're like miniature tropical oases, wandering around on the outskirts of the Gulf Stream, and so tropical creatures can survive in them. The butterfly fish must have been just a larva when it was picked up, not yet in need of a reef to feed from, and so it was carried into exile, borne most of the way from the tropics by the Gulf Stream and then transported for the last short part of its journey inside a warm-core ring. Then that ring reached Rhode Island and discharged its cargo into the murky waters of Narragansett Bay. Although they can't survive in the bay for long periods, small numbers of tropical fish regularly appear each summer, unwilling ocean fugitives doing their best to find the warmest water they can.

There are five main ocean gyres, and they all form in the same way. Two are the North Atlantic Gyre and the North Pacific Gyre, rotating clockwise around the northern parts of those two ocean basins. But south of the equator there are three: in the South Pacific, South Atlantic and southern Indian Oceans, all rotating anti-clockwise. The kelp carried from Japan to California after the last ice age was carried by the North Pacific Gyre, and the European eel juveniles were carried by the Gulf Stream across to Europe from the Sargasso Sea.[24] These gyres take years to turn, and the currents are generally only a few hundred metres deep, but these are the stately cogs of the surface ocean, moving heat from the equator towards the poles and providing the transport network that links the ocean basins with one another.

These great gyres form for two main reasons. First, the Earth is spinning, and the ocean is constrained by the land around it. But the nearby land can do more than act as a barrier to flow. Many surface ocean currents are driven by the wind, and the wind itself can be shaped by the land. The result can be huge open ocean currents, and humans will often opt for a free ride if those currents are going somewhere useful. But further back in history, it was often the other way round – humans created

---

[24] Incidentally, the gyre is what isolates the Sargasso Sea – one way of thinking about it is that it's the peak of that hill in the ocean, a safe haven where things like seaweed can accumulate while the gyre rotates around the outside.

'usefulness' wherever the currents took them. And in the early fifteenth century, the Chinese Ming Dynasty built a whole empire of maritime influence whose shape was dictated almost entirely not by where they wanted to go, but by where the ocean led them.

Treasure ships in a monsoon

In 1415, a group of mysterious and exotic creatures arrived at the court of the Yongle emperor Zhu Di, the powerful leader of the Chinese Ming Dynasty. In spite of the emperor's virtuous protests, his courtiers insisted that the appearance of these elegant aliens was entirely due to his excellent governance, a tribute to his wise leadership. From a practical point of view, some careful animal husbandry and a refusal to be defeated by the ludicrous definitely had something to do with it. But more than anything else, their arrival was due to the ocean, which had brought them all the way from Malindi, 5,500 miles away in present-day Kenya. They had been passengers in a giant fleet of 250 Chinese ships crewed by 27,000 men, known to history as the treasure ships. Over a period of twenty-eight years, this armada brought to China an influence that stretched far beyond its borders and was dictated by where the ocean currents took them. The fantastical creatures were just a memorable bonus. The Chinese identified them as *qilin*, a legendary magical creature with hooves, but in their territory of origin they were known as *zurāfa*, and today we call them giraffes.[25] The ocean currents that connected their home with China

---

[25] *Kirin* beer is named after the Japanese word for *qilin* (which means both the mythical creature, somewhat like a giraffe, which the Chinese invented before any of them saw the real thing, and the giraffe itself), although the creature on their logo is closer to the mythical beast than a modern giraffe.

aren't part of an ocean gyre, although they extend across a similar scale. The ocean engine is constrained in this region by the vast land mass of Asia, and so the blue machine turns differently here. The consequences have governed human exploration and trade in the Indian Ocean ever since the first humans set sail from its coastline.

If you turn the globe so that India is in the centre of your view, you will see a huge expanse of water to the south of it; this is the Indian Ocean. It's bounded on the left by the east coast of Africa, and on the right by a scattered collection of peninsulas and large islands which today are mostly part of Indonesia, Malaysia and the Philippines, stretching south-eastwards until they almost touch Australia. At the equator, the Indian Ocean is around 4,000 miles across, and since India protrudes southwards until it's only 7 degrees above the equator, most of this ocean is in the southern hemisphere. The giraffes that reached China in 1415 had survived months on ships, rolling all the way across the entire Indian Ocean from west to east, following a winding route through the channels between the mainland and the large islands, and then finally turning northwards and heading around the coast to the Chinese port of Nanjing. What matters for the ocean engine in this part of the world is the huge continent squatting on its northern side: the hills of India, the giant mountain chain that are the Himalaya and, north of that, land and more land until the Arctic Circle and the northern coast of Russia. It's this land that made it possible for the Yongle emperor's influence to stretch so far across the ocean.

The exact reasoning leading to the construction of the treasure ships is lost in the mists of time, but the outcome was the maritime version of an iron fist in a velvet glove: an impressive flotilla that exuded quiet menace while acting with extreme generosity. The emperor wanted the world to know that China was a force to be reckoned with, and he tasked one of his most dedicated and capable court servants, the eunuch Zheng He, with making sure the world heard the imperial message. Zheng He grew up in a Muslim family under Mongol rule, and was captured and castrated as a child with the specific aim of making him a dedicated servant of the imperial court. He grew into a physically imposing soldier with glaring eyes and a loud voice, but one who cared for the safety and welfare of the crews he took to sea with him. The seven long voyages of the treasure ships were to define his life. But this was not a fighting force tasked with defeating and taking direct control of the places it visited, or even with stealing the most valuable things it could find. In stark contrast to the European adventurers and mercenaries who would cross the oceans in the centuries that followed, the treasure ships were not named because they were collecting as much treasure as they could find. Instead they were giving precious treasure away: gold brocade, patterned silks, coloured silk gauze and porcelain, the most sought-after goods from the East. These were bestowed on the local leaders wherever the ships called into port, dazzling labourers and kings alike with the wealth and magnanimity of the Chinese. As long as the local rulers accepted Chinese influence, sent ambassadors back with the fleet to pay homage to the emperor (which looked impressive to the emperor's domestic audience) and generally played ball, the iron fist would stay safely out of sight. But when Zheng He wanted to make a military point, he had all the resources to do it. With this massive and richly endowed fleet, he could spread China's influence wherever the ocean took him. So where would that be?

Some of the countries the fleet visited were close by: the peninsulas and islands of Malaysia and Indonesia, all of which could be reached without leaving sight of land. But the real prizes lay further to the west, and getting there involved facing the open ocean. The expansive Indian Ocean sounds like the perfect place for an ocean gyre, but gyres can't straddle the equator, because the effect of the Earth's spin switches direction as you cross it.

That makes the northern Indian Ocean different from the northern Pacific and northern Atlantic; there isn't enough space between equator and the land for an ocean gyre to turn. Winds can still push on the waves to generate surface currents, but the ocean engine has to respond in a different way, particularly because the winds do something unusual: they switch direction.

During the height of summer, the colossal Asian land mass helps to drag a huge belt of weather from its typical position over the equator until it has shifted so far north that it sprawls across India from June to September. The northern Indian Ocean and any seafarers floating on top of it find themselves underneath one half of this wind system in summer, and the other half in winter. The cycle starts in the winter, when the 'normal' pattern for these latitudes is seen, and winds blow across India from the high Himalayan plateau and out over the Indian Ocean towards the south-west. But as the Earth rolls around towards the northern summer, and hot bright sunlight floods downwards from directly overhead, energy pours into the air via the ocean and land. This pulls an equatorial band of rain known as a 'tropical convergence zone' northwards until it has crossed the entire northern Indian Ocean and sits over the top of India itself. The winds on the other side of this rain band flow in the opposite direction, from the south-west towards the north-east. This is the south Asian monsoon, and every year the seasonal deluge over land that make it famous turns the dry, dusty hilltops of India into a lush green cornucopia of life.[26] Winds that go into reverse are terrifically useful for any voyagers who want to

---

[26] The standard explanation for the monsoon, repeated in many geography textbooks, is that the monsoon winds are caused by the land heating up more quickly than the ocean in the summer, driving a system of onshore and offshore winds. Scientists working on this topic today do not see this as the best explanation, partly because the timing of the temperature difference doesn't match the wind pattern. The updated view is that monsoons are a local part of a global pattern, although they all (there's more than one monsoon) have characteristics that depend on the local conditions.

venture far from home but still easily find their way back, and Zheng He was no exception.

The port of Semudera in Indonesia was described by Ma Huan, an interpreter who joined the fleet for three of the voyages, as the 'most important place of assembly in the Western Ocean'. Semudera was where ships gathered in autumn to wait for the right time to cross the open ocean. Once the wind had changed, it was time to go. But following the wind and travelling south-west from Semudera would have been catastrophic for Zheng He. The first land in that direction is 10,000 miles away across open ocean: the tip of South America. If that had been the only option, China's influence on the fifteenth-century world would have been limited to the ports of modern-day Indonesia, Malaysia and Thailand, with the possible addition of a foray around the coastline of the Indian Ocean. Fortunately for Zheng He, he was sailing on a spinning planet. Although the puffs and gusts of wind were pushing the wavy ocean surface towards the south-west, the Coriolis Effect meant that the water actually moved 45° to the right of that push, just like the ice that Fridtjof Nansen had observed in the Arctic. The winter monsoon winds blowing towards the south-west created a broad ocean current that went directly westwards. Similarly, the summer monsoon winds blowing towards the north-east created an ocean surface current that flowed directly eastwards.[27] The winds create a reversing east–west highway of water. All the fleet had to do was ride nature's conveyor belt.

Sailing west from Semudera took Zheng He's fleet almost directly to present-day Sri Lanka, 1,000 miles away and the gateway to India.[28] The speeds of the monsoon currents don't sound

---

[27] These currents are approximately 100 m deep, so very shallow compared with the depth of the ocean, but easily deep enough to carry a ship along.

[28] Zheng He was not discovering or exploring these routes, which had been well established as trade routes for decades or centuries, but he was making use of them in a new way.

impressive, reaching a peak of around 0.6 miles per hour (0.6 knots) in summer and 1.1 miles per hour (or 1 knot) in winter. But the average speed of the fleet on the open ocean part of the journey was only around 1.5 knots, or about 42 miles per day, so the currents were doing a big chunk of the work. Zheng He's fleet barely moved faster than a crawl, but its grandeur easily made up for that.

At the heart of the fleet were the ponderous giants – the sixty-two treasure ships themselves, now thought to have been between 117 and 134 metres long, and 48–54 metres wide.[29] In the world of wooden ships, these would have been titans at any point in history. Their shape was very different from that of most European ships: they were flat-bottomed and wide with asymmetrical masts, best suited to shallow coastal waters rather than the open sea. The rest of the 250-strong fleet was made up of various smaller ships, and the arrival of this multitude at any port would have been a powerful incentive for local rulers to be welcoming rather than dismissive or aggressive. Zheng He used Sri Lanka as his entry point to the ports along the Indian coastline, particularly the great trading port of Calicut on the eastern tip of India. On the fourth voyage and those that followed it, he ventured onwards to Hormuz in modern-day Iran, an even greater trading port which was full of foreign merchants getting rich on their cut of the valuable goods flowing back and forth. From the fifth voyage, Arabia and Africa were within Zheng He's reach as he pushed still further westward, bestowing the riches of China on the locals in return for

---

[29] The accounts from the time are limited and not completely consistent with each other, and there is still some ambiguity about the exact nature of the fleet and particularly the size of the ships. These numbers represent the current best attempt to reconcile the differing accounts. However, there is no doubt that the treasure ships were extremely large and the fleet as a whole easily big enough to be intimidating.

their tribute. Many rulers sent back gifts in return, including the unlucky giraffes, and these bobbed slowly back towards China on the reversed monsoon currents which gave each voyage a predictable schedule for departure and arrival. The treasure ships could ride on reliable currents for huge distances, in spite of being far from the most seaworthy ships ever made, as long as they stuck to the ocean engine's schedule.

Instead of an ocean gyre, the monsoon winds generate strong but reversing ocean surface currents both across the Indian Ocean and within the two great bays on either side of India. These connected together the human communities all the way along this huge coastline, forming a vast network that came to be known as the Maritime Silk Road, linking south-east Asia and China to Egypt and the Mediterranean for many centuries before the European Age of Exploration. These currents are driven by the wind, but the wind is constrained by the land and ocean working together.

The first six voyages of the treasure ships spanned the period 1405–22, and then Zheng He led the final voyage from 1431 to 1433. China had successfully stamped its mark on the trading routes of the Indian Ocean, without explicitly encouraging or controlling trade itself. But soon after the seventh voyage the Chinese leadership lost interest in the wider maritime world, and especially in the humongous cost of maintaining and using a huge fleet of long-distance ocean vessels.[30] Almost all shipbuilding stopped, and the flow of foreign tribute to the emperor's court dwindled rapidly. The monsoon currents continued to reverse twice a year, every year, and the ships of other nations continued to trade by following the paths that the ocean set. But nothing with the scale and ambition of the treasure ships would ever be seen again.

---

[30] The argument about the astronomical costs of these expeditions continued in China throughout the years of the voyages, but the emperor continued to prevail over his more financially squeamish advisers.

The ocean and the pyramids

It seems obvious that the reversing winds of the monsoon and their links to the ocean would have impacts on humans travelling in and near those winds and currents. But some of the consequences were felt much further away . . . even by those living in distant deserts, such as the pharaoh King Tutankhamun. He never even gazed on the expanse of the open ocean, and yet his luxurious lifestyle and the riches of his kingdom were possible only because of the faraway Indian Ocean and its monsoon winds.

The great civilization of ancient Egypt went on for an extraordinarily long time, marching onwards through the centuries from its establishment around 3100 BC until the defeat of Antony and Cleopatra at the battle of Actium in 31 BC. It bequeathed to us some of the most fascinating and memorable artefacts of the ancient world: the pyramids, the Sphinx, a rich language and culture, mathematics, medicine and materials science. But producing such luxuries required excess time and resources, far above and beyond what were required for basic survival. And what provided this additional bounty was water: the annual flooding of the Nile valley – the inundation – which carried with it rich and fertile soil and made food production easy. But all that water was a curiosity in the ancient world. The Nile valley rests in a hot, arid climate and gets almost no rain, and yet every year great torrents of water reliably came swirling down the valley to nourish the desert-dwellers.

The Nile ends where it flows into the Mediterranean Sea, but it starts around 1,240 miles to the south in the highlands of Ethiopia.[31] This area is known as the 'roof of Africa', a jagged expanse of ridges and peaks with altitudes ranging from 1,500

---

[31] This is where most of the water comes from, although a second contributing branch starts in Uganda, even further south.

metres up to 4,500 metres. A relatively short distance away is the African coast, and the point where the summer monsoon winds leave the ocean and whoosh across the land, carrying warm air which is full of water that has evaporated from the ocean surface. As this humid air is forced upwards over the Ethiopian mountain range, the water condenses and succumbs to gravity, to be dumped as a seasonal deluge of mountain rain. This trickles down through the upland soil and streams to join the Nile, producing a substantial pulse of water carrying nutrient-rich sediment. The ancient world had no idea about its oceanic origin, but could rely on this flood arriving in the Nile valley every August and September. The people living in the valley developed ingenious agricultural methods to take advantage of this regular inundation, but they would have made little progress without the raw materials provided by nature. The annual floods continued through the millennia until the construction of the Aswan Dam in the 1960s, which put the water release under human control and ended the natural cycle of annual flooding. But the luxuries of the ancient Egyptian world were only possible because part of the ocean engine reliably delivered water to a desert, and to people who never saw the huge ocean that made their lifestyle possible, giving them a land of plenty and therefore the capacity to develop their culture, riches and infrastructure.

We have looked at the anatomy of the great ocean basins, and the shallow and deep currents that shunt water around the engine, moving heat from the equator towards the poles. But the polar regions dance to a different tune. Once water becomes cold enough, ice forms, and that changes the working of the mechanism quite significantly. Even though the polar regions are both remote and relatively small, they have an outsized impact on the rest of the blue machine, and that's because the ocean anatomy in this region controls huge flows of the Earth's most critical currency: energy. Yet even when you're standing right in the middle of the action, that energy is almost impossible to perceive directly. Studying these subtleties is challenging, and success requires teamwork on a huge scale. The strong friendships which are common in the ocean science world are built on facing work like this together, on exploring both the ocean and our own relationship with it at the same time. Experiences like this often strongly influence the attitude of those humans to their work and its importance. So let's face the cold, and voyage to the North Pole.

The frigid north

We're at 88° 30′ N, creeping closer to the pole but going ever more slowly as we approach. From up here on the bridge, *Oden*'s snub bow looks as though it's sliding over the ice, but the trail of broken ice behind us tells of the uneven contest between the weight of the ship and the metre-thick ocean shell. Every so often, the stately forward progress stops as we meet a ridge, and the ship backs up perhaps 20 metres and then slides forwards again until the ice yields. The grey-white sky merges with the white-grey ice and we are a dot in the middle of it all, alone at the top of the world. The hard ice we found at 86°N has been replaced by a softer blanket without clear boundaries, still enough to resist the progress of the ship, but covered in melt ponds and slushy-looking. It's also drizzling. For a ship, this is the most accessible that the pole has ever been in human history. On foot, this melting summer surface would be completely impassable. Beneath the white icy shell lies a vast, dark ocean. The ship's sensors trace an imaginary silhouette as the water depth zigzags between 3.5 kilometres and 5.5 kilometres, indicating that we are sailing over great mountain ranges. But no sunlight has ever brushed those mountaintops, and they are some of the least studied places on Earth. As the ship breaks a path, we crack open a window between ocean and atmosphere, letting the gloomy light penetrate a few tens of metres into the black water, until the shell closes again behind us.

*

Planet Earth is home to an astonishing richness of habitats: steamy deep green rainforests, harsh lichen-covered ridges, flat grassy plains, friendly babbling streams and far more. But the starkness of the polar regions sets them apart: sculpted of water and yet jagged and unyielding, tranquil and yet treacherously untrustworthy, and shockingly white in a world where the messiness of existence demands every colour in the paintbox. This aloofness feels right, because the role of the polar oceans in the workings of the blue machine is unique; remove the ice, and the whole engine would work differently.

But the Arctic and the Antarctic have very different characters. Through nothing more than an accident of geology and the lottery ticket of the aeons, the North and South Poles of Earth present two perfectly contrasting examples of what a polar ocean can be. The continent of Antarctica covers the South Pole with solid land, and the Southern Ocean runs continuous eastward rings around it, connecting the global ocean by forming the entire lower edge of the Pacific, Atlantic and Indian Oceans. To get to the nearest ocean from the South Pole you need to trek 800 miles, and even though the land at the South Pole itself is only just above sea level, it's covered by ice that is 2,700 metres deep. The ceremonial marker of the South Pole, a metallic sphere on a red and white striped post,[32] is therefore nearly 3 kilometres above the continent. In contrast, the North Pole is not only in the middle of an ocean, but in a deep basin: the sea floor here is

---

[32] The ice is moving very slowly seawards at that point, by about 10 m per year, so there's a ceremony every New Year's Day when the post marking the pole is moved to the new and most accurate position.

4,261 metres below sea level. The Arctic Ocean is relatively small, almost entirely enclosed by Greenland, Canada, the United States, Russia, Iceland and Norway crowding in around it. Its connections to the rest of the global ocean are narrow. The nearest land to the North Pole is only 430 miles away, and yet in spite of its proximity to land and also to countries sufficiently well motivated and resourced to prioritize such things, the first expedition with an undisputed claim to reach the North Pole did not take place until fifteen years after Amundsen had planted the first flag at the South Pole.[33] The constantly shifting ice of the north poses a far more challenging obstacle course than the bitter, windy, rocky plateaus of the south. There is no permanent visible marker at the North Pole because there is no solid land to place one on.[34]

We are constantly hearing about the importance of the polar oceans, and yet they're the two smallest oceanic regions on Earth. The Arctic contains just 1.3 per cent of all ocean water, and covers only 3 per cent of the total ocean surface. It's tiny. So why does it get so much attention? Everyone loves a polar bear (from a distance), but bears are not the reason the icy top and bottom of the world matter for planet Earth. The polar oceans are the gearbox of the entire ocean engine, and they are also a critical lever constantly tweaking the energy budget of our planet. That's why *Oden* was nosing its way through the ice towards the North Pole. We were going to the heart of the machine, so that we could perch inside and watch it turn around us.

[33] And even then it was in an airship, so although flags were lowered, no humans set foot on the ice. Amundsen was one of the airship passengers.
[34] A Russian submersible reached the sea floor at the North Pole in 2007 and placed a Russian flag there, in significant contrast to the marker at the South Pole, where the metallic sphere is surrounded by a ring of all the flags of the countries that were the original signatories to the 1959 Antarctic Treaty.

'The most important thing to remember,' Professor Michael Tjernström paused for effect as he surveyed the scientists and crew squashed into the far end of *Oden*'s dining area, 'is that *longwave is king*'. This was the first of the informal evening seminars aimed at helping us all to understand the many scientific perspectives on board while *Oden* chugged northwards. Michael was referring to longwave radiation, known to most of us as infrared, and he was describing how the Arctic affects the overall energy budget of the Earth. Although we cannot see it, the Earth is constantly glowing with infrared radiation, and the Arctic is the largest plughole – the sink – for the planet's energy reservoir. Whatever energy gets up to on Earth – being taken up by plants in photosynthesis, keeping rivers and streams running, or helping you lift weights in the gym – it will almost all eventually end up as low-grade heat energy which leaches away invisibly into space as infrared light. If you want to understand the overall energy budget of the Earth, the income – energy arriving from the sun – is relatively straightforward. But the outgoings – the invisible infrared energy seeping away from everywhere – is where all the subtlety lies. The Arctic and Antarctic make a significant difference to the energy flow out of this plughole, and this makes them critical valves in our planetary life-support system. Up here, longwave is most definitely king.

On 13 August, after nearly two weeks at sea, *Oden*'s crew found an ice floe suitable for our science camp. It was very close to the North Pole: an irregular oval just over a mile across, wedged between many other floes, its flat surface speckled with bright turquoise melt ponds. *Oden* would stay moored to this ice floe for the next thirty-two days, basking in weak twenty-four-hour daylight. From the feel of it, I would never have known that the floe was anything other than solid land – it felt like any other icy landscape until you saw that everything that wasn't sky was ice, and that the orientation of the sun and the ship shifted unpredictably as the ice meandered across the top of the world and we meandered with it. Other large and small ice floes gently drifted and spun near by, forming a lid on the ocean that was almost continuous except for the irregular watery gaps where the icy jigsaw pieces didn't fit. The sky was almost always grey, and the whole landscape was quiet: no birdsong, only the slightest bit of noise from the lazy wind, just the sound of boots crunching on snow and the occasional muffled 'whoop' as the ship's multi-beam sonar probed the ocean beneath. If I had a moment alone, I could almost flick a switch in my mind away from the functional, busy, list-oriented organization of tasks for the day and get lost in a planetary perspective. I could mentally zoom out to see myself standing on the ocean at the top of the world, perched on top of a fragile icy shell only 2 metres thick as the Earth rotated around me,[35] tilted towards the sun's weedy rays, but surrounded by invisible flows of infrared energy as the planet balanced its energy budget. It was the best vertigo I've ever experienced.

As we have seen, there's an important pattern in energy's arrival on and departure from Earth. The regions close to the equator are places of net energy gain, and the top and bottom of the planet are places of net energy loss.[36] The consequence is that the entire ocean engine, assisted by the atmosphere, is constantly moving heat energy from the equator towards the poles, before the surplus flows away into the universe as infra-red light. But there is one obstacle that can slow down the outward infrared flood: clouds. Even with a modern icebreaker, it's not easy to get to the central Arctic, so we don't know very

---

[35] This is the one time when the world literally does rotate around you, although obviously it would be getting on just fine if you weren't there.

[36] All of these statements are averages over time and space, so there can still be local exceptions.

much about the clouds there.[37] But the atmosphere, the ocean and the ocean's fickle lid of ice all matter when it comes to energy loss and cloud formation. Studying this maze of interactions is not a task for any single individual. It requires a team. And my time on *Oden* was a powerful reminder that teamwork is not just what keeps Hawaiian canoes safely traversing the Pacific; it's what keeps science going as well.

And a team there certainly was, for *Oden* carried seventy-four people: forty-two scientists and thirty-two experts in logistics, ship operation, cooking, safety, weather and all the other practicalities needed to operate a floating steel village in complete isolation from the rest of the world for two months. The scientific expertise on board was very broad, covering meteorology, ecology, tiny atmospheric particulates called aerosols, sea ice itself, physical oceanography and ocean chemistry. All of us had separate research projects, but the environment around us was so complex that collaboration was critical. The ice was affected by the ocean, which was affected by the atmosphere, which was affected by ocean ecology which was affected by the ice. None of us would succeed without drawing on the expertise and hard work of the others. And none of the science would get done unless someone was feeding us, keeping the ship safe, and keeping an eye out for polar bears. We still focus on the lone genius science stories of the past: Edward Jenner as a pioneer of vaccination, Einstein's *annus mirabilis* papers and Marie Curie's persistence in studying radioactivity. Great scientists of the past may well have earned the title 'genius', but it's very unlikely

---

[37] This might make you wonder about satellite data in the polar regions. Very few satellites go right over the poles – orbits are commonly tilted slightly so that their northern and southernmost points are some distance from the poles. This can have advantages like keeping the satellite in a 'sun-synchronous orbit', and most satellite operators consider the loss of a small circle of coverage around the pole a small price to pay.

that they did their work in complete isolation. It's always a good idea to ask who was doing the laundry while the 'great minds' did their thinking.[38] Today all science is a collaborative endeavour, whether that collaboration is at the lab bench or in publicly challenging and exploring ideas to test their robustness. And this is especially true for ocean and polar science.

I had been invited along by Dr Matt Salter from Stockholm University, who in turn had been invited along because he's an expert in the tiny particles that are carried by the atmosphere, which have a considerable influence on cloud formation. Matt is exactly the sort of person you want to work with in challenging outdoor environments: a brilliant scientist, permanently optimistic and cheerful, who somehow managed to pack every last spanner and screw for an incredibly complicated set of experiments at the North Pole while forgetting to bring a hat. He had to borrow one – a white beanie with a black bobble – from a PhD student. Our joint project was to test the hypothesis that tiny bubbles in the watery gaps between ice floes could be responsible for ejecting tiny particles of ocean water into the atmosphere as they burst. Matt had a wooden floating platform with a large particle trap covering a square metre of ocean surface. I had a specialized submerged camera to look for bubbles in the water, and a collection of other sensors to measure what the water was doing. Our work site was on the opposite side of the ice floe to the ship, so every morning the team working at that site had a beautiful walk in the cold and quiet across to the patch of open water where we were working. The ice edge was sharp, at almost a perfect right angle to the flat top, so you could stand on the edge of the floe and look straight down into the gloom below.

Modern science is always keen to emphasize the most

[38] For the avoidance of any doubt, on *Oden* (as on all research ships), we were all responsible for doing our own laundry. And yes, ships do have washing machines on board.

advanced and shiny pieces of electronic equipment, often accompanied by scientists wearing incongruously spotless lab coats and in charge of small robots that extract samples with precision from perfect glassware. Doing fieldwork sometimes reminds me of that quote about the famous dancer Fred Astaire and his equally gifted but not-quite-as-famous partner Ginger Rogers; it was said that Rogers did everything that Astaire did, but backwards and in high heels. While doing fieldwork (the ocean science community still persists with this name, even though no one has yet discovered any fields in the ocean), we can't quite do everything you can do in a lab, but we do have to do it outdoors in all weather, perched on top of a very mobile fluid with cold hands, minimal equipment and no access to anything we didn't bring with us.[39] But this is the fun of it, the observation and problem-solving combined with physical hard work with a team of people who muck in rather than step back. Matt and I spent almost two hours one morning digging his floating aerosol chamber out of a 50-centimetre-deep layer of slushy ice that had filled the space around it, with scoops and sieves and much pushing and pulling to try to dislodge the platform. We were parked on top of the world and the only way to free that platform was to do the work ourselves, so we just got on with it. This is not the sterile science of the lab: it's the real world, and coming back with robust data involves drilling and digging and plodding and heaving – and watching: so much

---

[39] And then the walrus turns up. This one hauled itself up on to the ice to sniff at a small black sledge, mysteriously fat and happy even though it was a long way from the shallow coastal waters where this species normally survives by snaffling mussels and other marine molluscs. Then it flopped back into the water and swam around the corner to inspect a large buoy by trying to spear it with its tusks. After deciding that science was neither food nor enemy, it arched its back and disappeared back into the deep.

watching to see what the natural environment is really doing and whether it's about to throw a curveball your way.

For many of us who work at sea, this is the foundation of our relationship with the ocean. It's a dual view. On one side there's a mental model of the best understanding of what's going on, with tidy water layers and wind directions and conceptual atoms shifting from one form to another: the sort of stuff we read in textbooks and teach in lectures. And on the other is the observed physical reality, the wind and ice and waves that are messy and irregular, and make us constantly question our conceptual ideas. It's perfectly possible to be an ocean scientist without going to sea – there are lots of expert modellers and engineers who don't go into this environment themselves. But to me it feels essential that *someone* is there, in the clutter and beauty of reality, making sure that the computer models don't stray into the world of computer games. Nature is constantly surprising us, especially in the ocean, and the quickest way to find those surprises is often to put experienced eyes in a position where they can see the problem in its context. There is a lively debate in the oceanographic community today about the future of ship-based science, and whether the need to reduce our carbon footprint means that we should send robots instead of humans. I'm certain that the robots will be a huge help, but we scientists still need a human relationship with the ocean. This is not just about numbers and predictions, but about the kind of world we want to live in. Do we want to live in a sterile and tidy mental fortress where the science is as efficient as possible and our assumptions and our muscles go unchallenged? Or do we want to face the real world as it is, to take time to understand that we exist as part of nature, and to use that perspective to shape our science? I know which future I think is healthier and better, both for the planet and for us.

As our ice floe pirouetted and looped west and then east, before a sharp change in direction that spat us out of the tiny circle above

203

89°N and directly southward towards Svalbard, it was easy to take the calm grey surroundings for granted. But this environment is very strange by the standards of the rest of the global ocean, and the icy lid is as important for what it prevents as for what it enables. It stops light penetrating into the surface ocean, instead acting as a white reflector that sends what little visible light energy there is back out into space. It prevents the wind pushing directly on the water, so that any waves in the gaps between ice floes rarely grow to be more than ripples before they are stopped by another piece of ice. This keeps the ocean surface calm, and slows down the transfer of gases between the air and the water. Although, as Nansen observed, the wind does push the ice around, it can't push along surface currents as efficiently as in other parts of the world. Since the wind doesn't push directly on the surface, it can't mix up the surface waters easily and so thin layers defined by density can form undisturbed. I had a package of sensors which measured temperature and salinity; these I frequently lowered down on a rope, and they showed a clear surface layer only 30 metres thick, which was significantly fresher (with a salinity of around 32, when the average in the global ocean is 35) than the water below. This fresher layer comes from the sea ice melting as the summer goes on, supplying salt-free water right at the surface. As we have seen, the stratification here is dominated by salinity, not temperature. If it were not, sea ice wouldn't be able to form, because the cooler surface layers would sink and warmer water would take their place, preventing it ever getting cold enough to freeze.[40] Further down, far beneath the ice, there's water that has trickled in from the Atlantic and Pacific that circulates around the Arctic Ocean

---

[40] Incidentally, there is more than enough warmth in the deeper layers of the Arctic Ocean to melt all the sea ice. It doesn't happen because there isn't enough energy to mix it all up, so the warmth stays trapped safely down below. But it really shows how important the salinity variation is, and also how important mixing is.

basins before flowing back out into the Atlantic. Some of this is the densest water in the global ocean, formed by the ice squeezing out the salt, as we've seen, before it spills over the giant underwater waterfall between Greenland and Iceland, and drives the large-scale circulation of the global ocean.

As August turned into September, the sun started to touch the horizon on its daily spin around the ship, where previously it had only hovered above it, and the temperatures started to drop dramatically. We packed up our equipment when the open water started to freeze over, and I brought all of my kit back to the ship on 11 September. Matt's wooden platform and a meteorological mast were due to come back in the following day, but bad weather kept us all on the ship. On the morning of the thirteenth, the crew member who had gone out for a 6 a.m. safety check came back to report that the mast and the aerosol platform had gone. Matt and another colleague, John, went out to look and came back to report that the site where we had worked every day for the past five weeks had been mashed up by a collision with another ice floe overnight and was now unrecognizable. They spotted some pieces of wood in a floating pile of ice rubble, but could not retrieve anything. Their scientific instruments, the storage boxes and the last chances at calibration were all lost to the ice. It was a stark reminder of how lucky we had been – this could have happened on any day throughout the expedition, instantly ending all of our science plans. When the Arctic plays nice, it's a beautiful, awe-inspiring place to be. But that character can turn at a moment's notice. We were lucky not to have been there when disaster struck.

The relatively small Arctic Ocean has an outsized influence on the global ocean engine for two main reasons. The first is that sea-ice formation acts as a machine for generating cold salty water that spills out into the global ocean. This sets the nature of the deep ocean all around the globe, and is fundamental to the way that the whole ocean engine works. The second reason is that the white ice reflects sunlight back into space, preventing the feeble Arctic sunshine from warming up the surface of the ocean, which would melt the ice, creating more dark energy-absorbing ocean surface, which in turn would create more warming. The ice controls the energy flow in and out of the Arctic, and that means it acts as an important regulator of the global energy budget. The clouds, which limit the energy flow once it's left the ocean and is on its way through the atmosphere, are also affected by the ice because ice changes how and where heat flows through the system. These energy flow controls are very sensitive to the exact configuration of the system. The quantity of ice functions like a tap dictating the overall flow of energy away into space – turn the ice up, and the Earth stays a bit cooler. Turn the ice down, and Earth heats up a bit more. Those energy flows over the Arctic also influence how the atmosphere further south works, tweaking global weather far, far away from the North Pole. Those are the dominant influences, but the Arctic is a very complicated ocean basin, and it plucks and twists the rest of the global ocean and atmosphere in a variety of subtle ways. The many aspects of the polar region that we studied on *Oden* are all linked, along with many others, and the work of understanding this region involves picking apart all those influences and their connections. Because the Arctic is so hard to get to, there's a lot of work still to be done. But what we do know is that this distinctive part of the ocean engine has an extraordinarily disproportionate effect on the rest of the ocean engine.

The waters of the Antarctic – the Southern Ocean – are just as influential, but the same principles express themselves differently here. The dominant feature of this ocean is the 13,000-mile Antarctic Circumpolar Current, which flows continuously around the continent connecting the Atlantic, Pacific and Indian Oceans, even while the Southern Ocean water maintains its own character. Antarctica is surrounded by areas that act as sea-ice factories and therefore also creates a huge amount of cold dense water which finds its way down into the deep ocean. But the circulating waters around Antarctica are also places of upwelling, where nutrient-rich water from the deep can come up to the surface and provide the raw materials for a flood of life. And the ocean here snuggles around the icy Antarctic continent, its warmth or lack of it directly affecting the flow of fresh water from glaciers into the world ocean.

Ice – its presence, formation and loss – changes ocean water in distinctive ways. It can cause huge shifts in the character of a water mass in a relatively short time, creating types of water that otherwise wouldn't exist. This provides shape and structure to the rest of the blue machine, and gives the physical processes in the rest of the global ocean something to work with. That's why the ice in the two smallest oceans on Earth – the Arctic and Southern Oceans – matters so much for the rest of the planet. Up at the North Pole on *Oden*, we were literally only scratching the surface of the processes beneath us, and even so what we could see was both incredibly subtle and constantly surprising. It's a tough environment in which to operate, but it's also a scientific necessity.

The blue machine contains plenty of nuance not covered here that adds richness and further complexity to the picture sketched out in the first half of this book. But we have seen enough to understand the overall structure of the global ocean and how this dictates the way that the ocean engine turns. It's beautiful and gigantic, and never ever still. Now we have seen the shape of the engine, we can turn to what happens inside it: the messengers that travel through it, the passengers that are carried by it, and the voyagers who have the freedom to navigate from one part of the engine to another. They will show us why the blue machine matters and will lay the foundation for us to consider our own relationship with it, both present and future.

PART TWO

TRAVELLING
THE BLUE MACHINE

4

Messengers

LOOK UPWARDS ON A CLEAR dark night and the majesty of our galaxy stretches across the sky. It's familiar, overwhelming, and a constant reminder that our planet is part of something bigger: a universe. It's easy to take the night sky for granted, but we shouldn't. The rest of the universe is a very long way away, and the only reason we even know that it exists is that light travels from out there to us down here, flowing for years across the vast expanses of space through our atmosphere to find us on the ground. If those messages did not reach us so easily, our personal perspective on ourselves and on what it means to inhabit planet Earth would be completely different.[1] We are astonishingly dependent on such messengers, because otherwise we would live only in the world we could touch, without any way of knowing what lies any further away than we can reach. In our own lives, the main messengers are light and sound, and these create our view of our world.

The messengers shape the message; sound cannot tell you what colour a painting is, and light cannot tell you what a guitar sounds like. So when we ask what the ocean *is* and how it works, and how our perception of it may be biased, we need to

---

[1] Light is the first and the main reason that we know the rest of the universe is there. Neutrinos are the second, and gravitational waves are only the third. It's easy to forget how tenuous our connection to the rest of existence is.

understand how the ocean's messengers work. They flow through its physical structure, influencing their environment along the way. And every messenger carries energy, sometimes a significant amount, sometimes not, so following the messengers also tells us something about how energy moves. The main messengers in the ocean are the same as on land – light and sound – but the way they operate and their relative importance is very different. Because we humans are visual creatures, we will start with light.

Light

The defining image of the underwater world is that of a coral reef. We have all seen many versions of it: a bright scene with exotically patterned fish gliding past more subtly shaded corals, the beauty underlined by an epic royal blue backdrop. But although the underwater camera doesn't lie, it certainly has a flexible relationship with reality.[2] Light in the atmosphere can usually be trusted to illuminate everything equally, to convey information from many miles away reliably, to travel without transformation, to show us accurately whatever there is to see. Light in the ocean is a more fickle friend, easily diverted and diminished. The way it behaves is critical to the character of the ocean, but it can be hard to appreciate until the consequences are right in your face. And they can be very disconcerting.

Several years ago, I spent a month working as a scientific diver to help a friend with her research project in Curaçao. She was studying ways to modify the traps that local fishermen used in order to reduce their impact on the species that the humans

[2] In this case, almost always aided and abetted by bright photographic lights which accentuate colours that may not be visible in natural light, as we will see.

didn't intend to catch. This involved two or three hour-long dives every day around the stunningly healthy Curaçao reefs, opening and closing differing traps made from local natural materials while she counted what was in them and then let it go. There were sometimes pink and blue parrotfish and long green eels, and often very entertaining stripy butterfly fish which were so busy bustling around this new object that they never seemed to realize they were actually going in and out of a trap, navigating the narrow entrance as though it was just another hidey-hole on the reef. The one concession to scientific practicality were the zip ties, which had to be tightened and cut each time a trap was opened and reset, and which seemed to have unnecessarily sharp edges. I learned this the hard way a few days in, when one of these small plastic daggers cut the back of my hand. Bleeding underwater didn't bother me that much, but the blood itself was startling: a darkish *green* plume leaking out of my skin, with wisps that twisted and turned as they mixed into the water around them. I wondered whether I had temporarily turned into the Wicked Witch of the West. But beneath 10 metres of ocean, all of us would bleed green.

As sunlight travels into ocean water, it meets a liquid that is made up of constantly jostling water molecules, each one capable of vibrating by bending and twisting if you nudge it in the right way. This mêlée can absorb light energy much more quickly than the air does, and the distance that the light can penetrate depends on its colour. Light is a wave and each colour in the rainbow has a different wavelength, from 380 nanometres for extremes of violet to 750 nanometres for the opposite extreme of red.[3] All the detail that our eyes can see – the kaleidoscope of a Mardi Gras parade, the bright colours of a meadow in summer, the vivid stripes of the Grand Canyon at sunset – is carried by the tiniest variations in wavelength within that narrow range.

[3] One nanometre is one-billionth of a metre.

Those tiny nuances also determine how easily water can snaffle the energy carried by a light ray. Red light is absorbed very quickly, and almost two-thirds of it is lost within just a couple of metres. But blues and violets can travel much further, more than 100 metres before losing a similar proportion of energy.[4] You don't have to go very far into the ocean before all the bright and beautiful colours of our surface world start to lose their edge.

Down there with the fish trap, I was surrounded by a flourishing reef inhabited by a fabulous variety of life of all colours. But the limited illumination from above meant that the scene was tending towards the monochrome. The sunlight that reached down here had been almost entirely stripped of one side of the rainbow, so that the reds and oranges were gone and only blues and greens and hints of brown were left. As I watched the dark green blood dribble out of my hand, I realized that the darkness came from the pigments that should have been red. Something that is red will absorb the non-red colours and reflect the red back to your eyes. Here, the blues were being completely absorbed but there was no red available to be sent back out. However, blood does reflect back a bit of green light. We can't normally see it because it's overwhelmed by the plentiful red, but here where there was no competition the green was the only visible colour. Of course, if I had had a flashlight that could shine white light on my hand, the reds would have become visible again, because most of the light rays from the bulb would have survived the short journey from the flashlight to my hand and then my eyes. It isn't that things lose their colour at these depths, just that the natural light is only ever partial, and so the visible colouring is also only ever partial. Whenever you see reds and

---

[4] This is in perfectly pure water, but of course anything that's in the water – life, particles, pollution – will increase the amount of light absorbed. Also, this is considering only absorption and not scattering.

oranges on a photograph from a reef that is clearly more than a few metres below the water surface, you can bet your bottom dollar that the photographer had bright lights pointed at the subject to supplement the natural light.

But our visual loss is the thermometer's gain, and in physics, as in life, there is no such thing as a free lunch. The energy in the light doesn't vanish when the light is absorbed. Energy is always conserved, and the energy carried by the sunlight has just been converted into another form to continue its journey through our planetary system. In this case, whenever visible light is absorbed by water, that energy is converted into heat. This is what heats the tropical oceans, keeping the surface waters nice and warm. If the physics of water worked differently, and all the colours streamed through its vast depths untouched, as they do through the atmosphere, the ocean surface would be far, far colder.

It's hard for us, as a very visual species, to take seriously the idea that light isn't a great underwater messenger. But 'light' is more than just the colours of the rainbow, so maybe there's hope in the rest of the electromagnetic spectrum. Light is an electro-magnetic wave, an interwoven chain of electric and magnetic oscillations that travels in a straight line unless diverted. All light travels at the same speed in a vacuum – the famously impressive 'speed of light' – but the wave peaks can be very close together, a very long way apart or anything in between. The light that humans can see is just a tiny fraction of this huge spectrum, which stretches from energetic and damaging gamma rays, through X-rays, ultra-violet, the visible rainbow and then off to infrared, microwaves and radio waves. They're all the same thing, just squished or stretched in a way that means they carry more or less energy. So what happens when any of those other types of light hit the surface of seawater?

The answer is not encouraging for visibility. Water is basically opaque to almost all light. The colours of the rainbow and some ultra-violet light are the exceptions to the rule, even though, as we have seen, water vacuums them up very quickly. Outside this narrow spectral window, light is devoured before it's gone anywhere at all. This is why you can't get phone or radio signals underwater, why submarines can't navigate using GPS (which relies on signals from satellites that don't penetrate to any useful depth) unless they're at the surface, and why you can't use laser rangefinders to measure the shape of the sea floor. But there is a glimmer of hope for the die-hard optical optimists, because out at the very far ends of the light spectrum, things get better again. If you're very keen indeed, there is a way to send long-distance messages deep into the ocean using light. The downside is that to make it work you need to create light that has a wavelength almost as big as the Earth itself. It's a brute force approach, not one for the timid or the financially limited. So of course, the people who stepped forward to try it out were from the military.

An antenna the size of a planet

The experimental transmission of very basic underwater electromagnetic signals predated the invention of the radio by many decades, but fizzled out very quickly. In the 1830s, the development of the electric telegraph was proceeding in parallel on both sides of the Atlantic, with William Cooke and Charles Wheatstone producing a working 'needle telegraph' in 1837 in England, while Samuel Morse and Alfred Vail in the United States demonstrated their working system in 1838. On 24 May 1844, a 44-mile experimental telegraph line from Baltimore to Washington DC opened, and almost immediately proved intoxicating to the news junkies, as election news could now whizz from city to city without waiting for the train. With the concept proven but no confirmed funding to extend the system, Morse turned his attention to the smaller questions, like how to cross rivers. Stringing wires across a river was far from ideal, and so Morse and his collaborators stretched two long wires attached to sub-merged copper plates along both banks of the 25-metre-wide Susquehanna Canal, and discovered that when they connected one side to a battery, a current could be detected on the other side. The electric signal had been conducted by the fresh water of the canal, and Morse briefly speculated about the possibility of linking all coastal towns through water only, avoiding all that tedious mucking about with wires strung from posts. The laws of physics squashed that idea and wires remained dominant, leaving the underwater world in electrical peace for nearly 150 years. But it wasn't to last.

After the Second World War, the Earth's military minds were in no doubt that the ocean was not merely a 2D battleground for ships to wage war at the surface, but was also potentially the world's greatest hiding place. With the development of nuclear submarines and improved underwater life-support systems, submarines could potentially stay submerged for months while roaming the globe, and the ocean's convenient eradication of almost all light meant that they could operate in pure stealth mode. As long as they stayed away from the top 100 metres, they would be completely invisible. Of course, this possibility came with a catch. The world over, top brass very much like to be able to tell their subordinates what to do. The cost of being shielded from the prying of the light was lurking beneath an equally effective shield against all optical communication from the outside world. What was required was a bat signal to sum-mon the submarines when required, one that could be projected into the deepest depths of the ocean instead of the heights of the sky. And so, in 1968, the US Navy came up with Project San-guine, the largest 'radio' in the world.[5]

---

[5] Purists will point out that it didn't operate using radio waves, and therefore perhaps ought not to be called a radio. But 'radio' is just a name that humans have given to an arbitrary part of the electromagnetic spectrum, and the waves

The audacity of the original project was simultaneously breathtaking and eye-watering. The first problem the engineers faced was the scale of the wavelength required to do the job. There was a clear and unavoidable trade-off: longer wavelengths are harder to generate, but the longer the wavelength, the deeper into seawater it would penetrate.[6] The longest radio signals, with a wavelength of about 10 kilometres,[7] would penetrate only a couple of metres. Wavelengths of about 100 kilometres, the largest of the range known as very low frequency (VLF), might penetrate 40 metres. But to push below a depth of 100 metres, the wave would have to fall into the extremely low frequency (ELF) range, with a wavelength of around 10,000 kilometres. To put that in context, the equatorial circumference of planet Earth is only 40,075 kilometres. So just four ELF waves would make a belt that could encircle the whole globe.

The second problem was that to make an electromagnetic wave efficiently, you need an antenna that is a similar size to the wavelength you want to generate.[8] Project Sanguine had a very clear proposal to address this: they would make their antenna by burying 6,000 *miles* of wire in a rectangular grid covering two-fifths of the state of Wisconsin. The electronics would be powered by 100 power plants, and the Earth itself would form part of the antenna. This device, they calculated, would be able to send ELF signals around the world, and these signals would

I'm talking about here are the same physically but have much longer wavelengths. Since 'radio waves' is a familiar term, I'm going to use it anyway.

[6] The salt makes this all much harder because it makes the water conduct electricity really well. The task is ever so slightly more do-able in fresh water, but then submarines are much less likely to be hiding in fresh water.

[7] Bear in mind that the wavelength of visible light is a few hundred *nano*metres.

[8] 'Similar' in this case can be half or a quarter, but it needs to be on the same scale.

reach their submarines while they were hiding in the ocean depths.

The reaction of politicians, environmentalists and peace campaigners was immediate and unequivocal: get lost. The astronomical cost of the plan, along with the unknown environmental and health implications, forced the military engineers to rewrite it again and again, scaling it down each time. Finally, in 1981, Project ELF was commissioned, and it became operational in 1989. This one involved 'only' 80 miles of overhead wires over two sites, one in Wisconsin and one in Michigan, making it less power-hungry but also less efficient. They could generate wavelengths of 4,000 kilometres, reaching across half the globe.[9] The system worked because the Earth has a natural shell in which radio waves can easily travel, between the ground and the ionosphere (which starts around 60 kilometres above the Earth's surface). There is a constant natural rumble of very long waves in this cavity, as lightning strikes provide continual electrical kicks that then reverberate around the planet. Now humans could add their distinctive bat signal to this natural rumble. This craziest of all radio stations could finally broadcast to the world, and some of that signal would leak through the ocean surface to the hidden submarines.

The next downside was that as a radio station, it was rubbish. Using a narrow range of extremely long wavelengths meant that signals could only be transmitted very very slowly. Voice transmission was definitely out, and the best you could hope for was a leisurely series of zeros and ones. It apparently took several minutes to transmit just one three-letter code. There was also no way for the submarines to reply, since that would have required each submarine to have its own antenna that trailed out behind it for many tens of kilometres. So it really

[9] The frequency of this signal – the number of times the wave wobbles back and forth in each second – was 76 Hz. Remember that – it'll be relevant later.

was only useful as an attention-grabbing signal. A constant standard code was broadcast so that submarines could check it when they were within range, and if the landlubbers wanted a submarine to surface in order to receive a more detailed signal, they would transmit a new code to summon it upwards to receive the details of its new task via other means. These ELF signals are the best that humans have when it comes to using light as a messenger over any distance in the ocean, and the best that ocean physics permits. It's inelegant, to say the least, as well as expensive and limited in potential, and in 2004, the system was shut down. The US Navy reported that other communication methods had improved, and it was no longer needed.[10] Given the scale of the antenna, it's perhaps amazing that it was ever built in the first place. But the effort required really highlights how incredibly difficult it is to get light to travel for any distance through the ocean.

When we look up into the night sky, we're seeing light that could have travelled through the universe for hundreds or thousands of years before it reaches us. It has left a distant star, zipped through the vacuum of space, made it through the Earth's atmosphere and weather systems and all the way down to the ground almost untouched. We associate light with almost infinite messaging abilities, because we humans are lucky enough to perceive the richest part of what's available. But really, our confident assumption that we see everything that there is is based on an exception: such fog-free and far-reaching views are not typical of the light and matter in the universe. If that same starlight hits the blue machine, the fundamental nature of the ocean means that it will go no further.

But although visible light in the ocean is efficiently extinguished, it doesn't happen immediately. This is important for

---

[10] It wasn't the only one in the world, though. Russia, India and China all reportedly still operate their own systems.

the ocean engine and the web of life within it. Partial availability is always more interesting than the extremes of all or nothing, because it allows for texture, character and unpredictability. So light does still matter as an ocean messenger, just not because it lets you see everything.

The colour of water

Our society has agreed on the unspoken convention that water is blue, even though it mostly isn't. If you provide a collection of coloured crayons and ask a child (or even an adult) to draw water, what comes out of their cartoon tap, fills up a cartoon aquarium and falls out of the cartoon sky will be blue, even though no one has ever seen this. If you inspect the water in all of those situations, it doesn't have a colour (and if it did, you certainly wouldn't be drinking it or putting your pet fish in it). Small quantities of water aren't blue, and that's because when light takes only a short trip through it, it's barely affected at all. What goes in is what comes out. But the crayon colour of water isn't a complete lie. Blue really is the true colour of water; but it's only apparent when you have very large quantities to look at.

The vast open expanses of the huge Pacific Ocean certainly are blue, and the absorption of red light alone can't explain this. If sunlight hit the ocean and then simply travelled downwards until it was absorbed, the ocean would look black, because no light would ever come back out from it to reach your eyes. The extra step, the one that really shows us the colour of water, is the jostling that goes on when water molecules nudge the path of the light rays as they travel through slightly more and less dense parts of the molecular mob. This is an example of what physicists call 'scattering', which is exactly what it sounds like: light keeps moving, but it's effectively bounced around inside the water so that it takes a zigzag path. The open ocean looks blue to us mostly because of a two-stage process:[11] first, light that isn't blue is quickly absorbed, and second, the leftover blue light bounces around inside the ocean rather than sticking to an unalterable straight path, which means that some of it may find its way back out to our eyes. These two processes set the stage for the visual world of the whale.

Humpback whales are patient ocean travellers, with stubby cigar-shaped bodies that narrow to a knobbly snout, and two long flippers that give them their Latin name, *Megaptera*: 'giant wing'. They are popular with human whale-watchers because they frequently rest and play at the surface, slapping their tails and flippers with enthusiasm, and occasionally breaching by launching themselves into the air and then thumping back down with a mighty splash. The populations that live along the eastern side of the Pacific spend their days cruising from Mexico up to the Bering Sea and back, benefiting from the nourishing bounty of the cold polar waters but returning to the tropics to raise their calves. Wherever it is, when a humpback takes in a last lungful of air and leaves the surface behind to dive, it swims away from the multicoloured sunlight into the blue dusk of the water.

Out in the open ocean, by the time it is even one body length beneath the surface, all red and green light is gone, and the whale is swimming through a hazy pool of blue. Above, there is a fuzzy glow from the direction of the sun, but nothing above the surface is visible. The brightness above fades to complete blackness below, passing through all the shades of blue on the way, and it's not even clear how far into the indistinct surroundings you can see. Around the whale, spaced out in the vastness, are all the

---

[11] When the ocean reflects the sky, it will also look blue – but that's a reflection off the surface, nothing to do with the colour of the water below. And of course, there are lots of occasions when ocean water looks greenish or brownish because of what's in it.

dramas of a huge ocean: shoals of small fish, large predators like orcas, powerful migrating tuna and unseen hordes of pulsating jellyfish. But none of it is visible, all of it hidden behind nothingness. This is the effect of scattering in the ocean: the water itself hides ocean citizens from one another by diverting and muddling the light, letting no ray travel directly from one distant whale or fish or jellyfish to another. Each whale swims in a benign fog that always graduates from light blue above to the darkest black below, and even the best eye in the world could not see further through the murk. Out in the clearest ocean waters, a whale's foggy visual bubble might extend for 200 metres, allowing it to make out the rest of its social group; but close to the coast, where the water is full of life and particles which scatter the light even more swiftly, the visual fog presses inwards and the broad shape of friend or foe only looms out of it at the last moment.

But this is the visual environment that whales have evolved to navigate. Evolution, with no agenda and no emotion, will vary and prune each generation's physiology on the sole condition that such tweaks don't make the animal's prospects of reproducing worse. The visual priority is to manage in the dim light. The land-mammal ancestors of whales had eyes similar to those of humans, with plentiful rod cells to detect low light, and two types of cone cells to distinguish between colours (we humans have three). Over the last 50 million years, as whales and dolphins adapted to their aquatic home, all of them lost function in one of the cone cell types and many of them lost both. Humpbacks and the other baleen whales (blue, fin and sei, among others), along with a few of the toothed whales (like sperm whales) now rely on rod cells only for their vision. This means that they can see in dim light,[12] but without multiple

---

[12] They have other adaptations that help them too, particularly a blue-reflective tapetum lucidum, which is a biological mirror at the back of the eye that

cones they are colour-blind. They can technically see blue light, because blue light is all that surrounds them, but they have no way of knowing it's blue. The ocean could turn bright pink, and they would not know. And so all whales and dolphins swim in a pool of blue that they are completely unable to appreciate. They also have relatively poor visual acuity, since there's no point in being eagle-eyed; things are likely to disappear into the fog long before they're too small to see.[13]

When feeding, a humpback whale generally stays in the top 200 metres. Even blue light will eventually be absorbed completely, and by the time you get to their deepest feeding sites, only the faintest glimmers of sunlight are left. This is only 5 per cent of the average depth of the ocean, and everything below that is dark even on the brightest summer days. This is why the top

---

bounces light back through the optical sensors, giving it a second chance to be detected. Cats also have this layer, which is why their eyes shine brightly in torchlight. But in a whale, this layer reflects most strongly in the blue part of the spectrum, as you might expect.

[13] My own most memorable experience of the ocean hiding things from me came when I was scuba diving off the Californian wreck HMCS *Yukon* in 2008. This is a huge Canadian destroyer, deliberately sunk 185 m off the San Diego coast to give divers something to look at. Our boat was tied to a line that stretched from the surface all the way down to the bottom where the ship was. California water is full of life and is consequently a fairly murky green. We were told to go down the line and wait for the tour guide at a marker some way off the bottom. I descended into the gloom, unable to see anything except the milky green-blue around me and the darkness below. I got to the marker and waited. There was nothing to see. I twirled slowly around the line and then just where I had been looking a moment before, the stark steel lines of the huge ship and its guns towered over me, perhaps only 20 m away, a silent giant that had been there all along. The current flowing past the ship must have had patches of murkiness, and the water had just happened to clear. I hate the 'boo' moments in horror films, and my trust in my surroundings was torn apart, because how can something the size of a building be completely absent and then so very present? The rest of the dive was fun, but I still haven't forgiven that ship for scaring me so much.

layer of the ocean is warm and sunlit while the ocean below it stays cold and black: the surface waters snaffle all of the sun's energy, converting it into either the molecular building blocks of life or heat energy. The deeper waters are hidden away from sunlight, by the same processes that hide whales from each other, and so they are also hidden from the heat source.

The outcome of all this is that ocean water and light have a strange relationship. Light is critically important for the ocean engine, since it provides almost all the energy needed for that engine to turn. But that energy is only injected into the ocean when the light itself is extinguished. Surveying a grand vista of the ocean depths would be amazing, but any ocean where that was possible would be a poor sight for spectators, because it would have no way of absorbing light energy to fuel interesting characteristics. It's the catch-22 of the ocean: to really see the beauty and richness of the engine, you would need to turn the light absorption off. But if you did turn it off and light could flow freely through the blue machine, there would be no energy entering the system to fuel the beauty and richness of the engine. The cost of the beauty is not being able to see it.

So the default state of most of the ocean is darkness. But determined oceanographers have split even the absence of light into categories. Imagine dropping a large grain of sand into the ocean above its deepest valley, the Mariana Trench in the Pacific. If we get the size of sand grain right, it will fall through the seawater at 1 metre every second, about the same speed as dropping a feather on land. As the grain enters the water, it's surrounded by bright sunlight, but the sunbeams fade after the first few seconds. The grain is now sinking through the blue fog of the open ocean, and the light around it is rapidly becoming dimmer and bluer. After just over three minutes of falling, the sand grain is at 200 metres. Only around 1 per cent of the light from the surface ever gets this far, and to a human, this environment would already look as black as night. This marks the transition into what is called the 'aphotic' zone,[14] and only the tiniest glimmers may survive to these depths. The sand grain takes another thirteen minutes to get to the bottom of the twilight zone at around 1,000 metres depth, and there is true darkness. But the sand keeps on falling, drifting and twirling in the darkness for another fifty minutes until it reaches the average depth of the sea floor. In most parts of the ocean, the journey would stop here; but we are right above the Challenger Deep, so the sand grain keeps falling, through the abyssal and hadal zones,[15] until, three full hours after it began its journey, it finally settles into the deepest part of the ocean floor. Almost all of the ocean is dark almost all of the time.

Billboards of the ocean

Ironically, it's in the darkness that light really excels at the role we take for granted on land. Instead of doubling up as an energy source, it becomes solely a tool for signalling. And this is where the scattering comes in handy, because each animal can only monitor its own local fuzzy bubble for potential light signals, and so there's no confusing background to worry about. Long-range light signals are useless in the ocean, but over short distances light is an incredibly useful tool. The darkness certainly doesn't stop ocean life broadcasting its intentions. It's estimated that 76 per cent of species in the open ocean can generate their own light (known as bioluminescence), and we are only just starting to untangle the billions of flashes, flickers, pulses, patterns and distractions that are sent and received every second under the waves. Tiny organisms called bioluminescent dinoflagellates flash and sparkle when disturbed, creating

[14] Literally 'without light', so this term applies to everything below this depth.
[15] The hadal zone is the region from 6,000 m to 11,000 m, and it only exists in the deep ocean trenches. It's named after Hades, Greek god of the underworld.

plumes of light whenever the water they're in is set in motion. As they crawl across the sea floor, brittlestars flash green and may eject luminous mucus. Krill, which are plentiful in the Southern Ocean, generate light on their underside to help disguise them against the bright sky above. Jellyfish pulse with light to deter predators, and also to lure in prey. Ostracods produce a puff of bright illumination when eaten which lights up their predator from the inside, causing the gobbler to vomit up the danger and sprint for cover while the gobblee swims off to get on with its night. And more impressive still is the Humboldt squid, which has found a way to use the complex signalling that's effective not only in the bright surface waters but also down in the dark depths where you would expect such subtlety to go unappreciated.

For fishing communities along the coasts of Peru, Mexico and California, the Humboldt squid is a mundane and useful commercial catch at best, and a personal menace at worst. These muscular molluscs grow to more than a metre long, are aggressive predators that often hunt in large groups and aren't above a bit of cannibalism when the mood or the opportunity strikes. In addition to their eight arms they have two tentacles covered in toothed suckers for snatching prey, which can leave painful scars on humans if the squid is attacked or disturbed. But they spend most of their lives well away from humans, living at depths between 200 metres and 800 metres in near-perpetual darkness. At night, if they can avoid moonlight, they will come closer to the surface to feed, disappearing back into the depths at the first hint of sunrise. Their prey are mostly small fish and crustaceans, and the squid often congregate at the bumper feeding grounds, gliding around each other with their arms held out in front, carefully minding the feeding hierarchy and avoiding fights. Targeting the same fish as a much larger squid could easily result in the impudent opportunist being chosen as the alpha squid's lunch.

The puzzle is how all this activity is coordinated in the absence of sunlight. Other squid species are masters of intricate signalling, but their signals depend on external light illuminating their skilful performances. Just beneath their skin they have thousands of pigment sacs which can be rapidly expanded or compressed to display or hide the colour they contain, enabling them to change the appearance of every part of their skin independently in the blink of an eye. They can also change their reflectivity and iridescence, with some species capable of glorious technicolour displays. If you are lucky enough to spot a squid on a reef, pause to watch it carefully, because it will constantly be shifting its colour, splashing complex patterns across itself for the world (and particularly other squid) to see, or carefully blending in with its background to hide in plain sight. They can even be deliberately two-faced, with males displaying soothing mating approaches to a female on one side of their bodies while broadcasting very aggressive 'keep off' signals to other males from the other side. But changing your skin colour is only useful if those changes can be seen. Typical squid signalling is impressive, but it relies on reflected sunlight to carry the message. A Humboldt squid's eyes are huge orbs that can be up to 8 centimetres in diameter, but even the best detector is only useful if there's something to detect. As a Humboldt squid stares out at a depth of 600 metres into the sunlight-free blackness with these big eyes, what does it see?

As it cruises along, tentacles stretched out in front, pinpricks in the darkness will constantly be lighting up around the squid's body: sparkles and flashes from tiny bioluminescent creatures, all sending their own signals to warn, attract or distract. These aren't messages for the squid, just the background optical chatter of a deep-sea ecosystem getting on with its day. A very faint glow may betray the presence of nearby fish, as their progress through the water tickles single-celled dinoflagellates which automatically light up when disturbed to give each fish a ghostly outline. But now the squid veers slightly to the side to avoid the drama happening just a couple of metres away. Another squid has darkened its skin and then rapidly drained one whole side of its body of colour as dark patches swell around its eyes. These broadcast a clear intention: to attack a fish. This second squid darkens as it strikes and then becomes pale again as it starts taking bites out of the hapless prey trapped in its tentacles. Throughout this display its fin edges shift between pale and dark, showing its size. All of this happens in just a couple of seconds, and the first squid knows to stay well away. The skin signals were clear, in spite of the darkness. The display was fully illuminated throughout, but not by sunlight.

Inside the gelatinous muscle of each Humboldt squid, there are hundreds of tiny globules each about the size and shape of a grain of rice. Each one contains a supply of two chemicals, one called luciferin and one called luciferase, and when the molecular machinery inside each globule brings them together, the result is a burst of bright blue light. The illumination they produce is inside the squid, not directed outward, and so it bounces around inside the muscle tissue. The outcome is that the muscly parts of a Humboldt squid's body (which is almost all of it) can glow bright blue, with particularly intense patches from the areas that produce distinctive skin signals. This glow provides a backlight for the patterns on its skin, and when the dark pigment sacs expand they form silhouettes against the glowing muscle. Even in the darkness, then, the Humboldt squid can broadcast its messages clearly, and the whole group of squid can coordinate their behaviour and their movement accordingly. Researchers have identified connections between specific skin patterns and specific actions, and it even seems as though the order of the signals may be important for interpreting the message. This is challenging research to do, and there are many more questions than answers. But the sophistication of underwater light signalling is mind-blowing, as creatures hidden in the ocean broadcast their messages outward, for the eyes of an unknown audience swimming through the same visual fog.

The physics of light in the ocean implies a disconnected underwater world, one where only local events are perceived, and only the immediate matters. But this isn't necessarily a disadvantage. Think about how complicated the world would sound if you could hear clearly the conversations of everyone within a mile of you; we would find it incredibly confusing. Up here on land, sound doesn't travel far, and when it does, only the most general information is available, slightly garbled by the merging of separate noises over long distances. This is the situation with light in the ocean: it's good for local communication only. We humans struggle to understand the world beneath the waves because our primary messenger – light – is demoted to a secondary role in the depths. But the long-distance connectivity isn't lost, because there is an alternative messenger available to take on that role. When you go below the ocean surface, you enter the domain of sound.

Down in the ocean world, it's sound that tells the great stories, that provides the long-distance links to mates and between habitats. Sound is the messenger that connects the ocean together. So let's take a step into the world of acoustical oceanography and listen in on the ocean's long-distance communication system.

Sound

In the western world, Jacques-Yves Cousteau is widely considered to be the giant of ocean storytelling,[16] and generations

---

[16] But not the first. That was William Beebe.

of documentary-makers now stand on his broad shoulders. From the 1950s onwards, his films, books and TV documentaries brought something entirely new into households across the world: a view of the ocean from the inside. He co-invented the 'Aqualung' – the first modern scuba-diving equipment – in the 1940s with his collaborator Émile Gagnan,[17] and opened the door to a more personal exploration of the seas. In the 1930s, William Beebe in his bathysphere had revelled in seeing the delights of the ocean, but Cousteau and his team weren't limited to mere observation of such living treasures – they could poke them, tease them, follow them, pick them up and bring them back for scientists to study. And, most importantly, they could film them, using newly developed underwater cameras, so that the rest of the world could join the adventure. It is undeniable that those films reset our view of the ocean, converting it from a frightening theatre of war into a wonderland of natural delights. Cousteau's first film, based on his first book, burst into the world in 1956, winning the Palme d'Or at the Cannes Film Festival and also an Academy Award for Best Documentary Feature. This was the ocean's great debut on the world stage, watched by kings and queens, politicians and artists, tradespeople, office workers and wide-eyed children. So it is to the lasting bafflement of anyone who studies the ocean that this great and important work of art had a name that was deeply misleading: *The Silent World*. The ocean certainly isn't silent.[18]

[17] The word 'scuba' didn't come along until someone else invented the acronym in 1952: self-contained underwater breathing apparatus. Cousteau always referred to his equipment as the Aqualung, which was its commercial name, but an Aqualung is a type of scuba equipment.

[18] From today's point of view, the title is a minor discomfort compared to the film's content. It's still educational, but mostly as an eye-opening example of how the bounds of 'acceptable' change over time. In his later work, Cousteau showed much more concern for the natural world, but in this first film his

Few film-makers would have resisted the temptation to give the ocean such a dramatic and spookily alien character. But the other reasons why this title might have passed without much comment are much more interesting, because they're woven into the nature of underwater sound itself. The water, our own anatomy and our methods of exploration are very effective at hiding nature's subaquatic music from us.

It starts at the surface. The familiar sound of the ocean is breaking waves at the beach, waves that roll in from far away and roar as they tumble and splash, generating billions of bubbles which fizz gently as they burst. But these are surface events, not the sound of the ocean's innards. On a completely calm day, no sound from the ocean reaches our ears, and that's because the ocean surface itself acts like a double mirror. Sound coming from below is reflected back downwards and sound coming from above is reflected back upwards. The worlds of above and below are acoustically almost entirely separate.

It takes an enormous effort to squash water and make it a bit smaller by applying pressure.[19] If you shove on water, those molecules, instead of being compressed into a smaller space, push into the ones ahead of them, and those ones quickly adjust by shoving on the ones ahead of them, which shove on the ones ahead of that. The shove then travels forward through the water without the molecules themselves having to go anywhere. A regular sound wave under water consists of a travelling pattern of very slight squeezing and stretching as the water molecules

---

team's casual treatment of the animals and environments they encountered is utterly shocking by today's standards. On the plus side, it shows that attitudes can change dramatically, so there is hope for those of us who feel we still have much further to go.

[19] Even with the weight of 10 km of ocean on top of it, the water at the bottom of the Challenger Deep is only squashed by about 5%. And that is a colossal pressure, well above anything we can see or generate closer to the surface.

wobble backwards and forwards in the direction of the sound. But when a sound wave reaches the surface, the air is so thin by comparison that the shove hasn't got anywhere to go. There isn't really anything that all those densely packed water molecules can push on. And so the wave reflects off the ocean surface and travels back downwards. As for sound from above the surface, sound obviously *can* travel in air, but when sound from the atmosphere hits the ocean it's like hitting a concrete wall by comparison with the rest of the air. So it just bounces off and goes back upwards. Water and air are so different from each other that sound travelling in one can barely penetrate the other. You have to cross that border physically to have any hint of what's on the other side. Humans can do that quite easily by just holding their breath and jumping in.

Jacques Cousteau was an experienced free diver before he started experimenting with what became the Aqualung. Free divers train themselves to hold their breath for several minutes, so that they can leave the surface behind and swim around pretending to be fish until they run out of oxygen and have to resurface. Cousteau had literally immersed himself in the world of underwater sound for years, and it clearly didn't strike him as particularly interesting. However, we can probably forgive him that, because human ears can't do their job properly underwater. The human ear has three parts, which perform (very roughly) three jobs. The bit most of us think of as our ear, the curved cone that sticks out of your skull, is actually just your 'outer ear'. Its main job is to funnel sound inwards and take it inside your head. This conveys the sound to the middle ear, which is an air-filled cavity containing an ingenious mechanism for converting sound in air into sound in water. And then the sound can pass into the liquid-filled inner ear, which is where the hearing process actually happens. So when you listen, you're actually listening to sound in the liquid in your inner ear. But that sound can only get there by passing through the outer ear and middle ear, and they operate on air-filled systems. Now we can see the problem, because every air–water boundary acts as a two-way mirror for sound. When you dive into water, the sound in the water around you meets a system that only works in air, and the critical entry point also is full of air. When sound gets from the water to that boundary, it just bounces off. You could be literally surrounded by underwater symphonies, and none of it would go in.

Except ... we *can* hear underwater. A little bit. There's a bypass. Water is dense and almost incompressible, and sound will travel from water into other materials that are similar to it in those respects. Bone isn't a perfect match, but it's pretty close, and when you're underwater, the bones of your skull and your jaw are right next to the water. So this offers a route inward for sound that is blocked by the air barrier in your middle ear. Sound arriving at your body can travel directly from the water into your jaw or your skull, and then along the bone towards the inner ear, circumventing the air-filled outer and middle ears.

Next time you duck down and take a couple of swimming strokes entirely underwater, consider the new physical connection between your head and the water and pay attention to what travels across it. It's not a particularly efficient way of hearing, and it works much better for low notes than for high ones. The deep sounds are conducted towards your inner ear, and you might suddenly become aware of the rumbling of nearby traffic or the pumps hidden under the concrete pool floor (if you're in an urban swimming pool), or boat engines (if you're at sea). All of that sound was there in the water all the time, but none of it ever crossed the double-mirror surface into the air, so you had no way of knowing what you were missing out on. But you won't hear the high notes as easily – the splashing from the surface or someone tapping the wall on the other side of the pool – because these don't travel through the bone so well. Sound heard through your head takes on a jumbled, grumbly character, and it's less distinct. But it is there. Free divers can certainly hear some of the sound that's passing them by, so Cousteau must have known that there was some sound underwater.

But possibly the most important reason why Jacques Cousteau shied away from the sounds of the ocean in his documentaries was that his filming method involved creating vast quantities of one of the noisiest objects in the ocean: the humble bubble. As we've seen, it's very hard to squash water, and so sound can travel a long way through it, because if you squeeze water molecules together even slightly, they quickly push back on the other molecules around them, passing the squeeze on. But a gas bubble is very squishy indeed. The Aqualung was designed so that when a diver breathed in, a valve would release gas from his tank at exactly the right rate to fill his lungs. When the diver breathed out, that gas escaped directly into the water, breaking up into a plume of quivering, shimmering globules of air that wobble their way upwards towards the surface and the sunlight. At the moment when each air pocket breaks away from the rest of the exhaled gas, the almost-bubble will be distorted and stretched, but still connected to the rest by a short neck of air. When that neck of air snaps, the bubble pings back towards a spherical shape, rapidly squashing the air inside it. The squashed air then pushes back outwards on the water around it, expanding until it overshoots its original size. And then the surrounding water squeezes it inwards again – and so the bubble expands and contracts until the gas settles down.[20] The regular shoves on the water in this 'breathing mode' oscillation create a sound wave.

If you pour out a glass of water and hear a tinkling sound, or you hold an empty bottle underneath the surface of a pond and hear the glug-glug as it fills up with water and ejects bubbles of

---

[20] The expansion and contraction are only a tiny fraction of the total bubble volume, so sadly this process is too subtle to see directly with the human eye.

air, this is what you're hearing: the ringing of each bubble at its moment of creation. Big bubbles bellow deep tones and smaller bubbles sing at a higher pitch, and each one adds its note to the music. Each new bubble is loud, and Cousteau and his divers released many of them with each scuba breath, a cacophony of creation right next to the diver's head. There is no doubt that this would dampen any diver's appreciation of the natural music of the ocean.

The adventures of Cousteau and the crew of his ship R/V *Calypso* painted a dramatic picture of the life of an ocean explorer. Bobbing red woollen hats, casual cigarettes dangling from the lip, technical work on the back deck of the ship, minds focused solely on the tasks of preparing the bright yellow tanks of air for their forays into the deep: this was the life of those who could transform themselves from man to fish and back again. But as far as the ocean itself was concerned, these were almost silent movies. There were microphones on the cameras – you can hear the scuba bubbles in the back of many shots although there's no natural ocean sound. These documentaries are cloaked in human narration and orchestral music, conveying a human composer's interpretation of the drama: flute trills, haunting oboe solos, triumphal brass and climactic piano chords. The ocean itself never got a say. This celebrated debutante was mute, and the misconception that the ocean is silent was never challenged. The visuals were easily enough to win prizes and popular acclaim, but they only told half the story.

So when you remove the human soundtrack and unmute the ocean, what do you hear? What messages were those early ocean stories missing?

Haddock hubbub

In the chilly water of the Norwegian Sea, a haddock is defending a rocky patch of sea floor in the gloom. Haddock are medium-sized fish, perhaps 50 centimetres long, with a dark back, a silvery tummy and a distinctive dark blotch just above their pectoral fins. Down here, 150 metres beneath the water surface, light is scarce and they are hard to see. The long, much-indented coastline of Norway twists and turns northward from the North Sea for 1,000 miles, extending far past the Arctic Circle and up towards the white expanse of the frozen Arctic Ocean. It's up here, in the far north, that this haddock is staking out its place in the world. Its territory sits on the shallow continental shelf, where nutrients mix easily from the bottom to the top of the water and the weak spring sunshine is just starting to fuel this season's new life. The sea floor here is divided by neighbourhood tiffs, each fish jealously guarding the borders of its own domain. Haddock spend most of the year grubbing around in the sea-floor sediment for their food, gobbling snails, clams, worms and occasionally smaller fish. Then when spring arrives they move to specific spawning grounds and the males pick their territories within the chosen area, creating a supermarket of potential mates for the females to peruse.

The flirting starts with the male haddock swimming in a repeating figure of eight, tightly circling its patch. And then the sound begins, dull thuds in a long sequence, as the fish advertises himself to the gloom. From his competitors near by come more knocks,[21] each with a slightly different pitch, pattern and repetition rate, until the clamour can be heard from hundreds of metres away. This is the nightclub of the hopeful, and the females will approach from above, pausing for a closer inspection if a particular drummer appeals. Mixed in with the audio adverts are the signals of success, since as the courtship develops the drumming gets faster until it blurs into a continuous hum. Sound is being used to coordinate the whole party, acoustical

---

[21] This is the technical term for the sound. The scientific papers on this topic are admirably restrained in respect of the obvious 'knocking shop' jokes.

messages rippling through the water in all directions, organizing the urgent need to mate into the fertilized eggs of the next generation. It's an impressive racket, and it's only possible because each fish carries its own internal drum kit.

As we have seen, water is unforgiving when it comes to tiny mismatches in density. If you are more dense than the water around you, you will sink. If you are less dense, you will float. If you don't want to do either of those things, you only have two options: swim continuously to keep yourself in place, or adjust your density so that it matches the water around you perfectly. The bony fishes, which is almost all of them except for sharks and rays, take the second option. Just behind the head of each fish there are two linked sacs lined with a silvery layer of guanine crystals that keep them airtight.[22] The fish can inflate the sacs like mini balloons, continually adjusting the amount of air inside them so that they are always neutrally buoyant and neither sink nor float. This apparatus is the swim bladder, and its most important advantage is that it lets fish stay in one place without having to expend any energy. But once you've got an air-filled sac, you can use it as a musical instrument without affecting its important work as a buoyancy aid, so the swim bladder is also the haddock's voice box. These fish have a pair of 'drumming muscles' attached to the outside of their swim bladder that twang the bladder at high speed to produce the knocking sound. The twanged bladder pushes and pulls on its surroundings, sending a pulse of sound outward. And just as scuba gas bubbles are efficient at producing sound, so this internal gas bubble is also efficient and therefore loud. Not every fish

---

[22] This is the same guanine that is the G of the DNA letters A, T, C and G. Guanine crystals are very useful biological building materials, mostly because they're iridescent. Guanine crystals are also what make fish scales silvery, and they've been extracted for years to add a natural shimmer to nail varnish, shampoo and metallic paint. It's very versatile stuff.

species with a swim bladder uses it as a musical instrument, but both male and female haddock excel at this additional skill. The females create a slightly different knock, and they mostly use it to warn off other fish when foraging. These are critical messages which cannot be sent in any other way, and this species relies on sound to coordinate its busy social life.

The haddock is not an exception. The list of noises that different species of fish make includes cackling, croaking, chirping, mooing, puffing, muttering, squeaking, grinding, groaning and howling, and that's only a small fraction of the total. It's not all done using a swim bladder – fish can also scrape body parts (say, fins or bones) together to create a wide range of pops and cackles. We may not think that their whistles, honks and howls have the sophistication or beauty of birdsong, but that's entirely a matter of perspective. The nightingale may sing in Berkeley Square in London, but the toadfish is singing from underneath its West African coastal rock and it's just as determined to make sure that everyone in the surrounding area hears its broadcast. Fish don't have a full musical range in the same way that birds do, but they're sending messages using sound for exactly the same reasons. But of course, all this singing is pointless if no one can hear it. So we have to ask where fish keep their ears.

As the female haddock cruise through the water above the patchwork of male territories, the water around them is full of sound waves, each one a travelling nudge rippling through the mob of water molecules. When the sound waves arrive at the haddock, the wave barely needs to change – the water pushes on the fish skin, and the fish is mostly water so the wave just keeps on going into the fish. The haddock doesn't need to keep her ears on the outside, because the sound goes right through her. But just behind her head, the sound pushes on something different. Instead of pliant fish flesh, here it encounters a solid lump of calcium carbonate three times as dense as water. The push from the sound can move this lump, but more slowly. The molecules in the fish flesh are vibrating as the sound wave passes through her, but the dense lump – known as an otolith – can't quite keep up. This is the fish's ear,[23] hidden deep inside her body but still easy for sound to reach.

The otolith rests on a carpet of delicate sensing hairs, which detect the mismatch between the otolith movement and the chamber it rests in. The hairs form the sound detection mechanism: if an otolith is rattling about inside the fish, it must be because the fish itself is vibrating with sound and the otolith is lagging behind. The hairs send signals to the fish's brain, telling her how much movement the sound generated and how rapid the vibration was. And so the female turns downwards, tempted by the thuds from the closest male. The clever thing about this sensing mechanism is that she can tell exactly where the sound came from, because the collection of sensing hairs can identify the direction of the otolith vibration relative to the fish. Imagine putting a ping-pong ball on the flat palm of your hand. If you waggle your hand side to side, the ping-pong ball will also move from side to side, after a moment. But if instead you shake it backwards and forwards, the ping-pong ball will wobble backwards and forwards, and the otolith behaves in a similar way.[24] So the fish has no problem locating her chosen paramour among the clamour of the surrounding wannabes.

Once the noisy orgy is over and the haddock are back to their normal routine of gliding around just above the sea floor and keeping an eye out for lunch, sound becomes a way of tuning

[23] Haddock have three pairs of these; the number and arrangement differ for other species.

[24] This is a relatively unexplored area of science, but it looks as though the way the otoliths are shaped and the influence this has on how they vibrate may add a huge amount of subtlety to the fish's hearing ability. Otoliths are differently shaped and differently positioned in every species of fish – they're so distinctive that they can sometimes be used to identify the species even without any other part of the fish.

into the local environment. Fish swim inside a flood of sonic messages about their surroundings, some of which are useful and some of which are just part of the background. When the wind rises in the atmosphere, far above the fish, breaking waves at the surface generate billions of bubbles, more and more as the wind increases, and the tiny peeps from the formation of each new bubble merge to form a constant background murmur. Rain can also generate noisy bubbles, with an acoustical character that depends on whether the source is a big thunderstorm or slow drizzle. The weather is heard inside the ocean, rather than felt. Up here, where the Norwegian coast is exposed to the storms rolling in from the North Atlantic, the weather up above is constantly making its presence felt underwater. Next there are the other fish species, grunting, honking and wheezing as they fight, feed and hunt. Pods of dolphins will pass by overhead, chattering to each other in high-pitched squeals and whistles while probing their environment with the staccato clicks that they use for echolocation. And from the distance, pouring out of the darkness, there are the deep groans, growls and grunts from humpback or minke whales.[25] Almost every species will be selectively deaf to some of those sounds – fish hearing is extremely variable, so they hear only part of the symphony. But almost all fish can hear. It's the only way to find out what's going on elsewhere in the gloom. The sounds may be subtle compared with the dawn chorus of the land, but even up here in the frigid north, the ocean certainly isn't silent.

Our human ears prevent us from being natives in the world of underwater sound, but over the past hundred years or so, we've developed technology that lets us eavesdrop. As with many areas of science, humans went into it assuming that they pretty

---

[25] If you would like to hear any of these sounds for yourself, you can find plenty of recordings of both natural and artificial underwater sounds at https://dosits.org/galleries/audio-gallery/.

much knew what they were going to discover, and that refining their instruments would just make their existing simplistic picture clearer. And as usual, they were wrong.

Stumped by sound

In June 1942, off the coast of San Diego in California, a collection of small creatures were scooting about in the darkness deep below the ocean surface. For a brief moment they felt a buzz, enough to put them on high alert for a second. But there was no sign of nearby predators. They returned to their search for food, unaware that they had just set in motion one of twentieth-century oceanography's most stubborn whodunnits.

Humanity had first realized how useful it would be to spy on the underwater world thirty years earlier. The sinking of the *Titanic* in 1912, and the influence of submarines on First World War battles, had highlighted a significant blind spot in human knowledge of the world. The lack of any way to peer into the ocean was becoming very frustrating. A potential solution did exist, but it wasn't simple to implement. Underwater sound reflects off anything that has a different density from water, and a submarine definitely comes into that category. So if you could send sound into the ocean you could listen for the echoes as it reflected off acoustical obstacles. Those echoes would even indicate the distance to the obstacle, as long as you could measure the time between your signal being sent and the echo coming back. It's far easier to do this with sound rather than light, because sound travels relatively slowly underwater – faster than in air, but slowly enough to be easily measurable.[26] By the start

---

[26] In 1826, Jean-Daniel Colladon and Charles-François Sturm dipped a bell beneath the surface of Lake Geneva, and then rung the bell at the same time as letting off a charge of gunpowder. Ten miles away, the visible flash from the gunpowder was compared to the time that it took the underwater sound to

of the Second World War, scientists and engineers had rudimentary equipment that could do this job (which later became known as sonar), but it was tricky. They were sending sound out into the vast unknown of the ocean and trying to interpret what came back. And the submarines weren't the only things that made echoes.

In 1942, three scientists – Carl Eyring, Ralph Christensen and Russell Raitt – were working off the coast of San Diego on the USS *Jasper*, a ship that had started life as a yacht before it was pulled into the US Navy to help with wartime science. Their job was to work out what sonar could tell them about what was going on beneath their ship. Problem number one was that it looked as though the sonar couldn't even really tell where the sea floor was. A fuzzy echo was coming back from around 300 metres down, even though the sea floor was 1,300 metres below them. The real sea floor was often detectable below this strange layer, which was a persistent shadow on every single measurement. Reports started to come in from other early users of sonar who were seeing the same thing. Something down there reflected sound, but it wasn't a solid object. Whatever it was seemed to be everywhere, in every ocean, between 300 metres and 1,500 metres beneath the surface, and a long way off the bottom. It mattered, because that sort of foggy uncertainty could potentially hide any submarines that were below it. Sonar was still useful, but the puzzle remained. No one could think of a reason why temperature or any other physical feature of the ocean engine could cause such a thing. And then someone noticed that the layer moved – wherever you were in the world, as dusk fell, it came closer to the surface; and then just before sunrise it went

---

arrive, and so the speed of sound in water was measured for the first time. The result, in the cold fresh water of the lake, was that underwater sound travelled at 1,435 metres per second, which was three times faster than sound in air but slow enough to measure.

back down. It became known as the deep scattering layer, because scattering sound was what it did, but no one knew why. Nets sent into it came back with almost nothing: the odd shrimp, a few jellyfish, one or two small fish. Nothing that could block, or divert, a strong sound signal.

Debate about the culprit became an occupational hazard for acoustical oceanographers, particularly the biologists among them – it moved so regularly and so quickly that surely it had to be life, not a layer of ocean water. Large cross-linked colonies of small predatory animals called siphonophores can carry a bubble of gas at the top of their bodies – could the sound reflect off those? Lanternfish had a swim bladder, so that was a nice acoustical obstacle – could it be those? But the nets went down and came back without answers, while the submersibles also explored and came back empty-handed. Whenever anyone went into the layer, it seemed indistinguishable from the water above and below it. Sometimes the layer split into two as it rose, and sometimes it didn't. Sometimes it started as two or three layers, depending on what type of sound you used to look for it. And all around the world, the sound would show something going up and down.

It turned out that the culprit was a fish.

Lanternfish have been swimming around the Earth's oceans for more than 50 million years. This family includes around 250 species which are distributed all over the world. Their silvery bodies are teaspoon-sized, with big eyes and delicate fins, and they cluster in dense layers to avoid predators. They're an easy and abundant lunch for lots of larger species, and so they're skittish, rapidly moving away from any disturbance in the deep. They live up to their name, carrying rows of light-producing patches on their lower side and head, capable of emitting a blue, green or yellow glow to help camouflage themselves against the weak light from the ocean surface. Lanternfish are elusive, unbothered by the human world, part of the vast web of ocean life that is just getting on with the business of staying alive.

Small fish are vulnerable, and in the darkness of the ocean, hiding is survival skill number one. The closer to the surface you get, the more food there is, but also the more visible you become. And so lanternfish hide in the depths during the hazardous daylight and swim upwards to find zooplankton to feed on under cover of darkness. They have a swim bladder which throbs when sound goes past, stealing energy from the sound wave, and then losing it again as the pulsing sac re-emits that sound to its surroundings. The fish can't outrun sound, but they are incredibly sensitive to any other disturbances, fleeing if anything comes within 20 metres of them. This is why it took so long to discover their role in the deep scattering layer: by the time the scientists trawled through the layer with a net to find out what was there, the lanternfish were long gone. The first task of a lanternfish is not to be found, and it's very good at it. But they can't hide from sound.

The scientists could 'see' them, but they had no idea what they were looking at. The physicists and biologists kept looking and kept arguing, writing more papers and more rebuttals. Then, in the mid-1960s, a clearer picture began to emerge. Lanternfish are now known to make up a huge proportion of the deep scattering layer, although their skittish nature kept this secret from scientists for a long time. Millions of tiny swim bladders create an acoustical obstacle course through the compact layers of fish, as biology gets in the way of ocean physics.[27] But other organisms contribute too, and when two or more separate layers are observed, they're usually caused by different species behaving independently of each other. Any organism that swims

---

[27] We now know that they are only visible to certain frequencies of sound, and so submarines can't hide beneath their dense layers, because switching frequency will render them transparent.

from the depths of 1,000 metres up towards the sunlight acts as a link between the ecosystems of the surface and those of the deep ocean, an elevator for nutrient transport.

Around the world, this sound-scattering layer of life rises and falls every day. It's one of the most significant biological features in the global ocean, and it was only discovered because of its effect on sound. Had humans relied on nets and what they could see, we still might not know about the existence of so many lanternfish, a whole layer of life that was visible only with the right messenger. Learning to explore with sound opened doors to new treasure troves of oceanographic understanding. But what we missed was the opportunity to study an acoustically pristine ocean, because we were generating our own underwater sound long before we learned how to detect it. We couldn't hear it, so we didn't notice. But some ocean citizens certainly did.

Stories from a whale's ear

In the 1940s, the urgency of war had forced humans into the realization that the only effective long-distance messengers in the ocean are sound waves. Sonar, not searchlights, was the way to unmask the mysteries of the dense and fluid ocean. But sound is just a by-product of movement; anything that wobbles the molecules in liquid water will generate a gentle pressure pulse that ripples outward from the source. Haddock do this deliberately by drumming on their swim bladders, and the early sonar could do it deliberately by forcing a piston-like transducer to push and pull on the water. But the usefulness of sound as a messenger comes from its universality: *anything* that causes a shock or vibration underwater will send its acoustical influence outwards into the deep, whether intentionally or not. The sound-scape of the ocean is made up of the acoustical fingerprints of a thousand distant events as their ripples pass through your location.

And so the initial use of sonar was not the first time that humans had sent acoustical messages down hundreds of metres into the depths. It was just the first time we'd done it deliberately. Nature had been listening in on humans for years because sound travels a long way underwater and humans are hardly subtle. Sound may be ephemeral, here today and gone in an instant, but in some cases it can leave signatures in solid form too. Some of these are currently residing in jars full of alcohol in a London basement.

'You have to do a lot of work to get it out.' Richard Sabin is the Principal Curator of Mammals at the Natural History Museum in London, and he's holding a large glass jar containing a brownish stick with one pointed end, which is a bit larger than my thumb and swimming in colourless liquid. This is a part of a whale's anatomy which has to be seen to be believed, because surely no one would go looking for it unless they'd already discovered it by accident. 'It's a bit like getting a walnut out of its shell,' says Richard; 'you've got to go through a lot of muscle tissue to get to it, and then once you've found it you have to be very careful about peeling the tissue away.' At the end, instead of a walnut, you have a very unexpected prize: a plug of earwax. Whale earwax. Which is now stored here, behind a nondescript door in a huge maze of tall grey metal cabinets, pickled in alcohol for posterity.

Museum curators have to be hoarders by design, because their job is to be ready whenever someone knocks on the door and asks for a specific piece of the past. Even so, this plug seems like an odd choice when there are so many other parts of a whale that could be stashed away in museum basements. But whale earwax is one of those rare substances where evolutionary coincidence indulges scientific curiosity so generously that it feels like winning the lottery. Written in this brown gunk is the entire life story of the whale, starting at the pointed end and progressing down to the base. This is scientific gold.

It's not clear why whales should have earwax at all, although their extraordinary evolutionary path means that a lot of options were on the table. The ancestors of all whales were small land mammals with four legs and a long tail, and ears with a similar structure to ours. Around 50 million years ago, these mammals started to shift into the water, becoming over millions of years better swimmers and more adapted to underwater life. Their outer ears – the bits that stuck out from their head – shrank as the advantages of being streamlined grew. But the tiny tube that led inside to the sound-sensing parts of the ear remained intact. If the early whale ancestors dipped their heads below the water surface, sound could travel directly from the water into their skull and jaw, bypassing the tube but still reaching the sensing parts of the ear. And then, as they started to dive deeper, the outside of the tube closed over, protecting the sensitive innards from the water.[28] A whale's ear is now completely sealed off from the outside, accessible only to sound that travels through bone to get there. But in baleen whales (the gentle giants that vacuum up plankton, like blue, fin and humpback whales) the leftover tube is still there, still lined with skin and still producing earwax.

Richard puts the jar back in the cabinet and shows me a diagram of the bony capsule called the tympanic bulla which is attached to the whale skull and contains the sensing parts of a modern baleen whale's ear. The base of the tube is next to this bony capsule, and throughout the whale's life it keeps busy, producing mostly fats during the whale's feeding season, and a fibrous protein called keratin while the whale is migrating and

---

[28] If you go to the Natural History Museum in London and look at their life-sized model of a blue whale, you can see a tiny dimple in the skin, next to a white spot that's about a metre behind its right eye. That dimple is where the ear used to open to the outside world, and the plug of earwax is buried directly below that.

fasting. This discarded material has nowhere to go except up the tube, so it squeezes outwards until it is stopped by the outer skin, with one dark and one light stripe for every year that the whale is alive.[29] There is no outlet and no way for it to be recycled, so it just keeps building up. It's not clear why evolution hasn't discarded the tube, but scientists like Richard certainly aren't complaining.

'You know the way the police will take a hair from someone to determine whether they've taken drugs in the recent past?', he says. 'It's the same principle here.' This is a scientific treasure trove, but the real prize only comes when you combine the latest techniques of modern science with decades of careful work by museum curators. Fixed in the earwax gunk are tiny traces of hormones such as cortisol, which can be used to track stress, and progesterone, which can be used to track pregnancy over the long life of a individual whale – perhaps twenty or thirty years. Museums have old specimens, and careful records of their exact origins. Richard and his colleagues worked out that the Natural History Museum and the Smithsonian Institution in the United States together could assemble plugs of earwax that covered 146 years of history. And that means they could map out how the stress of the global whale population has changed over time. Pollutants are also stored in the earwax, so that the plugs hold a record of ocean health as well as whale health. These nondescript brown sticks with the consistency of waterlogged wood offer glimpses into the historical experience of ocean life: what was it *like* to be a whale in the past?

The answer over the last 150 years is: patchy, but overall pretty terrible. This is entirely because they have had to share a planet with humans. Once the cannon harpoon was invented,

---

[29] These alternating stripes of light and dark are the original reason why the plugs of earwax were kept, since it seemed likely that they could be used to calculate the age of the whale even before that was scientifically confirmed.

the blue, fin and humpback whales became easy fodder for commercial hunters. Industrial whaling had mini-booms before and after the First World War, but whaling really took off in the 1950s and 1960s, when many tens of thousands of whales were slaughtered every year. And the whales certainly knew what was going on. Their cortisol, the stress hormone, follows almost exactly the same pattern as the number of whales caught, reaching a painfully high peak that coincides with the whaling peak in the 1960s. But there's something else in the earwax too. Whaling largely stopped during the First World War, and especially the Second. The humans might have been having a bad time, but the whales could perhaps have had a rest from being hunted. And yet whale stress shows an additional clear peak in the early 1940s, during the period when whaling stopped. This is when transient sounds became embedded in the physiology of the whale.

'Whales and dolphins have become completely dependent on an acoustical world,' Richard says. 'Baleen whales are solitary individuals, communicating and trying to find a mate across hundreds or thousands of kilometres.' Sound is the only messenger these whales can use, the only way they can connect with other members of their species who are more than a few tens of metres away. And during the Second World War, humans filled the ocean with the sound of battleships, depth charges, torpedo strikes, submarines and plane crashes. Even if the whales weren't right next to the action, the sound would have travelled easily through the ocean, pervading this normally calm environment. The stress of all of that sound interfering with their daily lives was comparable to the stress of being chased by humans who had only death on their minds. We don't have a direct record of the volume of this war-generated sound, but we can see its effects. The whales' most effective messenger was drowned out in the racket, and it mattered. It's recorded in their physiology, written in the earwax for us to find and acknowledge. War created a vast amount of noise pollution, and although the pollution itself was fleeting, its effects weren't. This is a global phenomenon, consistent across whales of three different species, in both the Atlantic and the Pacific.

I ask Richard about that consistency, and he presents an interpretation that would not have occurred to me. 'Was the sound produced reason enough to cause this stress? Or are we seeing the distress that's being caused to whales in one part of the ocean being communicated by those who have survived to other whales? We just don't know enough about whale culture to answer that question.'

The thought of a global population of whales somehow sharing the news that there are threats out there, that vigilance is necessary, all because humans fancied killing each other instead of the whales for a few years, is chilling. But the clear message is that although we are not primarily sound-based creatures, most of the species that inhabit the ocean are. We may not prioritize sound, but we still create it and it still travels incredibly efficiently through the ocean. We may not care about filling the ocean with sound, but we are creating an acoustical fog that means ocean life has to struggle to communicate.

Curators like Richard see their job as connecting the preserved past to the potential of the future. It's not just about putting items on display to educate and entertain the public, but about protecting and developing our shared repository of planet Earth's history. London may seem a long way from the ocean, but the Natural History Museum holds a huge library of ocean history that is just waiting to be read, once we develop the tools to do so – perhaps in twenty or fifty or two hundred years' time. I find it incredibly exciting that so many undiscovered stories are right there, holding on to a past that we cannot visit, but that still has a lot to teach us.

As we have seen, sound is *the* critical messenger in the ocean. But it doesn't travel completely unimpeded. The ocean deflects, sculpts and absorbs it, and so even though sound can travel for thousands of miles underwater, the sound itself carries the imprint of its surroundings. And if sound travels far enough, it will carry a snapshot of the entire global ocean. This was the premise of one of the most audacious experiments ever undertaken in acoustical oceanography, one that took everything humans had learned about sound underwater and applied it to the study of Earth's climate.

'Wish I was with you, but glad I'm not'

In January 1991, at the beginning of the first Gulf War, the eyes of the world were on Iraq, Kuwait and the United States. With conflict dominating the news, relatively little attention was given to one of the other things happening that month: a strange project run by a bunch of oceanographers with the lofty aim of taking the temperature of the entire global ocean in one go. As the first bombs were dropped in Iraq, two US Navy ships were 6,000 miles from the fighting, heading away from human civilization and out into one of the most remote parts of the global ocean. Their destination was a bleak ice-covered volcano called Heard Island, which pokes through the sea surface 1,000 miles off the coast of Antarctica, about halfway between the southern extremities of South Africa and Australia. The seas here are cold and fierce, and the only links between the ships and the rest of humankind were crackly telephone lines and a couple of fax machines.[30] Although the big

[30] For those under the age of thirty, fax machines were how people sent pictures and documents over a telephone line before the internet came along. They were actually invented before the telephone, since telegraph wires could also be used to send the signals that would tell you whether each square on a

question was about ocean temperature, being able to answer it depended on a practical question about sound: could you send one sound through the *entire* ocean, a single acoustical message to the whole globe?

The ships may have been remote, but they weren't completely disconnected from the rest of the world, and so the plan was to use satellite telephones to check whether the messages sent through the ocean using sound had been received. It was very much reminiscent of the early days of email, when people would send an email and then ring up the recipient to check that it had arrived. But in this case, it wasn't the acoustical message itself that mattered most. The important bits were what the ocean did to the message along its journey, and whether it would even reach the receiver at all.

Even though the main focus of the news media was elsewhere, the world's oceanographers were definitely paying attention. Regular faxed updates from the ship came in the form of the 'Heard Island Science Daily', a 2–3-page document detailing their location, marine mammal sightings and the progress of the experiment. Roger Revelle, eighty-one years old at the time and one of the world's most celebrated oceanographers and climate scientists, sent the scientists on board the project's ships a fax just before the experiment began saying: 'YOUR MESSAGES FROM DOWN UNDER ARE WONDERFULLY INTERESTING. WISH I WAS WITH YOU BUT GLAD I'M NOT.' But the scientists weren't bothered about whether the eyes of the world were watching. What they cared about was

grid should be white or black. If you accidentally rang up a fax machine, you would hear the distinctive beeps that were how these machines communicated with each other, and it usually took several minutes to send enough beeps for a single page. Fax machines were everywhere in offices in the 1980s and 1990s.

whether the ears of the world were listening. And a lot of ears were ready and waiting.

The Heard Island Feasibility Test, for this was its name, rested on three foundations. The first was that human-caused climate change was very much on the agenda, and there was a lot of discussion about robust ways to track it. One useful measure would be an average temperature for the oceans, since if Earth was heating up, there was good reason to think that a decent chunk of that extra energy would end up inside the blue machine. So the search was on for a global ocean thermometer.

The second foundation harked back to 1960, when American and Australian ships had deliberately set off a series of large underwater explosions near Perth in Australia. The sound of the blasts had been picked up by hydrophones (underwater microphones) on the opposite side of the world, in Bermuda, an astonishing 12,315 miles away. The sound had taken around three hours and forty-three minutes to cross the southern Indian Ocean and then travel diagonally all the way across the Atlantic to reach Bermuda. But the oceanographers thought that this travel time was a bit odd. It implied that the sound waves hadn't travelled along a direct line from source to receiver, as might be expected. The ocean itself had apparently tweaked the path of the sound, and had maybe even created multiple paths for it. That suggested that you could learn things about the inside of the ocean by following where such sound signals went and how long they took to get there.

The third foundation was the most important of all, because this is what makes it possible for sound to achieve the ludicrous feat of travelling halfway around the entire planet while underwater. It had been discovered in the 1940s that the invisible structure of the ocean engine – its distinctive layers, with their varying temperatures and pressures – corrals sound, creating acoustical barricades that enclose an efficient long-distance communication channel. The ocean itself can act as a sound guide, so that the deepest notes can travel immense distances with almost no loss of signal along the way. It's an astonishing natural feature. The contrast with light couldn't be greater: as we have seen, even the biggest brute force attempts can only force very low radio frequencies a hundred metres or so into the water; but sound of the same frequency (around 60 Hz) can whoosh along a natural and ready-made acoustical highway for tens of thousands of miles almost untouched. If you're interested in long-distance acoustical communication underwater, you're in luck: the ocean has a built-in system for doing it.

Take an imaginary journey down into the Atlantic Ocean. At the surface, bright sunshine glints off choppy waves and there are flecks of white in all directions, as breaking waves leave brief bubbly signatures behind. As we have seen, as you sink beneath the surface the red and green light will quickly disappear, leaving you in a blue foggy pool of increasing darkness. The water around you has sucked away the light and converted it to heat, so this surface layer is a very comfortable temperature, around 28°C. This is the mixed layer, the warm lid of the ocean which we have already met. As you sink further down, the light fades but the water stays warm because the mixed layer lives up to its name: constantly stirred by winds, the water is a similar temperature throughout this thin surface layer. Even though little sunlight reaches the bottom of the mixed layer, this water still holds plenty of warmth because it regularly mingles with everything above. But as the last glimmers of light begin to fade, at perhaps a depth of 80 metres (a little under the height of the famous clock tower in London that houses Big Ben), we reach the bottom of the mixed layer and the water starts to change. As you continue to sink, it quickly becomes colder – much colder. Sink down a little further, into the intense cool blackness, and then let's pause in the depths, 800 metres beneath the surface, where you can't see anything except a few bioluminescent sparkles. You are deep inside the blue machine.

From down here, we can appreciate something that we couldn't detect above. The speed of sound has changed, and that's because sound travels faster in warmer water. At the surface, where the temperature was 28°C, sound could move at 1,542 metres every second. But down here, where it's 10°C, the sound speed has slowed to 1,504 metres every second, 2.5 per cent less. That may not sound much of a difference; but it matters, because in regions where the temperature is changing, the water can actually steer the sound just because of its temperature. If you imagine a line of people all holding hands and walking straight forward, the line will advance without changing direction. But if the people on the right-hand side of the line start walking faster, the whole line will swerve to the left. Sound does the same thing. If you were to ring a bell down here at 800 metres and follow the sound rippling out sideways, the top of the sound wave would be travelling slightly faster than the bottom. That will make it swerve away from the surface, and head downwards at an angle. Travelling sound waves bend away from higher sound speeds.

Let's follow the diverted sound and keep sinking downwards. Now you start to notice something else. The deeper you go, the more the ocean squeezes on you. The water above you is literally pressing down on you, and the deeper you go, the greater the pressure. But pressure also changes the sound speed, and if we go down to 2,000 metres depth at the same temperature, the sound speed has gone back up to 1,524 metres every second. The situation has reversed, and now the sound speed below is greater than the sound speed above. So the sound will bend back upwards again, until it meets the warmer water and swerves back downwards.

This is how the ocean traps sound without needing any reflective surfaces. The physical structure of the ocean creates an acoustical channel, and sound spreading horizontally will stay trapped within it. The axis of the sound channel is the depth of minimum sound speed, caught between the warmer ocean above and the higher pressure below. It's often around 1,000 metres down, but can be much closer to the surface. Sound generated in this channel will mostly travel sideways, but will weave up and down as it goes, crossing the axis repeatedly, unable to escape this narrow channel. The corridor is called the SOFAR channel (for Sound Fixing And Ranging, because that's what happens when very functionally minded people name things).[31] This is incredibly helpful for several reasons, one of which is that trapping the sound in a horizontal layer keeps the signal strong and prevents any of it being absorbed at the sea floor or the ocean surface. It's like a secret sound tunnel, and its depth varies depending on the ocean conditions.

So that was the context for the Heard Island Feasibility Test. And here again, Walter Munk pops up in the story, unafraid of big ideas and willing to take risks with bizarre-sounding schemes. It appears, he said, that long-distance sound offers a way to calculate an average of the hundreds of small fiddly ocean features *without measuring them all individually*. If you send sound across an entire ocean basin, the sound that arrives at the other end only gives you the big picture, with the effect of all the details smoothed out. And the speed of travel of sound depends critically on the ocean temperature. So maybe, if you could send long-distance sound signals and measure the differences in their

---

[31] Later scientists weren't afraid of revealing at least some sense of humour, though. In the Second World War, someone worked out that if something made a loud noise in the SOFAR channel and it was heard by lots of different receiver stations, the location of the source could be pinpointed by comparing the detection times of the noise in different places. Later on, it became apparent that it was more useful if you did it backwards – had lots of known sources sending signals and then something deep in the ocean could listen to all those signals and calculate its own position. The second system was the first one backwards, and so to this day it's called RAFOS, which is SOFAR backwards.

arrival times around the world accurately enough,[32] you could make a single global measurement of temperature and keep track of that over time. If the ocean is heating up, the sound signals will travel more quickly as the years go on, and the journey times will decrease. Maybe the planet's changing climate could be measured using sound.

It was a compelling idea, but no one knew whether it would work. The only way to find out was to try. So a large team was assembled, the US Navy was persuaded to provide a ship, and planning commenced in earnest. As momentum picked up, many countries volunteered to host listening stations in their local patches of deep ocean. And so, in January 1991, the ship M/V (motor vessel) *Cory Chouest* was loaded up with huge acoustical transmitters and set out from Australia with Munk in charge. They were accompanied by another ship, the *Amy Chouest*, which would monitor whales and dolphins to look for ill-effects, with the proviso that the experiments had to stop if it appeared that they were harming marine mammals.[33] The area of ocean next to Heard Island had been chosen because of its location – from here, there were expected to be acoustical routes into the Atlantic, Pacific and Indian Oceans – and also because the SOFAR channel came conveniently close to the surface in that area.

The sound to be transmitted was far more sophisticated than the loud bang which had been heard from Perth in 1960. The *Cory Chouest* carried ten underwater loudspeakers, each one

[32] Incredibly, 'accurately enough' meant to one hundredth of a second, across tens of thousands of miles of ocean. And it was achievable.

[33] This was a constant source of friction between the acoustical oceanographers and the whale-loving public, who quite reasonably thought that loud artificial noises in the ocean might disturb creatures who communicated mainly using sound. But as no one knew what the effects would be, the experiments were allowed to proceed.

about the size of a telephone box and designed to produce a main signal at 57 Hz (about one octave up from the lowest note on a standard piano). The nuances of the sound were carefully designed so that the signal would be rich in information. No one knew how far this sound would go,[34] but on the day before the tests were due to start, the engineers asked Munk whether they could turn the loudspeakers on for five minutes to check that they worked. He said yes, and several hours later a fax arrived from Bermuda asking what they were doing, because the sound had reached them and it wasn't time to start the test yet. The first question, of whether artificial sound could even travel that far, was answered before the experiment started. And then, on 26 January, the tests started in earnest. Sound waves flooded outwards from the loudspeakers into the invisible SOFAR channel, spreading out across the acoustical highway of the ocean for hours until hitting the continents finished them off. The messengers were on their way, and the question now was how much information they'd be carrying when they arrived.

The delight of the scientists grew as the calls started coming in – from Bermuda, Nova Scotia, Washington State, Ascension Island . . . Heard Island had been heard around the world. But their delight was tempered by the sad demise of first one loudspeaker and then another and another. The rough weather in this stormy ocean was taking its toll. The technicians worked around the clock to fix everything, but the last 'Heard Island

---

[34] You're probably curious about how loud they were, but there isn't an easy way to tell you. We're used to thinking about sound volume in decibels, but actually there's no such thing as '40 decibels' – that's not a valid unit. Decibels are a *relative* measure, so really, you should say '40 decibels, referenced to 20 micropascals', which is the lowest threshold for human hearing. In everyday life, the reference level is assumed. But the reference level in the ocean is different, and so decibels on land and decibels in the water aren't directly comparable. But the sounds emanating from the speakers were slightly louder than the loudest known natural whale calls.

Science Daily' of 31 January included a tongue-in-cheek hand-drawn plot of one set of data they weren't proud of: the miserable loss of almost all the loudspeakers as the days had passed. When the number left hit zero, the experiment was over, four days early. But the scientific point had been made. Almost all of the receiving stations had heard the globe-encircling sound. The signals still had to be analysed to tease out the details of the path each had taken, but there was no doubt that there was a wealth of information to harvest. And as long as the signals arrived and their travel times could be measured, it would be possible to track changes in the average ocean temperature.

The scientists were optimistic that a global network of loud-speakers and receivers would soon be established, monitoring changes in sound arrival times that could only be due to temperature changes. But it was not to be. Fear of potential harm to whales and other marine mammals, and the difficulty of making reliable underwater loudspeakers that could provide the deep sounds necessary, scuppered the biggest ambitions for measuring ocean temperature this way. And so things rested.

But science often goes in cycles, and in recent years, another version of this idea has come along, one that avoids both of those problems. Instead of making loud noises, a recent study has suggested that perhaps we could listen for the loud noises that nature makes regularly: the deep booms of distant earthquakes. In theory, identifying earthquakes and tracking how long the sound takes to travel to lots of different sites would allow changes in ocean temperature to be measured. It might even be possible for historical measurements of earthquake sounds to be analysed in this way, possibly opening a window on the ocean temperatures of the recent past. Perhaps one day, this 'seismic ocean thermometry' will become standard practice.

Long-distance messages

The distances that very deep sounds can travel through the ocean are astonishing. Next time you look at a globe, imagine a whale calling out from a spot in one of the great ocean basins, and imagine that sound travelling outwards across the globe, rippling away far further and far faster than the whale itself can move. The physics of the ocean means that a whale, one single animal, can broadcast its message across a significant chunk of the planet. It's not the case that all ocean sounds travel anywhere near that far, but every location in the global ocean is criss-crossed with acoustical messages, flooded with sonic imprints of the environment near and far. When it comes to communication, sound in the ocean often plays the role of light on land. This is where the big picture comes from, the reassurance that the state of the world today is as it should be – or the warning that there is something out there to worry about.

The ocean's messengers dance through the water, carrying information and energy from place to place without the water having to go with them. But there's another way to move: to surrender to the water and to exist as a permanent drifter, going wherever the ocean takes you. These travellers include some of the most critical components of the blue machine, which are also some of the most underappreciated. It is time to meet the ocean's passengers.

5

Passengers

Drifters

THE COAST OF TANZANIA FACES eastward, away from the massive African continent and out into the Indian Ocean. The beaches here are narrow sandy strips sandwiched between the turquoise water and a line of scrubby greenery that looks determined to overflow from the land. It's peaceful, basking in its remoteness, but that doesn't mean nothing of note ever happens here. In December 2004, two turtle conservation officers, Jumanne Juma and Saidi Jumbe, watched as an animal the size of an upturned kettledrum got a last push from the ocean waves and then advanced up on to the beach near the village of Kimbiji. Sea turtles often come to these beaches to nest, doggedly dragging themselves up the sand with powerful flippers that are ill-suited to moving on land. But this new arrival was no turtle. It was walking. The lower half of its shell and its feet were covered in huge goose barnacles, and as its determined plodding lifted it out of the waves, it became clear that it was a giant tortoise. It had drifted in on the ocean currents, a passenger from far, far away. Giant tortoises are very definitely creatures of the land and this one's journey was an accident, the consequence of unwittingly hitching a ride with a parcel of salt water on its way around the globe. Seawater is a convenient host for accidental drifters like this, because it's generally benign and has a huge carrying capacity. And these ocean passengers matter, because this is how the blue machine keeps our planet connected.

The tortoise was only one of the trillions of passengers carried by the ocean that day, each one slotted into a part of the engine and taken wherever that patch of water went. This is the ocean as a transport system, ferrying atoms, molecules and life around the globe, hiding some passengers from the land and air and revealing others as the engine turns. But these passengers aren't entirely passive. They are a fundamental part of the ocean engine. Some of them are alive, while others are just travelling fragments of Earth – individual atoms and molecules, or residue from physical processes. Now that we have seen the overall shape of the ocean engine and how it moves, it's time to look at who its passengers are and what they're doing. Even the Kimbiji tortoise was more than 'just' a passenger, because journeys like that one have shaped entire ecosystems.

On the other side of the globe, the Galápagos Islands sit in the eastern Pacific Ocean at the equator, right in the path of the fertile Humboldt Current. Their most famous inhabitants are the giant tortoises which gave the islands their western name (an old Spanish word for tortoise is *galápago*).[1] But these volcanic islands are relatively young, and only started to belch themselves into existence from the sea floor around three million years ago. They are more than 600 miles away from the coast of Ecuador, so it has never been possible to walk there from the mainland. And yet when the first western explorers set foot on them, they were full of tortoises. The question was how they got there.

Although they may appear to be unlikely long-distance ocean

[1] Although I reckon that the finches studied by Charles Darwin make this a very close race.

passengers, tortoises have several natural quirks that help them out. They are able to survive without food or fresh water for many months, because of their extensive fat reserves and slow metabolism. Their lungs keep them buoyant and their long necks act like snorkels, enabling a tortoise to keep breathing even in choppy seas. They live for a long time – a captive Galápagos tortoise called Harriet was 175 years old when she died – and consequently the need for a mate isn't urgent. Females can also store sperm, so a single female could establish a colony by herself. And they can swim, albeit slowly, so once they spot land (a tricky task from their low vantage point), they can head in the right direction. The odds are still stacked against any individual tortoise washed out to sea, but over long periods of time, one or two are going to be lucky. And they are also helped by the ocean.

In the vicinity of the Galápagos, the Humboldt Current turns away from the South American shore and flows right through the islands, sending any tortoise castaway in the same direction as any of its previously unlucky compatriots. So this is how the Galápagos islands got their tortoises: very occasionally, one or two got washed away from the South American coastline, survived the journey as ocean passengers on the Humboldt Current and established families when they arrived. That was enough to give a new species a start.[2]

All the evidence pointed in that direction before the Kimbiji tortoise washed up in Tanzania in 2004, but this was the first time that a successful conclusion to such an unlikely journey was directly observed. The Kimbiji tortoise had spent two months drifting all the way from Aldabra Atoll, a tiny island 460

---

[2] It seems likely that humans moved the tortoises around a bit in very recent history, but there is extensive evidence to suggest that these tortoises reached the islands long before humans were anywhere near them.

miles away that is part of the Seychelles.[3] The Seychelles are the other group of isolated islands where giant tortoises have established themselves by becoming accidental ocean passengers over very long distances. Tortoises wandered into the ocean on the shore of the African continent, and the ocean currents did the rest. The same mechanism operated at both the Galápagos and the Seychelles completely independently, to shift tortoises from the nearest mainland to the islands.

If an ocean can carry a tortoise, it's clearly capable of carrying plenty of smaller hitchhikers too. At the other end of the size spectrum, down at the scale of atoms, the ocean looks like an endlessly restless soup of water molecules jostling each other, with sodium, chloride and magnesium ions (the most significant ingredients of ocean salt) occasionally visible in the fray. But there are other atomic passengers dissolved in this crowd, each so rare that looking for them is like searching for a very small needle in a very large haystack. Scientists often quote their concentrations in parts per billion or even parts per trillion, which sounds impressive but doesn't necessarily help. Imagine a standard bath that's about one-third full of medium-sized sand grains. That bath would contain about one billion grains of sand. Now, if we imagine that those grains represent atoms or molecules, the vast majority of them water molecules, it's a bit easier to imagine the oceanic concentrations of other things. In that bath, there would be about 7 atoms of copper, 36 atoms of iodine and 72 atoms of lithium. That's all, in a bath of a billion. But they are there, being carried around the ocean, jiggling and jostling and occasionally bumping into things that aren't water, like life. But the ocean is very big, so the sum of each type of atom over the whole ocean is huge. And that can seem irresistible, particularly if it's a very valuable element and you are in

---

[3] It came from an island to the mainland, rather than the other way around, but still it proved the point.

dire financial straits. For a while, it looked as though Germany's economic woes after the end of the First World War could all be solved if they could only find a way to isolate a very particular type of needle in that haystack: gold.

Hope for an ocean gold rush

Fritz Haber was a German chemist who left a very complex legacy when he died in 1934. He unlocked the way to feed a ballooning human population when he invented a method for synthesizing ammonia and therefore artificial fertilizers (providing an alternative to the bird poo we considered in chapter 2). It's estimated that this invention alone supports half the global population today. But when he was given the Nobel Prize for this work in 1919 (because of the war it was officially the 1918 Nobel Prize), the Nobel Committee chose to overlook the wartime work that had earned him the title of 'father of chemical warfare'. For Haber had led the team that developed chlorine gas as a weapon and then personally supervised its devastating use at the battle of Ypres, defending his actions by saying 'during peace time a scientist belongs to the World, but during war time he belongs to his country'. The war ended with the 1919 Treaty of Versailles, which stipulated that Germany must pay for the damage the war had caused, valuing this at 132 billion gold marks. But Germany was broke. And so the country turned to its star chemist, who proposed an idea: perhaps he could extract enough gold from seawater to pay Germany's debts. At the time, it was thought that there were around 47 gold atoms for every billion water molecules, and Haber was confident that his electrochemical methods could filter them out. So he embarked on an ambitious project to survey the seas with the serious aim of plucking money out of the ocean. But at some point in the 1920s, he identified errors in the previous calculations and measurements. The key number wasn't 47 gold atoms per billion water molecules. It was 1 gold atom for every 50 *trillion* water molecules. To go back to the analogy of the bath full of sand, you would have to fill two 50-metre Olympic swimming pools full of sand to get to 50 trillion grains of sand, and just one of those grains would be gold. Picking those gold atoms out of the multitudes of water molecules would cost far more than the atoms were worth. The project was abandoned.

The first cohort of ocean passengers are these atoms and molecules that are dissolved in ocean water, rare members of a large crowd. They are present throughout the ocean, but for the most part in incredibly dilute solutions. They are integrated with the liquid water, so they flow along with the currents, and up and down with upwelling and downwelling, constantly shifting around the planet. Almost the whole periodic table is in the ocean somewhere, although sometimes in quantities so low that they are very hard to measure. But a subset are more important than the rest, because they provide the raw materials for the next cohort of passengers: life. Life doesn't just exist *in* the ocean, like an actor standing on a stage. It's an integral part of it, as atoms become part of living creatures and then disengage from them, with consequences that affect the whole ocean.

Drifting through life

One of the most pernicious myths about the ocean is that it's empty. It's true that if you scoop up a handful of ocean water from almost anywhere in the world, it *looks* empty: colourless, salty water, a liquid blank canvas waiting for a fish or a boat to give it purpose. It's a romantic idea that appeals to those who want to see a void with their own eyes, to give emptiness a material place to exist because it's just too hard to imagine nothingness. And yet the buoyant balloon of this beautiful concept was popped with a very large pin very shortly after the microscope was invented. Alexander von Humboldt wrote this in *Cosmos*, his sweeping treatise on the nature of nature:

> The application of the microscope increases, in the most striking manner, our impression of the rich luxuriance of animal life in the ocean, and reveals to the astonished senses a consciousness of the universality of life ... However much this richness in animated forms, and this multitude of the most various and highly-developed microscopic organisms may agreeably excite the fancy, the imagination is even more seriously, and, I might say, more solemnly moved by the impression of boundlessness and immeasurability, which are presented to the mind by every sea voyage.

The microscope had opened a door on the world of the plankton, the living passengers of the ocean. Their minuscule size makes it easy for a casual observer to dismiss them as insignificant bystanders. But the plankton form the bulk of the living fabric of the ocean, the web of life that is an integral part of the blue machine. The ocean certainly isn't empty, and it's vital for the whole Earth system that that's the case. But seeing the magnitude of what's hidden in plain sight takes a bit of effort. And von Humboldt's impression of 'immeasurability' only gets more intimidating when you consider that the plankton ecosystem can change over relatively short distances, and with the seasons and depth and even the time of day, and that this pulsing mosaic of microscopic life permeates the entire global ocean. Taking a single water sample from one place and carrying out a full census of what's in it is like taking one page from one book in the world's biggest library and counting how many times each word is used.[4] It's certainly informative, but it's far from the whole picture. To understand plankton properly, you ideally

---

[4] The Library of Congress in Washington DC holds this title, with more than 167 million books at the time of writing.

need to measure the patterns in detail, again and again, over the entire global ocean. This is not a task for the faint-hearted or the impatient. Many thousands of researchers study plankton today, but one project stands out for its persistence and steadfastness in the face of the apparently immeasurable. While governments, social movements and entire industries have come and gone over the last ninety years, scientists in a quiet corner of Plymouth in the UK have patiently persisted with one gargantuan and painstaking task: recording the story of the world's plankton.

Archiving the living ocean

'This one is fifty years old,' says David Johns cheerfully, 'but we've got some which are much older.' David is the Head of the Continuous Plankton Recorder Survey and we are looking at the armoured outside of a Continuous Plankton Recorder, or CPR, device. The metre-long heavy-duty metal casing narrows to a snout at one end, with a solid towing attachment just above it. At the other end, a square arch of metal sticks up from the body and protrudes backwards, open at both ends. A window has been cut in the side of the device, and it sports a plaque that reads: 'CPR No 138, Manufactured Oct 1970. First tow Feb 1971, last tow Mar 1987. Total distance towed 18,458 nautical miles.'[5] It looks as though you could drop it off the top of the Empire State Building and it would barely be dented on arrival at the bottom. Which is just as well, because it's designed to be chucked

---

[5] That's just less than the equatorial circumference of the Earth, which is 21,639 nautical miles. A nautical mile is 1852m, which sounds unnecessarily awkward until you see where it comes from: it's the distance of 1/60 (or a 'minute') of a degree of latitude. Basically, it makes sense if you are following maps plotted in latitude and longitude degree co-ordinates. Don't shoot the messenger.

off the back of a cargo ship travelling at full speed and hauled back in without the ship slowing down.

The no-nonsense solidity of the outside protects something utterly unexpected on the inside: a beautifully crafted mechanism made of shiny brass cogs and spindles attached to rolls of stiff mesh. It looks as though it came from a watchmaker's workshop, and in a way it did – this is old-fashioned watchmaking technology from the 1930s. It was designed in 1931, and such is the brilliance of that design that with the exception of some minor tweaks in the 1950s, the device has not needed to change since then. This is critically significant, because it means that plankton from ninety years ago can be directly compared to plankton now without worrying that the measurement technique might bias the result. Hundreds of these devices have now been trawling the seas for decades.

The ingenuity of the CPR is breathtaking. In the snout at the front of the housing there is a small hole 1 square centimetre in size. As the device is towed through the ocean, water flows into that hole and passes through the silk mesh on one roller, trapping any plankton contained in it. The mesh on the second roller closes down on top of it, making a plankton sandwich, and formaldehyde seeps in to preserve everything that has been captured. But the really clever bit is that there's an impeller inside the metal arch of the casing,[6] and the water flowing past also turns that impeller, which is geared so that it turns the mesh spindle and advances the roll of mesh about 5 centimetres for every 5 nautical miles travelled. If the ship speeds up or slows down, the impeller will turn more quickly or slowly, automatically advancing the mesh at the right speed. The whole device is

---

[6] An impeller has blades like a propeller, but instead of turning in order to push on the water, the blades are pushed around by the passing water in a similar manner to a wind turbine being turned by the wind. By counting the rotations, you can work out the speed of flow past the impeller.

a bit like a tape recorder, with the plankton from each spot trapped at a specific position on the mesh roller. It can be dragged behind a ship going at full speed for 450 nautical miles before the rollers run out of mesh, at which point it's pulled back on board. These devices are given to ships in the merchant navy which take the same routes across the ocean month after month, year after year. Cargo ships, ferries, liners and sometimes even large yachts all do their part as they chug across the seas, moving people and goods around the world, and snaring a few plankton as a bonus. It's a collaborative project, with particular routes funded by the British, American, Canadian and Norwegian research agencies. By the end of 2021, Continuous Plankton Recorder devices had been towed for seven million nautical miles, equivalent to 326 circumnavigations of the globe. The routes only cover a tiny fraction of the ocean, but the consistent sampling has produced incredibly valuable long-term data. The rolled-up samples all come back here to Plymouth, in the UK, to be studied and archived, and then the mesh is replaced and the devices are sent back out again.

When the mesh returns, battered by travel and stained slightly green, what exactly is it carrying? I'm a mere physicist, used to thinking about the extremes of time and space rather than the messy machinery of life, and trying to get my head around the plankton is a struggle. The more you zoom in, the more you find and the more complicated it gets.

The smallest thing that a human can see with the naked eye is around the width of a human hair, things that are 0.1 mm across. We'll take the example of one type of foram as the top of our scale.[7] This particular species, *Trilobatus sacculifer*, can be about 0.25 millimetres in diameter and it's a single cell

---

[7] Short for 'foraminifera', but everyone calls them forams. This one floats in the water column, but most forams live in the seabed. They were first noticed at the Great Pyramid of Giza by Herodotus in the fifth century BC, because a

surrounded by a hard shell which bulges out into several separate rounded chambers. Thin, jelly-like tendrils extend out of the shell; these help the foram to eat, move and excrete. If we imagine enlarging this up by a factor of about 20,000, it would be the size of a Range Rover. Moving down using the same size scale, at about the size of a modern Mini Cooper we find an entirely different type of organism, the dinoflagellate *Ornithocercus*. This group are commonly considered algae, and this single cell is covered in cellulose platelets but sports an intricate fan of delicate but rigid outgrowths, giving it the appearance of a cell with an elaborate headdress. Neither the foram or the dinoflagellate can make its own food from the sun – they scoff other plankton for sustenance. There are two broad categories of plankton: the phytoplankton, which are the sun-harvesters, and the zooplankton, which eat other organisms to survive. But because ocean life is complicated, there are plenty of organisms that do both, depending on the circumstances.

Carrying on down the size scale still further, at about the diameter of a large watermelon we find diatoms, such as *Thalassiosirales*, a single cell in the shape of a beautiful flat disc made of silica, covered with a honeycomb pattern and adorned with delicate spines poking out into the surrounding water. Diatoms are sun-harvesters and they generate a huge proportion of the Earth's oxygen.[8] Still smaller, about the size of a galia melon,

---

large type of foram (Nummulites) makes up the rocks used to build the pyramid.

[8] This oxygen is generally consumed again within the ocean as other animals use it for respiration. We often hear that 'the ocean generates the oxygen for every other breath that we take', but that's not actually true. It *is* true that almost half of all the photosynthesis on Earth happens in the ocean, and therefore also almost half of all oxygen is produced there. But it's also the case that most of that stays in the ocean, being consumed within the water without ever getting up into the air. The oxygen that we actually breathe is mostly there because of the long-term burial of organic matter, which frees up oxygen

are coccolithophores, stunning spherical cells covered in intricate circular platelets made of calcium carbonate. And even further down the scale, at about the size of a Malteser, there are cyanobacteria like *Prochlorococcus*, the most abundant photosynthetic organism on Earth. These are tiny globules, too small to have a complex structure, and the smallest known organism that can photosynthesize. And for the biologists who really like to be thorough, at around the size of a fragment of dried couscous, there are the marine viruses which multiply by hijacking the cellular machinery of bigger hosts to reproduce. Put a Range Rover next to a grain of couscous and you have some idea of the range of plankton sizes.

The Continuous Plankton Recorder focuses the larger end of this range. David tells me that the biggest thing they've ever found was a 30-centimetre-long pipefish,[9] which must have reached the hole in the CPR snout head on and then found itself wrapped around the rolls of mesh. The smallest plankton may be trapped in the mesh fibres, but may also be small enough to escape.

The real work begins when the mesh arrives back in Plymouth. At the headquarters of the Continuous Plankton Recorder Survey there's a long room with twelve microscopes, staffed by scientists in white coats, goggles and the obligatory purple latex gloves. They are identifying and counting what they find on the mesh squares, turning what David describes as 'plankton roadkill' into data.[10] The expertise in this room is incredible. Laboratory manager Claire Taylor stops her work to

---

because the buried carbon isn't around to react with it. Like most things on this very complex planet, the story of the oxygen we breathe isn't simple.

[9] Pipefish are long, thin fish related to seahorses, and when it comes to body shape, they really are true to their name.

[10] The mesh squares are all archived for future scientists to access, and all the data are freely available to anyone who asks.

talk to us, and I peer down her microscope to see tiny ocean beasties trapped in the silk strands. There's clearly a lot of variety and I ask her how many species she knows. I'm floored by her answer: eight to nine hundred – and that doesn't include all the various life stages of each species. We look at another sample, which resembles a small, slightly squished shrimp, illuminated by the harsh microscope light. David points to a bump on its fifth limb, a critical clue for identification. This is why humans are still better at this job than the machines; they can nudge the limbs to reveal important features, and recognize when a specimen is squashed, broken or caught at an odd angle.

Along with many other research projects, the CPR survey has helped to uncover the basic patterns of drifting life in the ocean. The biggest single feature is the spring bloom: a wave of life pulsing across the ocean as the seasons change. In each hemisphere, big winter storms stir up the upper layers of the ocean, mixing nutrients and moving them upwards towards the surface. When the longer spring days kick-start the seasonal flood of sunlight, the abundance of light and nutrients provide the raw material for an explosion of ocean life: first diatoms and other phytoplankton, then zooplankton to feed on the phytoplankton, then dinoflagellates, and then a secondary diatom bloom in the early autumn.[11] Although some species may dominate, plenty of others are present. There are no fixed rainforests in the ocean, but these huge waves of life are the oceanic equivalent because of their size and diversity. But their nature is very different: mobile, short-lived, and carried wherever the currents take them, likely to change within a few days if the nutrients or light or temperature shifts. There are also shorter-lived local blooms that pop up whenever the conditions are right, and disappear again when the

---

[11] This pattern of seasonal blooms doesn't occur in the tropics, where there's no annual cycle of winter storms, and where the day length doesn't change much throughout the year.

feeding frenzy has burned itself out. But the basic conditions for life – a good supply of nutrients in a place reached by sunlight – are often determined by mixing processes bringing nutrients up from the deeper waters, as we have seen. And where the plankton blaze a trail, they will fuel the fish and birds and bigger life forms as well, creating hotspots of life.

The ocean's passengers are therefore not evenly distributed around the engine that they ride. The ocean's equivalent of both rainforests and deserts have boundaries set by the components of the blue machine. Even the tiniest passengers – atoms of gold or iodine – display complex patterns of abundance, being more numerous in some places than in others. Where the physical conditions and the availability of energy and raw material are favourable, inert passengers will be built into living passengers – the phytoplankton – which will themselves alter the chemical signature of the water around them. The phytoplankton drift with the ocean currents, swelling and fading in number as the ocean's chemical makeup and the physical ocean engine shift around them. Without this vast web of tiny life there would be no penguins, dolphins or sharks. But no life at all would exist without the chemical passengers carried by ocean water, which provide the atomic building blocks that shape both the living passengers and the entire workings of the Earth's engine. The ocean's most important passengers aren't tortoises or humans, or the flotsam and jetsam that our society abandons to the ocean's whims. The real VIPs are too small for us to see, although the consequences of their existence are very visible and affect us all.

Our society is used to the idea that microbes are generally counted in millions and billions, but what do those numbers actually mean? To see how big the citizenship of the ocean really is, we need to start with our own relationship with ocean life. And because our species is adventurous, adaptable, curious and omnivorous, for many people that relationship starts at the dinner table.

Ocean bounty as banquet

It's 5.30 a.m. on a beautifully sunny spring morning. The bright blue sky is still and the pinkness of the dawn is caressing the landscape, but there's no birdsong to celebrate the start of a new day. The great glass monsters of Canary Wharf in London rise around me, skyscraping monuments to banking, law, financial regulation and telecoms industries. This district of London is tucked into a loop of the Thames estuary on the eastern side of the city, and the towering modern buildings were constructed from the 1990s onwards around two-hundred-year-old docks which had once been the landing point for the riches of the Caribbean. The towers come to an abrupt halt as I cross to the other side of the North Dock and back into an older world. Here there is no modern glass and steel, just a long, low, yellow-roofed building running along the waterfront and surrounded by a huge car park. This is the Billingsgate fish market, the UK's largest inland fish market, and the modern expression of an unbroken London tradition stretching back more than three centuries.

Dozens of small white vans are scattered across the car park, their back doors open, each one accompanied by a clutter of wooden pallets and white boxes. The sign at the entrance announces that the market is open from 4 a.m. to 9.30 a.m., Tuesday–Saturday. I find my way upstairs to the seafood school which organizes tours.[12] We don white coats, which seem to be

---

[12] I haven't eaten fish or any other seafood since I was a small child, and I was reassured by the availability of a guide as I walked into an alien culture. They're lovely people, helpful and enthusiastic, and I kept quiet about my personal detachment from the focus of their lives.

the uniform for everyone in the building, and step out on to the market floor.

The stalls are arranged along narrow aisles, each one packed with white expanded polystyrene boxes which leave no free space on the metal racks that support them. The hard green floor is wet from hosing and ice, and every few minutes there's a cry of 'Backs!' or 'Mind your legs!' from porters propelling trolleys of pallets loaded with white boxes, presumably hurrying out to the vans. In the white boxes, there are fish and more fish. What strikes me at first is how well sorted everything is, and how uniform: one box contains only red mullet of exactly the same size, the box next to it is full of identical lemon sole, and another box has apparently cloned monkfish. Our guide picks up examples to point out the condition of the gills, the sheen of the skin and eyes, the texture of the flesh.[13] We pass by tiger prawns and trout, sea bass and scallops, mackerel, dover sole and turbot. They are almost all whole fish, with the exception of shrink-wrapped hunks of deep red tuna and occasional smoked fillets. Further along, my jaw dropped when I saw the ranks of tropical fish: parrotfish, goatfish, barracuda and grouper, species I had never imagined anyone eating in the UK. These are fish I know from diving with them, and I imagined the goatfish rummaging through the reef with their chin barbels, and the barracuda cruising near the surface of the turquoise water as they hunt.

To me, every stall is a confusing kaleidoscope, and after a while I realize that my disorientation comes from the biological hotchpotch I'm looking at: there are open ocean predators, coastal bottom-feeders, quiet foragers, vicious territory defenders, small fish that live in huge shoals and the independent voyagers of the high seas, all plonked next to each other in their

---

[13] She was at pains to point out that the traders let her do this because she's part of the seafood school; normal punters are not allowed to touch before buying.

white polystyrene boxes and cushioned by ice. There are fish from the North Sea, the Indian Ocean, the coast of Chile, the North Pacific – fish that would never have been within a thousand miles of each other in life, meeting in death. It was like visiting a museum and seeing royal Tudor headdresses piled up next to Roman coins, cheap Victorian clay pipes, a stuffed dodo, an Inuit canoe, gloves worn on the moon by an Apollo astronaut, a tin of baked beans, a lump of concrete and the World Cup football trophy.

Our guide was knowledgeable about all these fishes' biology and their origins, and the sustainability of each fishery. The buyers here are both fishmongers and members of the public, and these fish have come from all over the world to pass through this corner of east London on their way to dinner plates all around the capital. For individual buyers, the fish are weighed and dropped into blue plastic bags, a treat for that day's dinner. One couple in my tour group decide on salmon for their evening dinner party, and stop by a stall with salmon large enough to spill out of the boxes, each one nearly a metre long. They're magnificent fish, sleek, with powerful bodies, each shiny black along the top, speckling down the flanks into a white underside. The stallholder weighs one of the biggest and most majestic, accepts the considerable amount of cash dictated by the scales, and then slings the fish in a black plastic bin bag and hands it over. There is no ceremony here. If you paid that much for a shirt in the posh clothing shops of central London, the cashier would fold it carefully, wrap it in tissue paper held together with a pretty sticker, and reverently place it in a branded shopping bag. Afterwards, I can't decide what I think about this contrast. The bin bag seems like a degrading vehicle for a beautiful animal, but perhaps there's a purity and honesty to it – the value is in the fish itself, and no amount of packaging can disguise or misrepresent its true nature, so why would you try?

This is how most of us meet ocean life: on menus and in

sandwiches, at the fishmongers or a market, jumbled up as 'fish' with a variety of familiar and less familiar names. If you ask where it comes from, it's just 'the ocean', as though fish spontaneously arise wherever there is a bit of empty water. But of course they don't: each market fish is the tip of a biological iceberg, the visible part of a much larger structure. These fish eat life that is smaller than them, and that life might be fed by something smaller still. So how large is the submerged iceberg of invisible life, the ocean biology that we can't see? Perhaps surprisingly, there is a reasonably clear answer to this, although biologists are still working out the exact reasons for it.

One of the first stalls we come to has a white box bearing two huge red snapper which barely fit inside it. Every white box is the same size, perhaps 75 centimetres long, 40 centimetres wide and 20 centimetres deep. Each red snapper is almost the length of the box, with dappled peachy pink scales and a big pink eye adorning a tall narrow body ending in a delicate tail fin. This pair probably weigh around 5 kilos each. So let's consider one of these fish sitting in one polystyrene box by itself. This one fish – the contents of that box – represents a chunk of ocean life that is between 1 kilo and 10 kilos in weight. Now mentally put another box next to it, this one with ten Dover sole in it, each one around 500 grams. The total mass of life in the second box is the same, but it's made up of ten fish each a tenth of the weight of the red snapper, representing ocean life from 0.1 kilo to 1 kilo. The next box along contains 100 sardines, and the box after that, 1,000 Atlantic prawns.[14] Then there's another box with 10,000 small rock-pool snails, each weighing 1 gram. So now we have five boxes, each one with the same mass of life in it, but containing

[14] This is fishmonger's language. To an ocean biologist, a prawn isn't really a thing, and neither is a shrimp. They're both common terms for aquatic crustaceans with ten legs, and their use varies across different English-speaking countries. There is no formal scientific definition of either one.

creatures each a tenth of the size of those in the preceding box. Recent research has confirmed that the basic rule of the ocean is that each time you go down the size scale by a factor of ten, the total biomass in that size category is the same, and there are just ten times as many individuals. This includes all forms of life: crustaceans, fish, marine mammals, bacteria, sea stars and everything else; you put each one in its size category, add it all up across the entire ocean, and the mass of life is the same in each category. The number for the entire ocean is 1 gigatonne (one billion tonnes, if you prefer) of life in each size category covering a factor of ten; there's one gigatonne from 0.1 kilo to 1 kilo, one gigatonne from 0.1 gram to 1 gram, and so on for the rest.[15]

It sounds obvious that there are more small things than big things in the ocean, but the next question is how far down the size scale this relationship goes. Next to our one big red snapper, we already have four more boxes of the things we can see: Dover sole, sardines, prawns and tiny snails. Each box weighs the same. But let's carry on down the size scales, filling boxes with the ocean life that we can't see, each one full of creatures that are a tenth the size, and ten times more numerous, than the box before. The answer is that there are *fourteen* more boxes; the last one (which is mostly full of bacteria) has 10^{19} individuals in it, ten million trillion living cells, that all together weigh the same as one big red snapper. This is the ratio of life of different sizes in the ocean. Every time you see one big fish, it's worth visualizing how large the biological foundation it rests on is: around eighteen times the mass of the fish you're looking at.

In a healthy ocean, this relationship of a fixed mass for every size category also carries on up to the largest whales, covering a total of twenty-three orders of magnitude in size. The top of the

---

[15] There is a small caveat to this: the relationship is most consistent in the top 200 m of the ocean, because there are fewer large animals in the deep cold midwater.

largest size category is a trillion trillion times as large as the bottom of the smallest category. It's hard to get your head around these numbers, but one way of looking at it is that with the best will in the world (and without a microscope), we humans are incapable of even seeing 61 per cent of ocean biomass.[16] Thinking that 'ocean life' is all about dolphins and octopuses misses a large part of the point. The invisible majority – the plankton – are drifting along as passengers and nothing else could live without them. They provide both the raw materials and the energy that can be rearranged to create everything else.

By the time the ornate green clock hanging from the market roof shows 7 a.m., the clearing up is already under way. Floors are being hosed down, and boxes of fish are being wheeled off to the massive freezers at the back of the building. Huge scoops shunt ice from box to box, money is counted, and by 7.45 a.m. the hall is stripped back to bare metal racks and cold concrete floor. I asked whether restaurant chefs ever came here to buy fish and whether that was why it closed so early, and our guide laughed out loud at the absurdity of the idea. Almost never, she said. Ninety-nine point five per cent of London restaurants want pan-ready fillets which are ordered from big commercial retailers, and that fish never passes through this market.[17] The implication is that chefs are drifting away from familiarity with the original fish, away from bones and sinew and gills and towards neatly packaged lumps of protein. As we went back upstairs, we passed a class learning to 'gut, shuck and fillet' – the basic skills involved in converting an intact piece of nature into food. The proportion of people who eat fish in the western

[16] That's 14 of the 23 size categories, expressed as a percentage.
[17] The specialists, like high-end sushi restaurants, are probably exempt from this generalization, since they still take considerable pride in the details of fish preparation.

world who can confidently do that must be low and shrinking with time. Even the market itself, the place where you can at least appreciate that fish has to come from somewhere other than a supermarket shelf, is being pushed to the periphery. When Billingsgate fish market was formally established by Act of Parliament in 1699, it was on the site of the city's original water gate, the place that connected the citizens of London to the goods from the rest of the world. Until 1982, the market was housed in a grand Victorian building on the original site; in that year it was moved to its current location, 2.5 miles further east, away from the centre of the city. At the time of writing, it looks likely that the market will soon shift again, another 10 kilometres east, even further from the heart of London. In addition, more than half of the fish sold in the market today is farmed rather than wild, and that proportion is increasing all the time. We may think that fish connect us to the ocean, but the drive for efficiency and low prices is pushing even the enthusiastic fish-eaters away from the visceral reality of an actual whole fish, which embodies the complexity of ocean environments in its physiology and condition.

I still remember discovering, probably in my early twenties, what a tuna really looks like. They are large, powerful predators, muscular and lustrous, kings of the sea. I could not imagine anything further from the small sad tins of tuna that had made up school lunches and picnic salads in my childhood. Even now, I feel affronted when I see those tins, which offer no acknowledgement that this is majestic ocean nobility, chopped and processed, reduced to a meagre standardized disc. This ocean bounty had been made unrecognizable and no one ever told me what it was. There are arguments to be made for eating fish, when it's sustainably sourced and especially if it's local, but surely we need to insist that we are honest about what it is, and the huge richness of the ocean that it represents. When you eat a fish, you are ingesting the energy and raw materials accumulated by millions of phytoplankton and zooplankton spread over a wide geographical area, and incorporating them into your own body. A past part of the ocean becomes a present part of you.

But we also need to be more vocal about the tiny ocean passengers that form the invisible majority of ocean ecosystems, the living web of life that is a critical part of the blue machine and critical for maintaining the Earth as we know it. A shift of perspective is required, an adjustment to include the idea that what we can't see is far more important than what we can see. In 1844, Alexander von Humboldt immediately understood the significance of his discoveries with the microscope: 'The abundance of those marine animalcules, and the animal matter yielded by their rapid decomposition are so vast that the seawater itself becomes a nutrient fluid to many of the larger animals.'

We have now met the two types of ocean passenger – the non-living (gold and the other assorted solo atoms) and the living – and we can imagine them drifting around the world on currents that they do not control, transported by the workings of the blue machine. They are both dilute and omnipresent, although their numbers vary hugely from place to place. But these passengers aren't fixed in their character, and although they are mostly confined to their respective patches of ocean, there is immense potential for reinvention during their journeys. There are fundamental principles that keep these biological wheels turning, some of which we humans have been fairly slow to learn and apply in our own society. And so we will start our examination of another crucial aspect of what these ocean passengers are doing on their journey around the blue machine by looking in the mirror at ourselves. How do we humans connect to the ecosystems around us?

What goes in must come out

There are golden figs hanging above my head, dangling from bright green leaves. But this is no hothouse. These are metal figs, sprouting from a bright red iron pillar, one of eight connected by intricate painted ironwork which flaunts pale flowers, twisting greenery and confident pride. The dramatic octagon and its stunning ornamentation are the centrepiece of a grand hall, stretching upwards and outwards, supported by red, cream and green pillars linked by high arches. This space is deliberately magnificent, a spectacle, and so it should be: it has a lot to celebrate. It sits inside a small complex of tall, handsome brick buildings, clearly designed with both function and aesthetics in mind. But if you want to visit all of this, you need to find your way out to a long, winding driveway hidden on the flat Erith marshes, next to the Thames and 15 miles downstream of London. This site was picked because it was remote, out of sight and out of mind for Londoners of the 1860s. The beautiful figs that adorn the Cathedral on the Marsh were chosen with a knowing smirk; their relevance lies not in their horticultural quirks or their culinary potential but in their usefulness as a laxative. These buildings are given their purpose by poo. The great hall was put here to house the Crossness Engines, and the engines were put here to haul the daily deluge of London's sewage away from the city.

Cities are built on estuaries for many reasons. The banks of an estuary offer rich fishing and hunting grounds, access to overseas trade, nearby fresh-water streams and a commanding defensive position. But two hundred years ago, they also offered a free and convenient waste-disposal system. By 1800, the busy metropolis of London was the home of many smelly and dirty workplaces: abattoirs, leather tanneries, coal-fired bakeries, glassblowers and whale-oil processing plants, as well as a million people, along with their waste and especially their sewage.

Old Father Thames took the mess away, but the city just kept on vomiting up more and more. The following half-century was one of astonishing scientific, medical and engineering advances, but one thing didn't change: the slop went into the Thames, and you had to hope that the tide didn't bring it back six hours later. The empire grew, the city grew, the industrial revolution grew, and the stink grew faster than all of them.[18] The rainfall from a tenth of the entire country wasn't enough to wash it all away, and by 1850 the Thames in London was a disgusting stew of filth, disease and death. Everyone agreed that something needed to be done, but fixing a problem of this scale was going to be astronomically expensive. It's a sad irony (hardly the only one in human environmental history) that instead of recognizing that the Thames had been doing the city a massive favour in the first place by taking their rubbish away for centuries, the river itself was blamed for being the problem now that it was overwhelmed. The situation became known as the Great Stink, and by all accounts that was a considerable understatement. The tipping point came in the hot, dry summer of 1858, when the House of Commons (which sits right on the bank of the Thames) concluded that the intolerable stench made many of its rooms completely unusable. After a debate in which Benjamin Disraeli described the Thames as 'a Stygian pool, reeking with ineffable and intolerable horrors', legislation was finally passed to provide the colossal amount of money needed to banish those horrors. And so began one of the most ambitious engineering projects in Britain's history.

The leader of the project was Joseph Bazalgette, a brilliant civil engineer and the chief engineer of the Metropolitan Board of Works, which was given responsibility for tackling the

---

[18] The invention of the flushing toilet didn't help either, as it added huge volumes of water to the limited existing system of brick sewers, making the overflows worse.

problem of the poo. The plan involved building 1,100 miles of new drains which fed into 82 miles of brick-lined sewers, all of which slanted downwards slightly to ensure that gravity could move the liquid muck away from the city while completely hiding it from view.[19] The project reshaped the Thames in London, since the new Victoria and Albert Embankments that contained large branches of the sewer became the new banks of the estuary (and still are today). The whole system took sixteen years to build, but it was a stunning success with hugely beneficial consequences for public health, dramatically decreasing cholera, typhus and typhoid disease in the city.

Yet brilliant though the engineering was, the new sewers didn't actually solve the underlying problem. They just moved it. The vast network of smaller pipes joined up to make bigger ones, and the Cathedral on the Marsh marks the end point of the biggest pipe on the southern side of the river (there was a separate network on the northern side, ending at Abbey Mills). The land that London sits on is relatively flat, so in order to keep the poo flowing, the sewers ended up at a point about 10 metres below the level of the Thames. The brick buildings at Crossness were built to house four powerful steam engines and the twelve Cornish boilers needed to generate enough steam to power them. These iron workhorses would pump the sewage upwards to the level of the Thames so that it could be released as the tide flowed outwards. The poo would then be taken off to the North Sea, to become a passenger travelling according to the whims of the ocean. Since the point of sewers is that they are hidden underground, the pumping station at Crossness was one of the few visible markers of this huge new system, and its grand opening in 1865 was attended with much interest by the cream

---

[19] It's a huge testament to the quality of the engineering that Bazalgette's sewers are still in excellent working order today, although upgrades and additions have been made where necessary.

of London society. The architecture and the exquisite ironwork matched the high profile of the project, and also their confidence that this would be a long-lasting contribution to the fabric of London.

After several extensions and upgrades, the Crossness engines finally ceased work in 1956. The boilers were taken away for scrap and the great engine house was left to rust. But in 1985, the Crossness Beam Engines Preservation Group came into existence, later becoming the Crossness Engines Trust. The engines are open to the public now because of this group's deep and stubborn refusal to leave the great engineering achievements of the past to rot.

We are introduced to the site by an enthusiastic volunteer who proudly admits to being obsessed with sewage (and it's probably a coincidence that some time later he mentions having a poor sense of smell). The men (and all the people I see are men) who are our guides tell the tales of thirty years of patient work, chipping away at rust, disassembling, repairing and renewing the great engine components, stripping and painting ironwork, and returning one side of the engine house to its former glory. The other side has been left as it was, cloaked in the grey and brown of age, rust and memories. One of the cheerfully painted engines is now running again, and it is hoped that another will be soon. Each engine pushes upwards on a 13-metre-long beam, a bright green arm that stretches overhead across a central pivot point. The other side of the beam pushes and pulls on a huge green flywheel, 8.5 metres in diameter and weighing 52 tons. In its working days, as each beam rocked back and forth it hauled sewage from two deep tanks upwards, each stroke of the engine shoving 6,800 litres of London's effluent up to the high reservoir ready to be discharged on the next tide. This was a true wonder of its age, and the Crossness Engines Trust deserves enormous credit for making sure that it's been preserved for us to appreciate. But it will be amazing to future generations that the superb minds capable of executing the original plan so elegantly apparently also missed its fundamental flaw.

The city was reborn out of the fug of the Great Stink, having taken a big step into the modern world now that the fetid problem had gone away. But 'away' was just the other side of the fresh/salt water divide. The raw sewage pumped out by the powerful engines was discharged untreated until the 1880s. Part of the problem with assuming it would be taken away by the tide was the tide turning and bringing at least some of it back. And it soon became apparent that 'away' wasn't all that far away after all, as fishing downstream came to a halt, confounded by the concentrated stream of filth. It didn't take long for the arguments to start, but things only changed because of one of the most horrific civilian disasters ever recorded in the country. In 1878, the passenger paddle-steamer *Princess Alice* was struck and cut in half by a coal ship near Woolwich, then a few miles downstream of London. Hundreds of people were thrown into the water, and few would have been able to swim. But the real tragedy was that the collision occurred near to the sewage discharge point on the north side of the river, about an hour after the second half of that day's sewage had been released. The black, fermenting sewage was a far greater danger to life than the river, and the recovered bodies were covered in ooze. The inquest found that many hadn't drowned, but had been poisoned by toxic fumes. These were grim and unnecessary deaths. As a consequence of this disaster, systems to treat the sewage were finally installed – but six 'sludge boats' were also commissioned to take untreated human waste even further away, out into the North Sea.[20] The thick brown slop

[20] Incredibly, this continued until 1998.

gave these boats their nickname: the Bovril boats.[21] Humanity couldn't rid itself of the temptation to just keep taking all the horrors progressively further 'away'. Which meant the ocean.

There is no doubt that Bazalgette's engineering saved thousands of lives, and immeasurably improved the health of Londoners. But the brilliance of the engineering is matched by the clarity with which it reveals the attitude that built the modern world: when we have finished with something, we just put it somewhere we can't see it and then we don't have to think about it any more. It effectively vanishes from the parts of the world we care about. And historically, the ocean is so disconnected from us in our minds that it counts as 'away', making it the giant dustbin of humanity. But the problem, just as the Victorians discovered, is that the ocean isn't very remote, and things we put in it may well come back to haunt us. We may see the ocean as the end of a one-way pipe, but that is not how nature works. So let's follow the journey of a particular set of passengers in the ocean engine, and take a look at what nature does with what comes out of the back end of a whale.

Floating fertilizer

The icy continent of Antarctica, squatting over the South Pole, is completely encircled by the Southern Ocean. Fierce eastward

---

[21] For anyone who wasn't born in the UK before about 1975, Bovril is a dark brown, thick, salty meat extract that was either diluted with water to make a hot drink, or used as flavouring for soups. My vague childhood memory of it is entirely confirmed by the sludge-boat nickname. I'm sure there are still enthusiasts who will stand up for Bovril, although it's got the sort of robust character that is perfectly capable of standing up for itself.

winds drive the surface of these cold waters around the polar citadel, forming a continuous and chilly carousel that separates Antarctica from the rest of the world. For those living at or just above the surface – from majestic gliding albatrosses down to fragile and wind-ruffled storm petrels – this is a fickle and violent environment where one moment's misjudgement can slam you into the choppy surface, instantly making you food for the fishes. But cross the boundary into the water and the character of the environment changes immediately. The immense, frigid ocean is calm, a placid habitat for the largest animal that has ever lived: the blue whale. As this giant cruises upward from the darkness below, the edges of its long, slim silhouette sharpen, and then more than 100 tons of blue-grey bulk announces its arrival at the surface with a blast of spent air and mucus from its blowhole. Whales are mammals with lungs similar to ours, and so they must connect with the atmosphere regularly in order to live, especially after a feeding dive. As the whale rests at the surface, weak summer sunshine glistens off its sleek skin, nature granting a calm moment of beauty as this ocean leviathan pauses in the business of life. And then a bright terracotta plume billows outwards in the water just below the whale's tail, a torrent of thin goo that drifts upwards and forms a patch several metres across at the water surface. This is whale poo, the unwanted leftovers of what has passed through the giant's gut, now being ejected and left behind. And one whale's trash is a whole ecosystem's treasure.

Ecosystems are often complex webs of feeding relationships, but the blue whale's corner of the Southern Ocean ecosystem is relatively simple. In the polar summer, when the daylight is almost continuous, the tiny phytoplankton at the ocean surface harvest sunlight, collecting the energy that will power the rest of the ecosystem. Normally, this would take a while to work its way up the size scales of ocean life, but the Southern Ocean is home to a very effective short cut in the form of Antarctic krill.[22] These crustaceans look a bit like small shrimp, 4–5 centimetres long and almost completely transparent, with beady black eyes. They have the unusual talent of using their six bristly front legs as a 'feeding basket', and as they swim through the water they are constantly opening this basket and then combing the contents into their mouth. Krill hang out in huge swarms, great bustling mobs that can be hundreds of metres long. They are the tiny vacuum cleaners of the Southern Ocean, filtering out living ocean passengers that are 10,000 times smaller than themselves and scoffing the lot. But they in turn are filtered out by the blue whales, which feed entirely on krill. During their feeding periods, an adult blue whale is thought to consume around 16 tons of krill *every single day*. The whales spiral downwards to find the swarms at a depth of around 100 metres, and then lunge-feed, unfolding their huge jaws to engulf a chunk of ocean. Some time later, the leftovers are squirted out into the surface waters.

The story of blue whales in the Southern Ocean is straightforward and depressing. When humans first explored these waters, blue whales were plentiful and the water was said to be so full of krill that it appeared red. What happened to the whales was tragically predictable: humans saw them as an easy resource and killed them, turning their slaughter into an industrial process, and by the time blue whales were given formal protection in 1966 only 1–2 per cent of the original population remained. But what happened to the krill was less predictable. The blue whale was one of their major predators, so you might expect a population boom. Instead, krill numbers declined by 80 per cent in the regions where whales had been common. The ocean giants aren't the only species that depends on krill for food; the vast

---

[22] They go by the excellent formal name of *Euphausia superba*. They are also what shrimp paste is made from.

swarms of this small critter feed hundreds of other species: seals, penguins, squid and fish. The disappearance of so many krill emptied the Southern Ocean buffet for almost everyone. So where did all the krill go?

The machinery of life is simultaneously robust and delicate. Evolution has found the molecular components which fit together to make reliable bodies, carefully positioning the right molecules in all living cells to do the very specific jobs that will keep the machinery running and reproducing. It's impressive and resilient. But some of those specialized molecules only work when they have very specific atoms at their core. In the case of the two most critical processes for life – photosynthesis and respiration – the molecular machinery requires iron atoms – not that many of them, but enough to give a few critical molecules their function. Without iron, life cannot live. And although the surface of the Southern Ocean is well stocked with all sorts of other nutrients, it is short of iron. Without iron, there cannot be phytoplankton, and without phytoplankton, there cannot be krill. And so we come back to the defecating whale, and the red plume of poo billowing out behind it.

Whale poo is liquid and it floats, and both these characteristics are amplified in importance by its rusty colour: for this rich liquor is full of iron. The whale scoffs krill down below and then brings those nutrients back upwards into the sunlight, so that the raw materials and the energy needed for life can mingle once again. The whales are closing the loop, keeping iron in the system.

The phytoplankton don't exactly eat the iron-rich poo, but they can absorb the released iron in the water in order to use it to build the molecules they need. This is the way the ocean does things – not as a one-way pipe but as an unending cycle. One living creature's discarded material is reorganized, rebuilt and reused by others near by. All life on Earth – you, me, blue whales, sloths, the oldest tree and the most fleeting phytoplankton cell – all of it is made from recycled material. There could be atoms in your body which were once part of Julius Caesar,[23] a *Tyrannosaurus Rex*, giant 2-metre-long millipedes from many tens of millions of years before *T. Rex*, ancient grasslands, exotic iridescent beetles, and smoke from the early industrial revolution. The atoms that we're made from are completely interchangeable. Any carbon or iron or oxygen atom is equivalent to any other, and they move from one configuration to another, dancing through combinations and places and time, with new surroundings at every step of the way. The ocean's atomic passengers are all constantly circling around, for the most part staying within the ocean but shifting their outward form as they ride the blue machine. In general, the supply of and demand for each type of passenger is balanced, and so the cycles just keep on turning.

The problem for the krill was that killing the whales broke a huge link in the Southern Ocean's chain of recycling. Without the whales to help keep iron near the surface, this critical element gradually sank away from the sunlight, depriving the ocean surface of a key component of life. And so the krill had less food, and their population declined. The fact that iron-rich whale poo floats rather than sinks helps the ecosystem go round, fertilizing the ocean that feeds the zooplankton that feed the whales. The ocean recycles, endlessly and out of necessity. There is no 'away', except for the tiny fraction of material that gets locked up in rocks for a few million years, and even most of that reappears in the system eventually.

Since no squid or whale has worked out how to launch itself (or anything else) into space, the Earth is what scientists call a closed system when it comes to atoms: everything is trapped

---

[23] Sam Kean wrote a whole book exploring this idea, called *Caesar's Last Breath*.

inside and so everything comes around again.[24] This is part of the reason why we don't see and don't appreciate the tiniest ocean passengers: they don't generally accumulate into large visible piles, because they're dilute and constantly being reused. Nature has always operated on the principle that our society is finally rediscovering: what you do is *start* with the poo. That's the concentrated source of the materials relevant to life. If you can build everything you need out of waste, you're never going to run out of raw materials. Nor do you have a waste-disposal problem – the only thing you might face is a lack-of-waste-to-dispose-of problem. Nature recycles, endlessly. Poo is one of the most important resources on Earth.

The one-way nature of Bazalgette's brilliant sewage system wasn't entirely invisible at the time. Edwin Chadwick had been rattling cages about the links between sanitation and health since the late 1830s, and his 1842 *Report on the Sanitary Condition of the Labouring Population of Great Britain* galvanized the growing argument about sewage and water supply systems. This was the first report to use proper medical data to address a social problem, and it generated significant public pressure for the government to act. He proposed that sewage systems should be directed inland rather than out to sea, and that human waste should be used to fertilize farms. Bazalgette ignored these recommendations, and the one-way approach to human waste was locked into the structure of London in brick and stone. His sewers are still doing the same job today, although sewage treatment plants now deal with what flows out of the other end. But at the same time that our tunnels to 'away' are hidden underneath London's streets, each one providing direct evidence of the naivety of our thinking about our world, there are also very visible monuments to the vast scale of ocean recycling standing

---

[24] There is a tiny bit of leakage into space, but it really is utterly minuscule in the big picture.

proud, right in our faces. The modern city of London rests, quite literally, on recycled ocean passengers which have been accumulated by the blue machine over millennia.

Towers of phytoplankton

I am no artist, but one of my favourite photos of myself shows me standing in front of a blackboard, drawing with a piece of chalk. At the bottom of the blackboard is a printout of the photo I'm copying, an image of a stunning phytoplankton cell called a coccolithophore. The original coccolithophore was only perhaps 30 microns across,[25] and the outside of the cell is made up of around fifteen delicately ribbed rigid discs that overlap to cover the entire sphere. It looks like a cluster of tiny ornate parasols. The chalk had been lent to me by the Royal Institution of Great Britain, and it was a real piece of the White Cliffs of Dover, which had arrived in their stores decades before collecting fragments of this national symbol was discouraged. What excites me about this photo is the beautiful circularity of the event: I'm drawing this coccolithophore *with coccolithophores*. Because that's what the chalk is. Around 100 million years ago, great blooms of these armoured phytoplankton multiplied in shallow seas, each sucking in the sun's energy and using it to gather all the elements of life from its surroundings in order to build its cluster of tiny hard shields and everything alive on the inside. The blooms grew so rapidly that the biological recycling mechanism couldn't keep up – grazers couldn't grow fast enough to eat all of the sudden bounty. And so many of the dying and dead cells became diverted passengers, escaping the surface waters by drifting slowly downwards, and accumulating on the seabed over millions of years to form a layer of tiny delicate coccolithophore

---

[25] That's 0.03 mm, which means that you'd have to line up about three of them side by side to reach the width of a human hair.

fragments more than 100 metres thick. As the aeons passed, these were compressed by the increasing weight of the layers above, squeezed together into the soft crumbly rock that we know as chalk. Once the land and the seas had shifted to bring these layers back to the surface, in places like the coast of southern England, humans picked the chalk up, and later dug it up because it was so useful. I love the idea that for centuries, classrooms have been a place where ideas and knowledge – language, mathematics, history, geography and science – were passed on to the next generation with the help of tiny ancient ocean life.[26]

Chalk is a temporary stopover for a particularly interesting ocean passenger: calcium. It's a metal, although you would never know it because it's so reactive that it's almost always found in compounds with other elements. Calcium is a fabulous raw material for construction, and the beautiful armoured discs that cover the outside of the coccolithophore are all made of the basic building material of the ocean: calcium carbonate. This is a hard and solid mineral whose ingredients are ubiquitous ocean passengers, constantly available to be plucked from the water to create biological sculptures. Marine snails, barnacles, mussels, corals, sea urchins, lobsters, cuttlefish and almost anything else you can see in the ocean that has a shell has built its solid structures from calcium carbonate.[27] Like every other type of atom on

---

[26] Today, the 'chalk' you buy for drawing on pavements or blackboards is far more likely to be gypsum (which is formed by evaporation, not from sea creatures). And most schools in the western world have moved on to whiteboards and dry markers anyway, perhaps more 'modern', but lacking any visceral connection to the natural environment. Perhaps at least some biology and geology should still be taught with chalk. At the very least, it would involve less single-use plastic.

[27] There is another building material available: silica. This is very important for a wide range of phytoplankton, especially the diatoms. However, almost everything that grows its shell from silica is too small for a human to see directly.

Earth, calcium shifts from place to place, recycled again and again.[28]

But this ocean passenger is also part of a much slower cycle than the nutrients recycled by the whales. Calcium erodes from the rocks on land and is washed into the ocean via rivers, and then it travels the ocean currents until a coccolithophore or something similar vacuums it up to create its shell. If it sinks to the sea floor to be entombed as rock, it will join the ponderous churn of plate tectonics for tens or hundreds of millions of years. Only once the rocks have been lifted on to land and eroded by the weather will the locked-up calcium become an ocean passenger once more.

The ocean accommodates two speeds of recycling: the fast and the slow. The biological churn is relatively quick, turning over in days or months or perhaps years. But the slower processes can take thousands or millions of years, as some ocean passengers have their journey interrupted by being stashed away in rock, effectively parked in a geological waiting room, until the slow machinery of plate tectonics spits them out again.

The outcome of these calcium recycling processes is gradual concentration. Calcium is one of the more common elements in the Earth's crust, but it's still dilute, making up only around one atom out of every fifty in crustal rocks on average. Once it's dissolved in seawater, it becomes even more dilute – there is less than one calcium ion for every five thousand water molecules. But then biology starts to accumulate this valuable resource. The tiny coccolithophores scavenge calcium atom by atom, as they build their beautiful platelets. Each coccolithophore spends its existence as a minuscule ball surrounded by a calcium carbonate shield, a concentrated speck of calcium-rich

---

[28] Bones also contain calcium as a critical ingredient, but in a different form called hydroxyapatite.

life drifting in a vast ocean.[29] But these passengers grow and drift in their trillions, tiny biological machines that are copied again and again and again. The calcium carbonate is heavy, making it more likely that the fragments will sink once the phytoplankton dies. And then, if the conditions are right, these fragments accumulate on the sea floor together, forming calcium-rich rocks like limestone.[30] The route from dilute to concentrated has been organized by trillions of tiny sea creatures, entirely by accident, each one only keeping itself ticking along. Ocean life created this rich resource, but ocean creatures are not the only ones to have discovered the benefits of calcium compounds as building materials. Humans piggyback on this slow ocean recycling scheme, creating our own calcium-rich building materials.

Concrete is an ancient invention. The majestic Roman Pantheon and the inner walls of the Colosseum were both constructed from concrete, and are still standing two thousand years later. Concrete is a mixture of cement – the 'glue' – and aggregate – the gravel, rocks and sand that provide the strength. It's a dream structural material for any architect: a strong artificial stone which can be cast into whatever shape you like. The era of modern concrete only really started with the perfection of 'Portland cement' in the 1850s,[31] but once architects and engineers realized the potential of the medium, particularly when reinforced with steel, the course of construction of the modern city was set. Today, concrete is the second most widely used resource in the world, after water, and it's estimated that *every*

---

[29] Shield-like platelets are the most common, but some coccolithophore species are far more ornate.

[30] Chalk is one relatively rare type of limestone, but there are many others.

[31] One of the first people to write down a specification for Portland cement was John Grant of the Metropolitan Board of Works, as he set out the standards required for constructing Bazalgette's London sewer system.

single year we make 1.4 cubic metres of it for every person on Earth. And there are eight billion people on Earth.

The view from the Billingsgate fish market, of the soaring trophies of modern capitalism that make up Canary Wharf, is dominated by the visible outer shell of glass and steel. But this is only camouflage for the concrete foundations and core. Although the concrete itself is usually hidden, the mere existence of such giant forms betrays its presence.[32] And in our roads, bridges, car parks, pavements and plenty of modernist buildings, it isn't even hidden. Concrete is only easy to make because ocean life has already done all the hard work of gathering and concentrating the calcium for us. The critical ingredient in cement is calcium oxide, or lime, which we get from limestone. And limestone only exists because tiny sea creatures collected that calcium, bit by bit, to build themselves from the sparse resources around them.

The slow recycling of calcium in the ocean chugs along over aeons. Left to itself, limestone will eventually erode and deliver calcium back to the ocean ready for it to go round again. Interrupting the cycle to filch this calcium for our buildings makes relatively little difference to the calcium cycle because those atoms are just shifted from being locked up in natural rock to being locked up in artificial rock. But the calcium has a bunkmate during its sojourn in natural rock, another atom that shifts from ocean passenger to rock and then back again. This bunkmate also dances around the faster recycling systems in the ocean, on the timescales of zooplankton and whales. To get from the calcium carbonate in limestone to the calcium oxide needed to create cement, the bunkmate is kicked out: carbon, in the form

---

[32] It is possible to build very tall buildings without concrete foundations, but no one has actually done it yet. And there are now types of concrete that don't have cement in them.

of carbon dioxide. And carbon is the ocean passenger which matters more than perhaps any other.

Carbon

Over the last two decades, the word 'carbon' has flooded into public discourse, as the shackles which imprisoned it in school science textbooks have been strained and then smashed to pieces. Carbon is a small atom, tucked into the top right-hand corner of the periodic table, identified by the symbol 'C'. It isn't particularly common – it makes up only 0.025 per cent of the Earth's crust – but it's now discussed as if it were a type of currency, with budgets and accounting and markets. Other atoms are spared this scrutiny; no one talks about global limits for argon, or the problem with using too much potassium. But carbon really is different. The humble carbon atom is a master of transformation, whose potential for reinvention and adaptability make it the essential building block for the living world. It travels the world in the shifting blue machine in a variety of guises, deeply embedded in both the biological and the physical systems of our planet. But recognition of its significance was slow to arrive, partly because it wasn't clear that its many facades all hid the same atom. The first and most dramatic steps towards that understanding were taken in what could be considered a very frivolous set of experiments.

Diamonds are not for ever

Paris in the 1770s was not a happy place. Although France was a rich and powerful nation, almost all the money and power belonged to the king and the aristocracy, continually funnelled inwards towards the court by a complicated and regressive system of taxes. Wages were stagnant and food prices were rising. The reign of King Louis XVI and his wife Marie Antoinette began in 1774 with attempts to help the peasant class, but resentment simmered about the extraordinary wealth of the nobility while so much hunger stalked the streets. The nobility, meanwhile, had the luxury of choosing their pastimes with freedom, and some of them took an interest in science. The French Academy of Sciences, a strictly hierarchical body established by Louis XIV, found itself preoccupied by a puzzle left over from the previous century: why did diamond vanish into thin air when heated up? Diamonds were hard and valuable, prized for their beauty and their resistance to scratches. How could such a thing just disappear? This brought the Academy around to the question of what diamonds actually were, and they set out to solve the mystery. A scientific team was assembled, including a young Antoine Lavoisier, who is now seen as the father of modern chemistry for his systematic approach to chemical experiments.

Their chosen method of study, although scientifically justified, could not have blown more of a raspberry at the hungry peasants if they'd tried. Their approach was this: they would acquire valuable diamonds, lots of them, and then in a careful, sophisticated and scientific manner they would incinerate them. The nobility and jewellers of the era were interested in the outcome of the experiments and offered to donate the raw materials, often attending in person to watch the scientists effectively burning their money. The scientists borrowed a huge round lens, 80 centimetres in diameter, to concentrate the sun's rays on to their diamond samples. And then they waited for the sunny days that would allow them to fry the diamonds with sunlight, in air, in a vacuum, deprived of oxygen, encased in porcelain, and finally in a glass bell jar that allowed Lavoisier to catch and analyse the gas that the diamond had become.[33] By keeping careful track of everything, Lavoisier was able to establish that

[33] For anyone who is worried about the safety of their diamond ring, please be reassured by the fact that to decompose a diamond enough in the normal

they were watching combustion rather than the diamond float-ing off as a gas, and that the gas produced could not be distinguished from the result of burning an equal quantity of charcoal. The implication was clear: precious diamonds were made of exactly the same stuff as cheap charcoal. Carbon was an atom that was two-faced, to an extreme degree. Today we are familiar with this idea, that carbon atoms can lock themselves into an extraordinarily robust crystal structure to form diamond – and that a different, layered arrangement of identical carbon atoms makes graphite, a soft, black substance with characteris-tics almost completely opposite to those of diamond. It seems that Lavoisier himself didn't quite believe the results, and he moved on to other projects rather than chase certainty in this one.[34] But the multiple personalities of carbon had been revealed. The same atom could combine with oxygen to form the gas car-bon dioxide,[35] or could take the form of either the hardest or softest materials known. And this was only the start of the long list of guises that carbon atoms can adopt.

We now know that the most important quirk of carbon atoms is their ability to form long chains while still having enough available linkage points left over that other types of atom can attach themselves and branch off from the main backbone. Each carbon atom has the potential to connect to four other atoms at once, and these combinations can be rearranged in lots of ways – strings, rings, branching structures, and even all of those connected together. Other atoms, such as oxygen and nitrogen,

atmosphere, you'd have to heat it to 900°C. Unless you are planning on throw-ing it into an erupting volcano, your diamond is safe.

[34] He was sent to the guillotine in 1794, officially for dishonesty in setting taxes, which was one of his other jobs, but in practice because he was too closely associated with the *ancien régime*.

[35] Lavoisier didn't call it carbon dioxide – oxygen itself was only isolated in 1774 in England – but he knew that it was the gas that made lime water cloudy.

can be included in the framework, and the potential for different rearrangements is dizzying. No other atom combines so easily to form so much variety, in molecules with so many different shapes and sizes. It's so keen to react to form molecules that free carbon is relatively rare in nature, and so this element is most commonly found in an embrace with oxygen, in a form which we know as carbon dioxide.

There is one other critically important feature of these chemical structures. To take carbon from the surroundings in the form of carbon dioxide, and to build it up into this smorgasbord of molecular types, exacts a cost. That cost is paid in energy, which has to come from somewhere before any carbon architecture can be built. This is what plants do with sunlight: they spend the solar energy that they gather by assembling complex molecules based on carbon, in the process we call photosynthesis. The flip side is that if you dismantle the carbon structures, whatever they are, and return the carbon atoms to a stable carbon dioxide molecule, you get the energy back. Any complex organic molecule, each one a mini-castle based on carbon, is therefore a store of two things: raw materials and also energy. Those little living factories that we call cells depend on these two universally accepted currencies.

Carbon atoms are everywhere. As I write this, if I were to pluck a million molecules out from the air that I'm breathing in right now, about 419 of them would be carbon dioxide.[36] On land, rocks generally don't contain very much carbon, but the soil is full of it, and if organic matter builds up without decomposing (for example, in a peat bog), huge stores of complex

---

[36] It varies a bit depending on the season, whether you're in a well-ventilated room and how many people are breathing around you, but that's the global average. And the obvious and very serious caveat, as we shall see, is that this number is going up very quickly over time because we are still burning fossil fuels. When I was born, it was 335 parts per million (ppm).

carbon molecules can accumulate. All life is built from molecules with a carbon frame – about 18.5 per cent of a human body is just carbon atoms. But if there's all this carbon around, it sounds as though we can't possibly be short of it. And if life doesn't need carbon, it isn't forced to use it. So why is anyone making a carbon budget?

While carbon atoms are obviously critical for life, and woven through all of biology, they have some significant physical effects too. To see what those are, we need to zoom out a long way, far outside the Earth's atmosphere, further away than the moon, out into space to where we can look down on the whole solar system.

Out here, in the dark and the silence, we can see the slow rotation of planetary clockwork. The sun dominates the scene, making up 99.86 per cent of the entire mass of the solar system. The planets circle around this behemoth, held in their orbits by mutual gravitational attraction. From our vantage point out here, the most astonishing thing about this sight is the contrast between the dense spherical star and its planets, each one a crowded spinning zoo containing trillions of trillions of trillions of atoms, and the almost complete emptiness in between them. The laws of physics have organized each ball of atoms into solids, liquids or gases, shaped like mountains, volcanoes and atmospheres. But each one is an isolated island, spinning away in the darkness. Streaming out from the sun is light, which carries energy. Light energy that hits the planets can be taken in and turned into heat or chemical energy, and this fuels each planetary engine. But light energy can also leach away, mostly as infrared light. Each planet is like a bath for energy, which floods in from the sun, the bath's tap. Infrared light leaches away into space in all directions, acting like the bath's plughole. The question for each planet is: how much energy is in that reservoir – in the bath? For a stable planet, the energy flowing in must be balanced by the energy flowing out. That way, the level of the bath stays the same, even though energy is coming and going all the time. Most of the energy is held in the bath as heat, and so the overall measure of the amount of energy in the bath is temperature. A full planetary bath has a higher temperature than a half-empty one. And this is where one of the carbon atom's other eccentricities starts to matter.

The rate at which energy arrives at each planet from the sun is set by the size of the planet and its distance from the sun. Those don't change, so the solar tap adds energy to each planet at a fixed rate. But the planetary plughole changes size depending on the planet and its atmosphere. If the atmosphere helps to trap heat energy on the planet, the plughole is effectively smaller, and the energy bath has to fill up much more before the outward flow matches the tap.[37] That means that the planet gets hotter. There are a few gases that trap heat energy really well, the most significant being water vapour and carbon dioxide.[38]

The planet Venus has an atmosphere full of carbon dioxide, which accounts for 96 per cent of the total atmosphere. The planetary plughole is tiny, so Venus kept accumulating energy and kept getting hotter until eventually it reached today's painful temperature of 464°C, where the temperature at which energy flowing away into space finally balanced the solar input.

---

[37] The rate at which energy flows away from a hot object increases as the object gets hotter. So as the planet warms up, the outflow will increase again until it balances the inflow. When it catches up, the planet will stabilize at a higher temperature.

[38] It just so happens that the bonds between carbon and oxygen atoms in a carbon dioxide molecule are the right size to make the molecule vibrate as infrared light passes by, absorbing the light and then sending it out again in a different direction. This makes carbon dioxide a bit like an erratic baseball player – the balls are always thrown from the same direction, but the batter will deflect them in all directions randomly. So infrared light doesn't travel straight through carbon dioxide unimpeded, and at least some of it is redirected back downwards, reducing the amount that escapes into space.

This is an extreme example of the greenhouse effect at work, when the planetary plughole for energy is almost completely clogged up.

But things on Earth are very different, even though the early history of these two planets was relatively similar. If the Earth had no atmosphere, or an atmosphere that was only oxygen and nitrogen, energy would flow away through the plughole really easily and the average temperature of our planet would be around −18°C, rather than 15°C as it is now.[39] We need a bit of greenhouse effect to keep the average at 15°C because otherwise the Earth would not be habitable. It's only taken a little bit of water vapour (between 0.2 and 4 per cent of the whole atmosphere) and carbon dioxide (0.04 per cent) to clog up the plughole enough to give us a planet that isn't frozen solid. And now we come to the kicker. The plughole is extremely sensitive to the amount of carbon dioxide in the atmosphere.[40] Adding even a little bit more will have an outsized effect. So we need to pay attention to how much carbon is in the atmosphere, because that sets the planetary temperature.

The atmosphere is only one reservoir of carbon atoms. There is also carbon in the ocean and the soil and life itself – there is *loads* of carbon about. The multiple guises of carbon are everywhere. Carbon is constantly shape-shifting, from carbon dioxide to sugars to proteins to gas, and from land to ocean to life. But the amount of it in just one place – the atmosphere – determines the temperature of our planet. This is why the

---

[39] −18°C is a commonly quoted figure for this calculation, but it rests on an unlikely assumption: that the reflectivity of the planet is exactly the same in both cases. In reality, if the Earth cooled down as far as −18°C, it would be covered in far more ice than it is now. This would reflect away even more of the incoming sunlight, and so the equilibrium temperature would be even lower.
[40] And other greenhouse gases like methane, although carbon dioxide is the one with the largest effect at the moment.

budgets matter. There is only a fixed number of carbon atoms on Earth, and a fixed proportion of that total takes part in the constant cycling across life, ocean, atmosphere and land. The only way to really understand how many of those carbon atoms are in the atmosphere and are likely to stay there, in the one place where they have huge consequences for planetary temperature, is to keep track of all the carbon atoms everywhere else. Where are they going, how fast are they going to get there, and when will they float back into the atmosphere as carbon dioxide? The location of all those other carbon atoms really matters, and I'm sure it will come as no surprise to you now to find out that the vast majority of them are in the ocean, passengers riding the blue machine.

But before they can ride the machine, they have to get into it. And the only way to directly monitor the border between the atmosphere and the ocean is to be out there when the transfers happen. Sometimes, the ocean and atmosphere exchange gases in a calm, orderly way. And sometimes, there are dramatic events that temporarily let passengers flood into the ocean, almost overwhelming our attempts to keep track of them.

When the ocean takes a deep breath

'But I don't want to go among mad people,' Alice remarked.

'Oh, you can't help that,' said the Cat: 'we're all mad here. I'm mad. You're mad.'

'How do you know I'm mad?' said Alice.

'You must be,' said the Cat, 'or you wouldn't have come here.'[41]

The chief scientist is standing on the bridge of the ship, looking out over the sea and grinning as if all his Christmases have come

[41] From *Alice's Adventures in Wonderland* by Lewis Carroll.

at once. The bosun is scowling at the radar and muttering about this taking a year off the life of the ship. The wind speed display says 65 knots (75 mph), with a maximum measurement for today of 90 knots (105 mph). This is what we came out here for.

It's the autumn of 2013, and the American research vessel R/V *Knorr* is our home for six weeks. We are bobbing about in the North Atlantic, hundreds of miles from the nearest land, and sticking to an area that anyone else with a choice is avoiding. The ship is a giant steel can that roams the boundary between the ocean and the atmosphere, rolling and pitching as the waves slam into it, but safely glued to the water surface. There are eleven scientists and twenty-four crew on board, and we are all here with one purpose: to watch the ocean breathing. We are right underneath the North Atlantic jet stream, the atmospheric freeway that barrels eastwards high in the atmosphere, pulling powerful swirling storms along beneath it.

The chief scientist is grinning like the Cheshire Cat because he's been looking forward to this storm ever since it first appeared on the forecasts. And it's turned out to be even bigger than the forecasts suggested. As wind gets faster it stirs up the surface of the ocean, bringing new water into contact with the atmosphere. In addition, breaking waves generate bubbles which have a huge surface area, and are temporary parcels of the atmosphere trapped in the ocean. All of this mixing and stirring and contact means that the gas molecules in the atmosphere have plenty of opportunities to dissolve into the ocean, to cross the air–sea boundary from the free-wheeling molecular jumble of the atmosphere into the dense reservoir of the blue machine. From our point of view, a big storm is just the ocean taking a deep breath. And we have a very rare opportunity to measure what happens in some of the harshest conditions the ocean can throw at us.

My colleagues are measuring the atmosphere, the gases and the waves, and my job is to measure the bubbles. I've got a big yellow buoy covered in bubble detectors that looked such a monster when I first saw it on the dockside in Southampton. But out here, its 11-metre-long hull appears small and fragile. It was put over the side yesterday and it's still out there, recording the underwater details of the maelstrom as we pass through the eye of the storm and back out again. I'm not even sure what the data will look like, because there are so many bubbles that the sea surface has gone completely white. Now that the storm is here, and all our scientific instruments are switched on and collecting data, there is little for us scientists to do except ride the waves until the wind and the seas die down.

The ship is bouncing and swaying, and it's impossible to concentrate on desk-based work. Most of us are napping, or up on the bridge watching the spectacle. I am firmly in the second category. I could spend hours up there. By lunchtime the swell has built until the regular waves are around 12–14 metres high, and the ship is pointed right into the endless oncoming ridges of water. As each one goes past, the bow dips into it and then rides right back up, occasionally crashing right through the wave if it can't keep up with the ocean. The surface is covered in trails of bubbles being blown by the wind, and you can see the bubble plumes left by previous huge breakers sitting under the surface. It's as if the surface of the ocean is being blown away, snatched violently from the top and sides of a mountainous waterscape that never stops moving. And all around us, invisible gases are being captured by the ocean: carbon dioxide, oxygen and nitrogen molecules that hit the water surface and stick, and then are carried downwards before they can escape. At the same time, some molecules are leaving the ocean and floating upwards into the atmosphere.[42] But today I'm leaving thoughts about the gases to my colleagues, because all I care about is the

---

[42] The net outcome (whether it's an overall input of gas into the ocean, or a net output of gas into the air) depends on the balance of these two processes. In

bubbles – the speed at which the ocean creates them, the tiny processes in the top few centimetres that determine their fate and the way they change the ocean surface. I can see more bubbles now than I've ever seen in my life at one time. This, out here in the middle of the North Atlantic in a huge storm, is the bubble physicist's dream day out.

It turns out that, from a personal if not a scientific perspective, you can have too much of a good thing. A few days later, the heavy rolling and pitching is beginning to take its toll, even on the most cheerful of us. Sitting at the breakfast table, everything slides at once, and as soon as you hear that noise you reach for your plate with one hand and for the nearest stable object with the other. A few unlucky forks always escape, and a second after the sliding noise starts, there's a crash as everything in the kitchen arrives at either the wall or the protruding lip around the kitchen work surfaces. You keep holding on, waiting for gravity to be trustworthy again. The first mate, who had thought until a second earlier that he was about to step into the canteen, is frozen in time like a character in *Crouching Tiger, Hidden Dragon*. The pause only lasts until the ship rolls back the other way and then he arrives at the coffee machine rather more quickly than expected. Nobody stops eating, cooking or talking while all this is going on.

Last night was hard on us all. No one slept. Floating is a very relaxing state to be in, but it tends to be spoilt when it only lasts for half a second before you thump back on to your bunk. My cabin is close to the middle of the ship, so I'm mostly spared the floating, but I spent the night sliding down the bed and back up again and sometimes also from side to side. It occurred to me for the first time that bedsheets should be made with a measurable tread, so that you can choose the grip appropriate for the

---

this part of the North Atlantic, the average net flow of carbon dioxide is strongly into the ocean.

conditions. Only the enthusiasts are at breakfast, or the ones who gave up on trying to sleep hours ago. The steward cracks another egg on to the griddle, and waits as the ship saves her the bother of tilting the pan to spread it out. Wonderland, and the world Alice saw through the looking glass, have nothing on this.

Almost everything is strapped down except the scientists themselves, but the rolling is constant and even the most carefully confined objects sometimes work loose. That evening, I'm trying to read in the ship's library, and Trivial Pursuit zooms along the floor past me. When things fall on the floor, it's frequently easier to leave them there, to save them the bother of falling down again. But then you have to listen to them all playing bumpercars as the ship rolls with the waves. A few seconds later Trivial Pursuit is making its way back to the port side, accompanied by Pictionary. My tired brain notes that the games are playing games. Humpty Dumpty really wouldn't last very long out here, with waves slamming into us from every direction and no king's horses and no king's men to try putting him back together again. Half an hour later, the *Oxford Astronomy Encyclopedia* arrives at my feet, chaperoned by a chair and a small guitar. Time for bed. But the storm and the injection of billions of new ocean passengers arriving from the atmosphere carried on outside, oblivious to its uncomfortable human spectators.

Follow the carbon

Seafarers have looked out at stormy seas for centuries, understandably preoccupied by safety, thoughts of home, and their ability to sleep and eat while being thrown around like rag dolls in a tumble dryer. But in every case, all around them, the ocean was breathing deeply. The huge ocean surface is in constant contact with the atmosphere, and so gas molecules that bump into the surface may well stick to it and be carried downwards, while gas that starts off in the water may find itself in contact with the air and drift upwards to join the atmosphere. In calm seas, this process is slow, but when the wind rises, the breathing process speeds up; it can be perhaps fifty times faster in a big storm than on a very calm day. Our North Atlantic expedition may have been uncomfortable at times,[43] but every single one of us thought that it was worth it for the data and the understanding that we brought back. To be able to measure directly what was happening at such high wind speeds was really valuable, and the periods of the highest winds provided the most interesting data.

Gas molecules are always travelling in both directions, but if the concentration is higher on one side of the boundary than the other, the breathing process will tend to even out the difference. We can now draw maps of the ocean breathing in and breathing out carbon dioxide, maps that cover the entire surface of the global ocean. In the northern North Atlantic the ocean is predominantly breathing in, taking carbon dioxide from the atmosphere. But in the tropics – the warm waters near the equator – the ocean tends to be breathing out, and giving carbon dioxide back to the atmosphere. There is a seasonal pulse to the breathing process, which depends on water temperature and weather as well as what the ocean brings to the surface. This means that there is a regular transit of carbon atom passengers into the ocean from the atmosphere and back again, but it happens differently in different places.

We have seen that carbon atoms are distributed throughout the land, atmosphere and ocean, but they are not distributed equally. The amount of carbon in the land (including soil and

[43] If I'm honest, most oceanographers I know who work in these conditions (including me) love being at sea, and thrive in this sort of environment. Those who get seasick easily obviously suffer more than most, but if you stay out in rolling seas, it tends to wear off.

land plants) is about 2,000 gigatonnes. The amount in the atmosphere is 875 gigatonnes. And the total amount of carbon stored as travelling carbon passengers of the ocean dwarfs both of those – 37,700 gigatonnes,[44] around fifty times as much as in the atmosphere. So is there any danger that a significant chunk of that could come back up to inhabit the atmosphere? This is not something to worry about in the very short term because, as we have seen, the ocean is a huge engine rather than a well-mixed pool. The surface layer of the ocean, the warm, mixed layer that floats on top of the cold, dark water below, only contains about 670 gigatonnes of the ocean carbon. The rest is in the depths. If carbon gets down into the deeper ocean layers, it may be an ocean passenger for hundreds or thousands of years before next coming into contact with the atmosphere. And only if it's in contact with the air can it affect the carbon levels in the atmosphere, that very sensitive number that dictates planetary temperature.

Physical processes get carbon into the ocean, but the important question is what happens to these passengers next. Of course, the carbon will be recycled in the upper layer on short timescales, incorporated into life and then ejected as life's business goes on. But where else could this carbon go? There is a way to transport carbon atoms down below the warm surface mixed layer, the ocean's lid, and it requires both physics and biology. We've already met it: the tiny and fragile sinking detritus of life, which is collectively know as marine snow. The zooplankton poo that sinks out of the upper ocean layer and down into the depths is a very significant component; poo is very much the VIP lane when it comes to ocean passengers, helping its contents bypass the normal limitations and speeding up their passage to the next destination.

---

[44] All of these are minuscule when compared with the carbon that is stashed away in rocks – estimated to be more than 60,000,000 Gt – but this mostly stays parked in the rock and doesn't interact with living Earth systems on the timescales that humans care about, so we'll ignore it here.

Modern human history has not looked favourably on excrement. As we have seen, Victorian London was prepared to pay a huge sum of money to the best engineers of the day to make it vanish. Of course you don't want it piling up outside your door, not only because of the disease risk, but also because the pile will never stop growing.[45] In spite of this, the humans who have helped other humans by collecting it and manually moving it somewhere else – night soil collectors or gong farmers in the UK, dunny men in Australia, untouchables in India – have historically been looked down on. They were preferably to remain unseen and definitely to remain unsmelled. But of course, things are different in the ocean. In modern ocean science, collecting poo isn't a niche activity; it's an essential task for following carbon around the blue machine. The poo collectors are respectable, enthusiastic and important scientists, and understanding the biological structure of the ocean requires that they do a very good job of it. It's a bonus that the poo itself turns out to be fascinating.

The poo scientists

Professor Stephanie Henson is a principal scientist at the UK's National Oceanography Centre in Southampton. We are sitting in her office discussing the excretory habits of salps, common gelatinous and transparent tube-like creatures which are a few centimetres long and float around the surface ocean scoffing phytoplankton. 'They do poos which are quite enormous, considering how small they are,' Steph says; 'you get these big blobs, but the way they're processed in the ocean is quite different from krill faecal pellets, which are really compact.' The dense krill faecal pellets are feasted on as they fall through the ocean,

---

[45] Given our basic biology – as a sentient structure built around a tube connecting a mouth to an anus – it's arguable that there should be a third certainty of human life added to the list with death and taxes: poo.

but the salp excreta aren't nearly as popular with the scavengers in the dark depths and so may be more likely to reach the ocean floor intact. We spend a while speculating about how the biology of a fragile jelly-like salp allows it to produce any significant poo at all. But the popularity of falling faecal matter with the ocean's fussy eaters is critically important, because it's not just nutrients that are drifting downwards inside these pellets – it's carbon, tumbling away from the atmosphere all the way to the sea floor unless something gets to it and eats it first. The same goes for any other detritus from the surface: half-eaten zooplankton, expired phytoplankton, gel-like dross: they hold nutrients, yes, but packaged in a frame of carbon.

The question that matters for carbon accountants is how much of the carbon floating around the ocean escapes the part of the reservoir that's connected to the atmosphere. As we have seen, the raw material of life is rapidly recycled in the upper ocean, atoms shifting from water to life through other life and back to the water within days or months or years. Most of those atoms just cycle around near the surface, where they can make their way back into the atmosphere at any time. But the upper ocean leaks. Steph's group at the National Oceanography Centre are tasked with measuring that leakage, and just like any other tiny loss from a huge system it's a challenge to record accurately. There are generally two types of experiment when it comes to ocean science: the extremely simple and the frighteningly complex. Poo collection currently falls into the first category, but may be about to leap into the second.

The established tools of the trade are sitting outside on the concrete dock while they are prepared for their next foray into the deep. Each one is a giant yellow plastic funnel a metre high, its mouth open to the sky. They look like huge upside-down traffic cones. At the bottom, the neck of each funnel is stuck into a circular turntable. Corinne Pebody, another member of the NOC team, shows me how it works. Underneath the turntable, sample bottles are arranged in a ring, each one full of a preservative dissolved in very salty water, and the system is programmed to turn every two to four weeks to bring a new bottle to the mouth of the funnel. The funnel itself, with its turntable and bottles, is then suspended in the deep ocean to capture whatever falls into its big yellow maw. It's kept in place by glass spheres floating above it and a large weight on the seabed 2,000 metres below it. Amazingly, the ocean environment is so calm at these depths (typically 500–3,000 metres) that the high density of the preservative liquid is enough to keep it in the bottle, even while the mouth of the bottle is open to allow detritus falling into the funnel to be caught. This marine snow-catcher can be left hanging in the ocean for a year, moving on to a new bottle when the program dictates, and at the end of the year each bottle contains whatever fell into the funnel during its designated period. The bottles are tiny compared with the cone, perhaps the size of a home pepper-grinder, and Corinne says that after two weeks at a depth of 3 kilometres, whatever has fallen in typically only takes up the bottom 1–2 centimetres. I look up at the wide mouth of the funnel and try to imagine how meagre the fall of marine snow at these depths must be, if that's the entire yield from a few weeks. The upper ocean clearly doesn't leak very much. Corinne's job is to count, weigh, sort and analyse whatever turns up in the bottles. Perhaps it's a strange view of the ocean world, sifting out clues that tell the tale of slow and dilute ecological dramas from their fallen dregs. But Corinne thinks it's great.

Upstairs in the lab, she shows me what she has found in the bottles. The first microscope slide is full of a dense grey-green-brown fluff which wobbles when I poke it with a pipette, as if it's full of jelly. This is phytoplankton which died at the surface, sank as lots of tiny fluffy pieces and then aggregated into this big lump at the bottom of the sample pot. But that doesn't explain the jellied consistency. Corinne rummages around on the other side of the room and comes back with a small glass vial that seems empty. But when I look more closely, I see that it's actually home to a completely transparent snail shell. This is a pteropod, a free-swimming sea snail, and Corinne explains that some pteropods (particularly the ones known poetically as 'sea butterflies') exude mucus webs out into the water around them to catch food. If they're disturbed, they'll cut their gelatinous fishing net loose and this sticky web then drifts away, continuing to catch marine detritus and glue it all together. That's what I'm looking at under the microscope: a pteropod's jelly web that has bumped into and captured millions of tiny particles, gluing them together into larger pieces of marine snow. Big lumps will fall more quickly, so the presence of these natural gels in the water speeds up the leakage into the deep.

The second microscope slide is far less impressive, with just a dusting of the gelatinous phytoplankton clumps. The rain of carbon into the deep ocean isn't constant, with variations that depend on the time of year and the location. For the creatures of the dark, calm depths, these marine snow remnants are their only experience of the seasons, as the slow dribble of food from above fluctuates with the availability of light and nutrients at the surface. Also following that rhythm is the poo: pillow-like wads from the salps, round pellets from the pteropods, long sausages from copepods and a full selection box of visual variety from all the other zooplankton.

Steph tells me that in general, about 10 per cent of the carbon accumulated by biology in the ocean's mixed layer, the surface water, sinks down as far as 150 metres. And then perhaps only 1 per cent of that makes its way all the way to the bottom of the ocean without being eaten by something else. The carbon doesn't vanish if the mucus webs or the poo are eaten, because as organisms eat their food they break down the large carbon-based molecules and turn them into carbon dioxide. We do the same, which is why we breathe in oxygen and breathe out carbon dioxide. Once the carbon is back in the form of carbon dioxide, it will dissolve in the water and it won't sink any further. Instead, it will be carried along by that water parcel as a different category of carbon passenger.

The churning of ocean ecosystems produces very little carbon waste overall, and down at 3,000 metres, where the National Oceanography Centre's marine snow-collectors sit, the leftovers really are tiny. But this leakage of the ocean's carbon passengers adds up, a constant drip-feed across the seasons and across the global ocean, spiralling downwards through the darkness until they hit the bottom. And then what?

In the deepest parts of the ocean, most of it will still become food because the sea floor has its own citizens. The sea cucumbers, crabs and tiny fish that Adrian Glover's cameras observed on the vast plains of the Clarion-Clipperton Zone in the eastern Pacific are fed by the paltry dregs of the ecosystems above. Food is so scarce in the deep ocean that very little goes uneaten, and the carbon that reaches these depths will be released back into the water as dissolved carbon dioxide. Those molecules are then locked into the ocean engine, whose movement down here is dominated by the slow, density-driven thermohaline circulation that keeps the ocean turning on very long timescales. The carbon passengers and the water will slide sideways around the deep ocean until they come across a region of upwelling. It could be hundreds or thousands of years before these carbon molecules are carried back up to the surface, keeping them away from the atmosphere and land-based life for a while at least.

There is another option for carbon apart from this endless cycling, one that on ecosystem timescales is the equivalent of locking the storehouse door and throwing away the key. If the detritus of phytoplankton and zooplankton falls in large quantities near the mouth of a river, it can get covered over with silt and mud very quickly as the river outflow dumps tiny fragments of worn-away rock from the land. The covering of silt leads to a suffocating environment with hardly any oxygen, one where most life can't survive and therefore can't gobble up the organic goodness. As the mixture of mud, silt and carbon-based delicacies builds up and is squashed by the weight of the muddy layers that follow, the carbon passengers are locked away to become rock. Only a tiny fraction of the tiny fraction that reaches the seabed will suffer this fate, but its removal from the Earth's living system is almost permanent. This part-organic rock has been accumulating for hundreds of millions of years of the Earth's history, and over the entire globe, it carries a vast amount of energy that was harvested by phytoplankton at the surface of the ocean, locked away in a carbon framework tiny bit by tiny bit, and buried in the sea floor.[46] The vast majority of it stays buried. But some of it turns into something else.

For thousands of years, humans have been finding sites where black sticky liquid oozes out of the ground. Sometimes it flows like honey, and in other places it is so viscous that it hardly flows at all. It was useful for waterproofing in Bronze Age civilizations. The ancient Egyptians took it from the Dead Sea to embalm their mummies. It's mentioned in the *I Ching*, an influential Chinese text from 2,500 years ago, and indigenous peoples in North America used it as glue to attach the stone heads of their arrows to the shafts. The Japanese used it as lamp fuel. This natural oil was useful, but it was messy and supplies were limited. Throughout the nineteenth century, efforts to extract it by drilling were stepped up, largely because of the demand for kerosene for lighting, and it quickly replaced whale oil in lamps. And then the internal combustion engine roared into the world, and oil

---

[46] This buried store of carbon is indirectly responsible for the existence of free oxygen in our atmosphere. If all this carbon had been consumed to create carbon dioxide – CO_2 – there would be significantly less oxygen left over to fill our atmosphere. Its burial makes the separation almost permanent: excess carbon down below and excess oxygen up above.

extraction ballooned, reaching around 70 million barrels every day at the start of the twenty-first century. All of this black gold came from the fraction of a fraction of the ocean's total carbon store that had been locked away in rocks, slowly accumulating for millions of years. The fuel that powered every gasoline car and plane you've ever been in, the raw materials needed for almost every piece of plastic you've ever touched, the bitumen on the roads, synthetic clothing, lubricants, sporting equipment . . . all of this was made from carbon stitched together with solar energy that was harvested by phytoplankton in the ocean millions of years ago. Buried, heated, pressurized, transformed and trapped, carbon buried in ocean sediments could turn into oil and gas,[47] and once they found it humans released the energy and the carbon alike. Burning converts the long carbon chain molecules back into carbon dioxide, which is then pumped directly into the atmosphere. The buried accumulation of aeons was flooded back into the atmosphere in a matter of decades. The delicate balance of the atmospheric carbon budget was demolished.

Now we come back to carbon as an ocean passenger on a global scale. Let's zoom out to watch the Earth rotate beneath us, the blue machine glinting in the sunlight. The water we can see touches the atmosphere, and carbon dioxide is constantly moving back and forth across the air–sea boundary. In the coldest regions, carbon dioxide generally flows into the water, and near the equator it flows outward. The warm surface layer of the ocean, the mixed layer, tumbles and churns as the wind pushes on it, mixing it up. This water has loads of carbon dissolved in it,[48] billions

---

[47] Coal comes from land-based deposits: trees and plants.

[48] There is an important caveat to this simplistic description of 'dissolved carbon', which is tricky to understand but doesn't change the big picture. If you're a glutton for punishment, here's the overview: once CO_2 dissolves, it has a chain of possible chemical forms to exist in: carbonic acid (H_2CO_3),

upon billions of carbon molecules jostling around. They are carried by the currents and disturbed as fish swim past.

This is why my colleagues and I were funded to be out in the huge Atlantic storm. The amount of carbon that is taken up by the ocean and then potentially taken downwards and temporarily out of reach of the atmosphere is not only hard to measure, but has the potential to change depending on ocean temperature and how the ocean engine turns. This is a delicate planetary balancing act; the ocean is huge and the critical processes are sporadic and nuanced.

If we peer deeper, below the surface layer, we find even more dissolved carbon, held away from the surface in the cold middle levels of the ocean and augmented by microbes that gobble the rain of marine snow falling from above and breathe out carbon dioxide. The depths of the ocean are full of dissolved carbon, upwelling back to the surface when the ocean engine pushes them up, and sinking to the depths when the density of the water drags them down. The ocean engine is pushing and pulling this carbon around, and whether that carbon rejoins the surface waters or remains isolated in the ocean for hundreds of years depends on how the blue machine turns.

Floating, swimming and drifting around in this giant pool of wet carbon is life: the carbon-based complex molecules that make up the bones of fish, salp cells, bacteria and whales. This living carbon mostly swims around the surface waters, but when it dies it may join the slow rain of marine snow into the depths.

---

bicarbonate (HCO_3^-) and carbonate (CO_3^{2-}). The reactions from carbonic acid to bicarbonate, and from bicarbonate to carbonate, are reversible, and so what actually happens is that the populations of these three ions constantly adjust depending on the situation. The whole system acts as a pH buffer, and is important for keeping ocean water at a relatively constant pH. The detailed consequences have made many students of oceanography cross-eyed with confusion.

When a big storm mixes nutrients up into the surface waters, living cells grasp the dissolved carbon to build themselves and multiply, shifting the passive dissolved carbon passengers into wriggling life. Overall, the gobbling and swimming and multiplying and sun-harvesting make up a living pool of carbon that leaks into the depths as it dies, constantly and slowly taking carbon away from the atmosphere.[49] The pattern of life in the ocean depends on these flows of organic carbon – carbon that has been packaged up into complex molecules and therefore provides both fuel and nutrient supplies. Without sunlight in the depths, these carbon-based scraps are the only source of food. If you understand those journeys, you can explain the distribution of ocean life. And there is also plenty of new technology that can help, if we use it well.

Across the corridor from Steph's office, a gaggle of young researchers have gathered to show me what they've been working on. Dr Sari Giering is very clear about why she cares so much about this device. 'This will mean that we can actually do our job. Because our job isn't counting dead things, right? Our job is to make sense of the ocean and the ecosystem.' The focus of all the attention is a black tube about the length and width of a cricket bat. When it's in the ocean, water can flow through a

---

[49] Of course, this is part of a cycle too. Carbon that has been buried in rocks is naturally replaced in the atmosphere by volcanoes, which belch out carbon dioxide sporadically and slowly. But this process is far, far slower than the human release of buried carbon as we burn fossil fuels.

If you're wondering whether this gets us out of the problems of climate change, the answer is no. The ocean is already doing us a massive favour by taking up about a third of all the extra carbon dioxide that we humans are putting into the atmosphere. We would have whooshed past the Paris climate targets many years ago if it hadn't been doing this. But it's not clear that the ocean can continue to do us that favour at the same rate in the future, so we may be losing one of our best natural allies. Basically, it's still up to us to sort the problem out, and we must do it sooner rather than later.

small window at one end, and everything that falls through that space will be photographed by a holographic camera.[50] It can see everything from 0.02 millimetres to 20 millimetres in size, and it will be able to record every detail of the marine snow as it falls. It won't see plankton roadkill or squished goo – it will see the falling pieces of fluffy marine snow and the faecal pellets, the creatures that are feasting on them and what happens to them as they fall. They won't get as much taxonomic detail as from the studies of dead things, but they will watch the life of a marine snow particle as it really is. Sari tells me that they are recruiting experts in machine learning to identify what's in the images automatically. They need it, because this device will generate an unimaginably large torrent of data, many many terabytes, and certainly far too much for an individual human, or even a team of humans, to sift through. 'Google can do this with cars – identify what they are automatically – and we just need to do it with what we can see.' She's passionate about the difference that this could make to our understanding of carbon, and of the ocean and our planet. Her view is that the technology is progressing more slowly than necessary, constrained because the brightest and best often see a career with companies like Google as more attractive than one trying to understand the planet. 'But we have smart people in this building who gave up a career in banking to be here, and we just need to persuade more of them to come and do a really cool project like this and make the world a better place.' Sari ends on a wistful note: 'We don't have the money that's available to Google. But we make the tiny steps that we can.'

The invisible passengers of the ocean shape our whole planet, but they are not the only travellers in the blue machine. Swimming and sailing through and past the passengers, there are

---

[50] This will produce not a 3D image, but a very detailed set of 2D images. Maybe 3D holographic cameras will be used for this in the future.

humans and animals that actively and deliberately move around the ocean engine, using its features and its character to navigate, to survive and to explore. Now that we can see the outline of the blue machine and how it works, we can see why their migrations are worth it. They are the most visible connections between the ocean and everything else, and what they do has shaped civilizations and ecosystems alike. It's time to meet the ocean's voyagers.

6

Voyagers

THE MESSENGERS AND PASSENGERS OF the ocean are not in control of their destinies; they must go wherever the physics of the blue machine dictates. Their perspective is always local. But the dynamic internal architecture that they create offers a wealth of possibility to anything that can actively move around inside or on top of the ocean. Instead of having to live a compromise, constantly trading off the advantages and disadvantages of a single environment, a voyager can move between realities at will, benefiting from them all. It's the voyagers who will show us why the scale and intricacy of this beautiful engine matters, as they travel through its contrasts.

As land-based mammals, we ourselves are always ocean voyagers because we are not ocean natives. We can only ever pass through or over its waters, on our way to somewhere else. But we often do it with a very limited perspective on our surroundings. Evolution has forced voyaging animals to tune into every nuance of their surroundings, because correctly interpreting every clue is a matter of survival. We humans have the skill and intellect to excel at this, but our intuitive abilities have largely been swallowed up by the modern world. Today, technology, communication and the scientific method have given us the greatest privilege of a voyager: a global perspective. We still often manage to be mostly blind to what it offers, because we hide behind screens and numbers, because we have lost the habit of observation, and because we don't know what to look for any more. But maybe we can have the best of both worlds, if we re-examine ocean voyaging and why it matters.

Have anus(es), will travel

Whether you're a commuter on the number 32 bus or a migrating olive ridley turtle, the process of actively travelling from A to B is often time-consuming, expensive and/or dangerous. Short cuts are rare and the business of life can't be conducted remotely, so turtles and commuters alike must accept these disadvantages and travel anyway. But evolution can be relied upon to produce an eye-opening exception, and one of the most extreme is the marine worm *Ramisyllis multicaudata*. This slender wriggler completely outsources all of its voyaging to its anus – or, more accurately, anuses. *Ramisyllis multicaudata* is a very strange worm indeed.

There are many advantages to a reclusive lifestyle, particularly on a bustling coral reef full of predators and surprises. *Ramisyllis* can be found with its head so deeply buried inside a specific type of sponge that it pokes out of the base. The sponge is a solid water filter built of tubes and pores that incorporates just enough life to qualify as an animal, although it lacks any real organs and barely moves. The juvenile worm parks itself inside the base of the sponge and spends its life growing upwards and outwards through the protected internal channels. But the thing that really sets *Ramisyllis* apart is that, instead of being satisfied with one head and one tail like most of the rest of nature, it branches as it grows. Its nervous system, digestive tract and everything else splits and splits again to give it one head and *hundreds* of tails which all grow outward through the channels of the sponge until they reach the outside. Every tail ends in an anus, and the tails that reach the surface, hundreds or thousands of them, poke out of their holes and crawl around on the red porous surface of the sponge, constantly exploring their surroundings.[1] This complex branching structure means that the worm can never leave the sponge because it has grown to fit the internal maze of channels. And anyway, the water outside the sponge is dangerous, full of hungry predators and swirling currents. This worm certainly isn't getting out and about in the world to meet a mate, and so it has a problem. Its DNA must travel if the species is to continue.

Other species deal with this problem by chucking out eggs and sperm into the water and leaving them to manage as ocean passengers. Not *Ramisyllis*. After some time spent poking around on the top of the sponge, the end of each tail starts to change. The anus seals up, and a little segment at the top end grows eyes and a primitive brain. The gut atrophies and the muscles reorganize themselves, and DNA is packaged and prepared. And one day, this little autonomous gonad, called a stolon, breaks away from the tail (leaving a new anus behind) and swims off to the surface to get on with the business of mating. Its only role is to voyage upwards and hunt down another little stolon of the opposite sex.[2] So *Ramisyllis* stays exactly where it is, safe inside the sponge, while tens or hundreds of these little DNA-filled part-worms actively voyage towards the light to search for a mate and then die. *Ramisyllis multicaudata* takes 'having your cake and eating it' to an extreme but logical conclusion.[3]

This sounds like an extremely complex strategy, but a journey

---

[1] It's not entirely clear whether anything ever makes it all the way from the worm's head to be discharged on to the reef. In all the worms examined to date, the digestive tract has been almost completely empty, although it appears to be in full working order. The worm may feed by absorbing nutrients directly from its surroundings inside the sponge, but that's not clear either.

[2] All the stolons released around the same time by one worm are of the same sex.

[3] The other worms in this family also almost all do this, but they stick to having one head and one tail at a time.

from one part of the ocean to another – even though it's only a few metres in this case – can be so dangerous that the worm itself can't afford to make it. That journey is also so incredibly valuable that it's worth this massive outlay of time, energy and refashioned anuses. The first lesson from all this is that we are often extremely narrow-minded in our definition of what's 'normal' when it comes to life strategies. But the second is that the larger ocean life is, the less it can afford to take its chances as a passenger. Active voyaging across and through the blue machine puts the voyager in control, and this control comes with a lot of choice. The endless combinations of temperature, salinity, nutrients, trace metals and other passengers, arranged in great sheets or filaments or layers or isolated parcels, are not randomly positioned but they can be fickle, coming and going as the days and seasons pass. *Ramisyllis* only makes a short one-way journey, so let's move on and meet one of the more intrepid voyagers, a forager. Foraging is the search for energy stored in a form that is concentrated enough and plentiful enough to feed on, and the key to successful foraging is to know where to look.

A penguin's commute

It's December, nearly midsummer in the southern hemisphere, and a female king penguin has just transferred her egg to her mate. The egg will rest on his feet, protected from the harsh weather by a fold of feathered skin, until she returns from her trip out into the rough waters of the Southern Ocean. She stretches out her neck, turns her sleek grey back towards the sun, and waddles slowly through the crowd of this year's statuesque parents-to-be and last year's rotund brown chicks. She is moving to the shore on the Crozet Islands, grass- and lichen-covered volcanic specks that sit halfway between Madagascar and Antarctica, 1,400 miles from both, bleak and cold. Food on land is almost non-existent, so life here must look to the ocean for the fuel to stay alive. Almost as soon as the first waves brush her feet, she bellyflops into the water and is transformed from awkward shuffler to elegant torpedo. Underwater she can fly, with all the benefits of a fish-shaped body plus the additional advantage of two powerful wings that offer more agility than a single tail. She can spin, twist and dive with ease, a nimble predator who has a couple of weeks to fatten up before it's her turn to fast on top of her egg once more. All of the Southern Ocean lies before her.

Although life is everywhere inside the blue machine, the real bounty is extremely patchy, as we have seen. Penguins hunt mainly using sight, but they lack the high vantage point of the albatross or the petrel. This penguin has to be mere metres from her prey before she can see it, and she has no way to spot a big shoal of fish from a distance, or to see the other predators that might suggest an oasis of life in the watery desert. The water that surrounds the Crozet Islands isn't empty of food, but in order to make sure both she and her chick survive, she needs a feast – and she needs it quickly and efficiently. The slim pickings around here aren't enough. Hunting randomly in the vast ocean is too risky, partly because it would take so much energy just to search, and so the only way to survive is to know where to go. And she does, because the blue machine has a predictable structure. As soon as she's in the water she sets off southwards, coming up to breathe and then diving a few metres beneath the waves to fly through the water directly towards her target. To a human, her destination just looks like water in the midst of a lot of other water. But around 250 miles south of the Crozet Islands, the physical architecture of the ocean engine creates a huge feature that's a reliable biological treasure trove. This is what keeps these king penguins alive, and it is worth the great voyage they take to get there.

The waters immediately around Antarctica are frigid, partly

because this is a cold place and partly because a supply of cold water from the depths finds its way up to the surface here. This creates a shifting pool of icy water that surrounds this great white continent, where the surface water is below 2°C and only the hardiest survive. But as you go north from the Antarctic coast, you find the Atlantic, Indian or Pacific Ocean,[4] each one a huge basin that extends to the equator and beyond. Their surface waters are warmer, perhaps around 8°C at the latitude of the Crozet Islands, and the destination of this penguin is the great boundary where the warmer and colder waters meet: the Polar Front. This narrow belt of water – perhaps only 20–30 miles wide – stretches all the way around Antarctica, and it's the oceanic barrier that keeps these chilly surface waters separate from the sneaking upper warmth of the rest of the world ocean. As we have seen, these huge regions of warmer and colder water don't just meet and mix. Where great water masses come together, they brush shoulders in an offhand way, each on its way to somewhere else, wherever their masters of density, wind and spin are sending them. The result is these buffer zones, which oceanographers call 'fronts', and whose great snaking contours map out the shape of the ocean engine. The Polar Front is where cold northward-flowing water meets the warmer saltier water of the sub-Antarctic, and it's full of life.

The penguin ducks and dives southward, surfacing to breathe in the chilly air and then diving into the quiet calm below to travel onwards, progressing in almost a straight line and covering about 40 miles every day for most of a week. She doesn't feed much on the way, and the water she's in changes little, remaining at 8°C and relatively salty. But after 250 miles, the shift comes rapidly as she enters a different component of the blue machine. Over a few hours, the temperature drops to 2°C,

---

[4] Depending on your longitude at the time.

the water becomes ever so slightly fresher, and she knows that she has found the bustling metropolis that occupies this border territory. Now her behaviour changes. One deep breath of air sustains her dive all the way down through the surface mixed layer, until she can feel the water around her cool even more. Now she is 100 metres below the surface and she can see her prize: a flickering throng of lanternfish. They are fast, but she is agile and experienced, and she can take perhaps four fish before it's time to return to the surface. After a rest, she dives again and again, snatching the fuel she needs from this ocean metropolis. She will spend five to six days here, feeding during the day and resting at night because it's too dark to hunt.

Ocean fronts are full of life. Their character varies a lot, but they are places where water is mixed and sometimes dragged upwards, and where the advantages of two different water masses are both available to their citizens. An accumulation of nutrients means there are fewer limitations on existence here, and sometimes life is swept into a front because it won't follow its original water mass down into the depths. The phytoplankton grow, the zooplankton can stuff themselves, lanternfish and squid will find their way to the feast, and then the bigger predators – seals, penguins and albatrosses – will undertake long voyages just to join the banquet. Fronts are home to significant species richness and diversity, and most importantly they are *predictable* in both time and space. The Polar Front barely moves, because the vast water masses that it borders are just too big and too slow to change. It's a tiny fraction of the total area of this ocean, but it's always there to provide a consistent food supply. This is worth voyaging for, and this is what ocean predators do. They can't see great accumulations of food from a distance, but they can locate the physical places in the ocean engine where the conditions are right for food to grow and live.

After a few days, the penguin sets off back north on her

commute home. She has collected the energy supplies necessary to keep her alive while she incubates her egg for the following two weeks, and then she will undertake the journey again and again as the chick grows. This is a highly specialized feeding strategy, and it's extremely effective. Ocean fronts are just one of the types of ocean feature that provide sustenance to larger ocean predators, and there are plenty of others. Some are seasonal, so although they're not always there, both ocean predators and humans can plan their lives around them.

Following the followers

Our mental image of ocean voyaging is often one of battling against the odds on long journeys, of isolation, of enduring extended periods without friends and family, and of facing the mental and physical consequences of complete disconnection from the land. But both animals and people navigate the features of the ocean in far less dramatic fashion all the time. A seal will heave itself off a sandbank to go fishing. A flounder will wiggle a few metres further up an estuary. Together, these short, apparently random trips resolve themselves into parts of an extensive and sophisticated dance. Multiple species pulse and swirl in giant patterns choreographed by the blue machine and the seasons. And when human voyagers step into this choreography to join the hunt for food, they must dance to the same tune. The shifting structures of the ocean often leave their fingerprints on the smaller journeys of humans, on land as well as at sea, if you know where to look. So we should not be surprised to find out that a nondescript copepod – a slightly lumpy member of the zooplankton with dramatic antennae – is the reason that a rebellious community like the 'herring lassies' of the early 1900s even existed. Although I suspect that the herring lassies themselves would have had a few choice words for anyone who questioned their hard-won independence.

Up in the cold waters of the Norwegian Sea, in the southern part of the triangle between Iceland, Norway and the Shetland Islands, *Calanus finmarchicus* spends a quiet winter hiding from the world. We met this common copepod in chapter 3 in the North Atlantic, helping move nutrients downwards as its poo tumbles into the depths. Its teardrop-shaped body is only 2–4 millimetres long, but it has room for a sausage-shaped oil sac that can take up 20–50 per cent of its entire body in the autumn. As the winter approaches, it sinks a few hundred metres downward in the ocean and enters a partial stasis, shutting down much of its biochemical machinery to drift passively in the darkness. Above it, winter storms agitate the surface waters, bringing cold nutrient-rich water upwards and deepening the relatively warm mixed layer. But the winter is dark, so there is no fuel for life to take advantage of the bounty. *Calanus* is better off in the depths, away from hungry scavengers, eking out an almost-life by slowly consuming its oil store. The planet turns, the spring comes, and suddenly there are light and warmth, and up at the surface phytoplankton can bloom. It's not clear how *Calanus*, far below the sudden rush of biological activity, knows that it's time. But it knows, and vast hordes of this sleeping copepod switch themselves back on, paddle back up to the surface and breed. Their billions and billions of larvae gorge themselves on the phytoplankton smorgasbord which has been fuelled by the sun, and they start their growth cycle. They stay near the surface, but the ocean engine is turning and now that they've risen into the strong surface currents, the last tendrils of one branch of the Gulf Stream carry them southwards, through the gap between Scotland and Norway and into the North Sea. This army isn't marching – it's drifting along with the current – but as it advances it's doing something else: harvesting energy from the minuscule phytoplankton and packaging it up in its new and growing body. The North Sea is shallow, only 90 metres deep on average, and continental shelf seas like this are overloaded with nutrients, and therefore full of plankton. *Calanus* stuffs itself silly until it becomes a convenient package of protein and oil which is big enough to interest something else: the Atlantic herring.

A herring does not hang out by itself. These small, silvery fish gather in huge schools, which offer at least some protection against predators, and when they're not breeding their focus is food. Given the choice, they will seek out *Calanus finmarchicus* because it's such a rich source of nutrients, and as *Calanus* drifts south the herring move with them, to feed and then spawn. The copepods don't make it easy, darting away as they sense a fish approaching, but when a whole school goes by the continual evasive action tires them out quickly. Eventually, one of the herring will be rewarded with a fatty mouthful, when a copepod becomes too tired to jump. The herring store oil in their soft tissues as they grow, re-packaging the oily nutrients from *Calanus*,[5] and a herring is big enough to attract even larger predators: seabirds, seals and cod. The whole circus – the swarms of *Calanus*, the schools of herring, and a shifting rota of hanger-on predators – moves southwards along the coast of Scotland and then England as the season progresses, all the way from Shetland in May to the coast of East Anglia by December.

But of course, such a large food source did not go unnoticed by the humans on the land which surrounds all the action. For the past thousand years, the human history of the North Sea has been played out on a stage constructed largely from herring.

---

[5] Oily fish are famous for being a source of omega-3 fatty acids, a nutrient that has received increasing attention from health-conscious humans. But the oily fish don't make them. The herring steal them from the oily *Calanus*, and *Calanus* steals them from the diatoms they eat. The diatoms and other algae are the original source. So it's completely untrue that you need to eat fish to get enough omega-3s in a human diet. It's perfectly possible just to go straight to the source, although then humans need to do the work of extracting a concentrated enough dose themselves.

These small fish were scooped out of the water in their billions all around the North Sea coast, becoming a valuable addition to human diets. Herring provided the resources and the wealth to found and build Great Yarmouth in England, Amsterdam in the Netherlands and Copenhagen in Denmark, as well as an impressive collection of smaller towns and villages. From the 1500s to the early 1800s, the Dutch were the masters of the salted herring trade, raking in a mountain of money as they roamed the North Sea, preserved every herring they caught, and then sold what was now an incredibly valuable commodity. But the fish shifted and the technology moved on, and by the 1800s the Scots were poised to take over as the new human plunderers-in-chief of the huge schools of herring that passed down their eastern coast every year. Although herring were plentiful, making money out of them was not a straightforward process.

The problem was the oiliness. The same feature that made herring a rich supply of energy had a downside: once dead, the herring would spoil very quickly. By the late 1800s, the trade was more formalized, and a standard existed to reassure the buyer that their fish had been properly preserved and would not rot: the Scottish Crown brand. This required that the herring were gutted, salted and packed within twenty-four hours of being pulled out of the water. But the men were on the boats, doing the fishing, and so it fell to the women to take care of the catch. In fishing communities, the women would prepare the nets and the boats, take the catch, take responsibility for the money and the sale, and were generally treated as equal partners in the whole process. It was a hard life; every member of the team needed to pull their weight, and earned respect for doing so. But this was the height of the Victorian era, when most of society took the view that working women were barely to be seen and definitely not heard, and that their place was firmly in the home darning socks and cooking. The Scottish fishwives, living with something approaching equality in their remote coastal villages, were regarded as an oddity, to be visited on occasion by tourists who radiated smug horror at what they regarded as the immorality and sheer absurdity of the situation.

But as *Calanus* was carried south, the herring moved south, and so the Scottish fishing boats had to follow along. The fishing vessels could easily voyage through the cold North Sea to find the fish, but if the herring were to be gutted and packed within the twenty-four-hour window, their only option was to take their catch into the closest port. As the Scottish fishing fleet grew, railways were built, and options for travel and lodgings were scaled up. And so by the year 1900, another annual migration was well established: that of the 'herring lassies'. Curers, those responsible for making sure the fish were packed, would recruit women from Scottish villages to gut and pack the fish. And then, as the herring boats moved on from Shetland in May, passing through towns such as Wick, Buckie, Aberdeen, Scarborough, Great Yarmouth and all the way down to Lowestoft in December, thousands of herring lassies mirrored their journey on shore. In an era when women just did not travel, never mind on their own, and certainly not in large independent cohorts, these numerous teams of women would set out from Scotland, each with a wooden trunk to carry her belongings for the season, and arrive in the southern ports ready to work.

The hours were long and hard, often at least twelve hours per day, as they could not stop until the entire catch was packed and salted in the waiting barrels. Everything was done outside at long tables on the beach, whatever the weather, and they were paid by the barrel, so they had to be fast. Working at speed required incredible skill: there are reports of women gutting sixty fish per minute, while others carefully packed the fish so that they would not spoil. It was a dirty, messy job, and the splattered clothes and stench of herring would have been enough to mark the women as outsiders, even before the broad

Scots dialect, the loud banter with the fishermen, and their obvious confidence and pride in their work. These women had a voice and a presence that could not be ignored, far from the coy invisibility that the age expected. They kept their living quarters clean and were regular churchgoers, but they also liked a good party and weren't afraid to socialize. They were industrious and independent, with a strong work ethic; they earned and spent their own money, they did what they wanted, they travelled from port to port across the country, and they did it all without asking any men permission for anything. This was a degree of liberty sixty years ahead of its time. No one else had the skill or the will to gut and pack the herring, and so if the trade was to exist, society had get used to the her-ring lassies, or at least tolerate their presence. In spite of the hard work, joining the herring lassies was a good option for the women from the Scottish towns and villages, one they chose and enjoyed: a yearly adventure, an escape from their own tiny village and parental control, and a strong sense of comradeship with a cohort of other women who automatically became their friends. Being able to wave two fingers covered in fish guts to the established English patriarchy was just a convenient bonus.

For most of human history we have been voyagers just like any other mammal that ventures out to sea. The herring lassies were ultimately led by the biological needs and evolution of *Calanus*, and human society had to adapt its cultural patterns as well as its physical patterns in order to benefit from what the ocean had to offer. For a few decades, the herring lassies stamped their mark on human culture, having adapted to the niche that the natural world offered. But it was not to last. The First World War brought it all to a screeching halt. The year 1913 was a bumper one for the herring harvest, but the war stopped the fishing and it barely restarted afterwards. Ninety per cent of the market for British herring had been in Russia and Germany, so demand collapsed after 1918, although there were still some herring lassies until the start of the Second World War.

But the connection to the ocean is clear: as the physical ocean engine turns, and ocean life finds its place within that engine, human lives too are structured around the way that the engine operates. Human voyaging around and across the ocean also structured communities on land, through trade, wars, migration and exploration. You don't have to look very hard to see its traces almost everywhere.

Life in the ocean has a shorter list of concerns – survive, eat, mate – but just as much or more sophistication in how the voyagers travel around the ocean to attend to them. Without complex social structures or physical infrastructure to lift the burden of existence, every voyage has to be a success. So the most successful of these voyagers need to be able to navigate beyond the relatively fixed and predictable internal ocean architecture, to take advantage of the transient parts of the engine that are more risky to get to, but offer a far higher reward. And humans certainly aren't the only voyagers out there chasing herring.

Hunting the swirls in the big blue sea

To be an 'apex predator' is seen as an accolade, the status of being so dominant in your environment that you have no non-human predators and therefore nothing to fear. An apex predator goes where it likes and eats what it wants, competing only with others of its species for the richest pickings. On land, we know and revere many of these kings and queens of all they survey: the eagle, the lion, the wolf, the polar bear and the snow leopard. In the ocean, the orca (also known as the killer whale) is the most famous holder of this crown, capable of making even a great white shark decide it's got better places to be. But easily holding its own in this category is the giant bluefin tuna, a majestic and muscular ocean citizen which not only hunts with the best of them, but is also one of the great ocean voyagers. Bluefin tuna remain faithful to the same spawning grounds, continually returning to either the Mediterranean Sea, the Gulf of Mexico or sites along the North American coastline to breed in relatively warm waters. But once this critical task is done, they turn to face the open ocean. And, unlike many ocean creatures, they really do have almost complete freedom of the seas, and the ability to go pretty much wherever they want. It's a rare advantage.

What sets the bluefin tuna apart is anatomy and physiology that has been characterized by admiring researchers as 'the most efficient engine created by nature'. An adult bluefin is typically around 2.5 metres long, and its streamlined torpedo-shaped body is a rotund chunk of solid and incredibly powerful muscle which narrows almost to a point at its snout and the base of its tail fin. The body is stiff and tight, but that huge bulk of muscle connects efficiently with the slender crescent-shaped tail, and this is what drives this fish through the water, giving it astonishing acceleration and agility.[6] It's almost like a ship with a propeller at the back, except that it's far more manoeuvrable. The muscles have a rich blood supply, and this fish is partially warm-blooded to keep those muscles warm and effective. A bluefin tuna can maintain an internal body temperature of 28°C even in water of only 7°C water, an impressive feat. They are also extraordinarily efficient at converting energy from food into movement, and this is combined with an endurance that can take them round the world. It's been estimated that an adult tuna can

[6] It's this particularly powerful muscle, evolved for endurance, full of energy-supplying mitochondria and myoglobin, that is so prized for sushi.

easily cover 60,000 miles in a year, more than twice the circumference of the Earth.

Every year, a stream of adult bluefin tuna come out of the Strait of Gibraltar, the slim conduit connecting the Mediterranean Sea to the Atlantic, their dark backs and shimmering silver bellies hiding them in the deep blue of the open ocean. But in spite of their bulk, their fat stores are depleted after the long migration to breed and they need food. It takes a lot of fuel to keep this vigorous hunting machine going, and the tuna must choose the best-quality food the ocean has to offer. In the Southern Ocean, the weaker king penguin had a limited range and timescale, so it needed to forage somewhere predictable, somewhere it could reach on the first and only possible attempt. But the tuna is far more powerful and can afford to search. And it's worth it, because out there in the North Atlantic, there are mobile oases where a hungry tuna can get fat on herring, mackerel and squid. They don't know exactly where to go, but they do know where to start the search.

On the other side of the Atlantic, the great Gulf Stream barrels northward along the coastline from Florida to North Carolina. It's carrying warm water from the tropics, and as it reaches the knuckle in the coast at about 35°N it curls away from the land and flows eastward out into the cooler Atlantic. It's a current about 100 kilometres wide and 1 kilometre deep, but it has no fixed path. As it meets the cold Labrador Current flowing down the Canadian coast it starts to meander. These wobbles are a breeding ground for an extraordinary ocean feature: itinerant spinning islands of water, which can last for weeks and months. It's worth taking a moment to really appreciate how strange this is from a terrestrial perspective. On land, our distinctive geographic and ecological features – forests, grasslands, hills and ponds – are fixed in one place. We draw them on maps, and we use them as landmarks. But imagine that the pretty woodland just outside your town, with its woodpeckers and toadstools and squirrels, drifted slowly around the landscape. You wouldn't know where it was from one weekend to the next, and you'd have to hunt it down every time your dog needed a walk. And then imagine that after a couple of years it vanished altogether, but another two very similar ones popped up near by. That's pretty much how these spinning islands of water work in the ocean. They are nomads. But if ocean voyagers can find these oases, they can benefit from everything that this stray water is carrying.

The Gulf Stream can behave a bit like a river on land, in the sense that once a slight wiggle in its path emerges, that wiggle will get bigger. So if the eastward path of this huge flow of water develops a little bump, northward and then back southward, the wiggle will deepen slowly over time, growing into a loop that arcs outward and then curves back towards itself before carrying on to the east. In some cases the neck of the loop may get so narrow that it joins up with itself and the loop is cut free from the rest of the stream. But the northern side of the Gulf Stream sits on a boundary between cold water to the north and warm water to the south, and this means that there are two possible outcomes. If the loop pushes northward, the water on the inside of the loop is warm water, and the flow is moving clockwise around the loop. If this breaks away, it forms a clockwise-spinning island of warm water entrapped in the cold water to the north. This is known as a warm-core ring, and this is what the lost butterfly fish that we met earlier got stuck in. But if the loop pushes southwards from the main Gulf Stream, then when it pinches off, there's an anticlockwise-spinning island of cold water released into the warm water to the south. This is known as a cold-core ring. They are both types of mesoscale eddy, the formal oceanographic term for spinning islands of water of this size. These warm- and cold-core rings are typically 100–300 kilometres in diameter, and the distinctive patch of water at the core of each can extend down to 1,000 metres below the surface.

These liquid carousels keep spinning as they travel and this keeps them intact, with the water on the outer boundary of the ring making one circuit every couple of weeks or so. The original character of that water – its temperature, salinity and tiny passengers – will stay more or less the same, as it's effectively insulated from the water around it. The result is an intact area of water surface about the size of the island of Ireland, drifting across the ocean and leaving its parent current behind. The meanders in the Gulf Stream hatch about twenty-two warm-core rings and thirty-five cold-core rings every year, all of which go off, free to drift wherever the physics of the ocean takes them. The exact locations of where they'll end up aren't predictable, but the Gulf Stream eddy hatchery is a very good place to start looking. And although the spinning water mass mostly stays separate from the water around it, there's nothing to stop the ocean's voyagers swimming in and out of it; there's no wall. For any ocean voyager who can find these dramatic ocean features, the door to this lost chunk of water is wide open.

The bluefin tuna that have just emerged from the Mediterranean cruise across the Atlantic, a trip of 2,500–3,000 miles. They swim in and out of the warm surface layer, following a similar path to that of the fully grown European eels on their way back to the Sargasso Sea. But there is little food out in these huge expanses. After crossing the great desert that occupies most of the middle of the North Atlantic, they emerge on its north-west corner, where the Gulf Stream meets the Labrador Sea. Now it's time to feed. They can hunt anywhere, but they prefer the warm-core rings, the clockwise-spinning islands of warm water that are found drifting in the cold just north of the Gulf Stream. We know that their warm-blooded nature makes these fish capable of dealing with cold-water hunting, so perhaps this seems counter-intuitive. But it's a very effective strategy, because this isn't just an island of water – it's an island full of all the passengers that the water was carrying before it split away. There is also mixing at the edges and vertically, enough to bring all the benefits of mixing water masses that we've seen before: access to the nutrients and passengers from both. And so the warm-core ring is a mini-town of water floating in water, aggregating marine organisms and allowing them to flourish. There's significantly higher species diversity inside the eddy, small fish which attract bigger fish, and then the apex predators who arrive to lord it over the rest. Larvae and juvenile fish are transported around by the eddy, keeping these vulnerable young in a relatively benign environment as they grow. The rings attract turtles and seabirds as well as predatory fish, and they can even be observed from space, both because of their temperature, and also because they leave a signature in the shape of the water surface.[7]

The tuna have the endurance to search until they find this rich hunting ground, and then they are likely to stay, because what hungry fish would leave a feast? Tuna don't have great hearing, partly because their ear bones need a huge amount of padding to resist the great forces as they swerve and dart while chasing prey. But they do have excellent vision, and so they hunt mostly during the day. They'll eat several different species, but what they're really after are the herring, small, fatty prey that will fuel the tuna's huge muscles. The herring are here because they can feed on the plankton that have accumulated in the ring.

It's all one big happy bonanza, but it won't last for ever. Eventually, the warm core merges into its cooler surroundings, mixing everything it carries into that water. Its heat and salt are now added to the colder northern waters, so the existence of the ring has transferred a big chunk of heat energy northward. All eddies eventually fade away, but back where the Gulf Stream

---

[7] The specifics vary, but a warm-core ring will be around 5–10 cm higher at its centre than the edges, and a cold-core ring will be 5–10 cm lower, so there's either a hill or a depression.

meets the Labrador Current, more are hatching as the ocean engine keeps turning. The tuna will find another warm-core ring to feed in, navigating the structure of the ocean engine to find other promising hunting grounds. Or perhaps, if the tuna is fat enough and it's the right time of year, it will begin the great voyage back to the Mediterranean, to spawn once again. Bluefin reach sexual maturity at four to nine years old,[8] and they can live to be ten to fifteen years old in the wild, possibly many more if they're lucky. They will seek and find these warm-core rings year after year, and it's worth the 6,000–mile round trip to get there.

The cold-core rings that spin southwards of the Gulf Stream are also hotspots of life, but with a different character. In these rings, the effect of the spin is to cause an upward flow of water in the centre which brings nutrients from below up to the sunlight, allowing phytoplankton to thrive and attracting the smaller zooplankton that can feed on them. The cold-core rings drift southward, taking nutrients and food to the Sargasso Sea.

And the amazing thing is that these mid-size eddies are not the only ones. Similar spinning rings are found all over the global ocean, around three thousand of them at any one time, covering a few percent of the total surface area. They pop up and disappear again after weeks or months or (occasionally) years. Those formed by the Gulf Stream are particularly distinctive, because they occur at a boundary with such a large temperature difference and a strong flow, and they're a particular size and depth. But the global ocean is peppered with these temporary spinning islands, each shifting around heat, salt and life, and mixing water upwards and downwards. They come in a range of sizes and strengths, and most of them are moving almost directly westward, although

---

[8] It seems to vary for different populations of the same species, and the populations are separated by where they spawn, even though they tend to feed in the same places.

some are carried in other directions by the larger currents around them. They form the weather of the ocean, a slower version of the weather up in the atmosphere, and they add another rich structure to the ocean surface.

The voyagers of the ocean are constantly navigating between and around these features, according to their preferences. In the north-west Atlantic, yellowfin tuna prefer the cold-core eddies. Swordfish prefer to hunt on the outside, away from either type. Landscapes are fixed, but oceanscapes are fluid and dynamic, with mobile and unpredictable details inside reliable larger-scale patterns that are set by ocean physics. The ocean's voyagers aren't crossing uniform spaces as they navigate the huge ocean basins; instead, they are swimming in and around these dynamic patterns.

Human voyagers also used to be aware of these patterns, able to feel the nuances of the ocean environment as they crossed its surface. Even as we developed compasses and clocks, every sailing ship rode the shifting surface of the blue machine. Although the humans on board could not have seen the great churning below, they would have felt its influence and its changing character. Humans developed extraordinary skill at navigating through that environment, although their destinations were usually on land rather than at sea. They voyaged for the same reasons that tuna do: to get from one ecosystem to another, to benefit from access to a range of global ecosystems rather than being chained to one, to live richer and longer lives by benefiting from variety. But the focused pursuit of better, wider and quicker access to all those ecosystems led to a trade-off. At the pinnacle of the Age of Sail, in the mid- to late nineteenth century, humans were perhaps as skilful at feeling their way through the dynamic natural features of the ocean as a tuna is. But the relentless push towards modernity also drove a fundamental change in our human relationship with the global ocean. The place to see what we lost is the last great sailing cargo ship of that era.

The great disconnect

It's a beautifully still and cloudless summer day, and the streets of Greenwich in London are filled with the murmuring of relaxed humans who don't have any particular place to be. They're here to bask in both in the sun and the dramatic baroque architecture of the Old Royal Naval College, which looms over the bank of the Thames here. Perhaps they'll drift up the hill to stand on the Greenwich Meridian Line,[9] and to contemplate a past when time was uncertain and position even more so. It's the perfect holiday scene. But it's not particularly relaxing where I am, 26 metres up in the air, with nothing but the taut rope I'm standing on between me and the paved plaza below. I'm holding on to a steel rail that runs along the top of a thick horizontal wooden spar, and this juts out from the great mainmast of *Cutty Sark*, the world's last surviving tea clipper. Now suspended above her dry dock, with a commanding view over the Thames, this ship surveys the modern world that overtook her. She represents the pinnacle of humanity's ocean voyaging before it was decoupled from the natural world, and serves as a flag-bearer for a lost era. Humanity won huge advantages from the revolution that left this ship behind, but *Cutty Sark* is here partly to remind us that we lost something too.

Built in 1869, *Cutty Sark* was designed to be the fastest ship of her day. She has the sleek lines of a yacht, with a sharp and almost vertical prow that enabled her to cut through the water with as little resistance as possible. But the belly of the ship bulges slightly to provide extra cargo space for the valuable tea

[9] The Meridian Line is the official split between east and west, the line chosen to represent zero degrees of longitude. Every single GPS position you've ever used, or that your phone or your satnav has ever used, is a measurement made relative to this line.

she was expected to bring from China to London, creating room for ten thousand tea chests at a time. The tea trade was all about speed, and the only thing that came close in importance to the rapidity of the voyage was the amount of freight that you could bring back. When fully loaded, the packed cargo would have occupied the entire inside of the ship up to the main deck. Rising up from the main deck are three huge masts with giant cross-pieces – called yards – stretching outwards well past the sides of the narrow ship. I'm perched on the outer edge of one of these. I try to imagine what it would have been like to be up here when the ship was rolling in stormy seas, how close the end of the yard would get to the waves,[10] and the feeling of swinging over the ship and back again on this inverted pendulum. I can't get over how much the world would have been moving around you, and how much it probably would have felt as though the ship was trying to throw you off.

I spend a moment admiring the terrific view across the river, but then get distracted by the elaborate network of rope that braces and connects the masts: the rigging. From the ground it looks almost haphazard, the frantic overkill of a nervous camper as a storm approaches. But from up here, I can see how systematic the intricate rope structures are. My route up here was to climb the 'ratlines', effectively a rope staircase that adorns the 'shrouds', the ropes that keep the masts in place. A fully rigged working ship would also need an additional web of lines to shape and position the sails. In her heyday, this one relatively small vessel, only 65 metres long, would have carried *18 kilometres* of rigging. I'm up here with a harness and a helmet, being watched by two rope experts and with full confidence that someone somewhere has put a lot of thought into writing risk

[10] If you've ever wondered what a 'yardarm' is, it's that – the final end section of the yard, to which the sails don't quite reach.

assessments for this activity.[11] But when *Cutty Sark* was racing back to London with her precious cargoes in the 1870s and 1880s, the crew would have been constantly climbing up here and then back down, in bare feet, in rough seas and with no safety gear, to deploy, adjust and fix the canvas sails. Holding on with both hands at all times would not have been an option, because there was work to do. Back then, sailing required a very physical connection to the ship – pulling, adjusting, gathering, tidying – straining the human muscle which continuously resculpted the ship to meet the demands of nature.

And that is why this ship matters: because of the relationship it brokered between humans and the rest of nature. A sailing ship on a voyage is a collaboration, not an object. No sailing ship will go anywhere without its crew. By the time *Cutty Sark* was built, humans had been tweaking sailing vessels for centuries, learning by trial and error how to use a ship as a harness which could hitch humans to their environment in a very specific way. When you stand back and really look at *Cutty Sark* from the side, it becomes obvious that it's just a link in a chain – a minimalist connection between the cargo and the wind. This ship has a composite design with wooden planks laid over a slender steel frame, a delicate shell compared with the bulk you'd need to get the same strength from wood alone. Everything above the hull would have been dominated by the gigantic sails, up to 3,000 square metres of canvas held aloft to grab the wind. Humans were confined to the three cabins perched on the main deck, a thin living veneer on top of the cargo, and when they stepped outside those sleeping/living spaces, they stepped directly into the elements. Big swells could come right over the side, and one crew member wrote of the ship temporarily becoming just three

---

[11] Full disclosure: I really am confident about that because I have the great privilege of being a trustee of Royal Museums Greenwich, which owns the ship, and so I've read and discussed all of those risk assessments.

sticks (the masts) sticking out of the water, as the ocean engulfed everything beneath. The crew lived and breathed the ocean. This immersion in the natural environment was necessary to feel their way through the ocean engine, and they would have been either consciously or subconsciously aware of every nudge to the rudder, every additional sail added, because they would feel the ship shift under their feet in response.

In 1885, *Cutty Sark* set the world record time for travelling from Sydney to London, completing the voyage in just seventy-three days. This achievement was a triumph of human-plus-ship collaboration, as the captain and crew skilfully rode the nuances of the currents, wind and waves purely by making subtle changes to the shape of the ship – by altering its sails and its rudder position to shift how it sat in the great machine of atmosphere and ocean that was turning around it. The wind could exert a gigantic force, if the sail shape and ship orientation were tweaked to connect efficiently with the machine of nature. This is the skill of sailing. *Cutty Sark* was entirely at the mercy of the ocean, and when they met the 'doldrums' (a belt of persistently calm weather that girdles the Earth near the equator), and the wind dropped away to nothing, that dependence became clear. But when the wind was there, the collaboration was stunning. The top recorded speed achieved by *Cutty Sark* was 17.5 knots, or 20 miles per hour, which is fast even by today's standards. The crew had to understand and feel all of the ship's quirks to get the best out of it, and it was art as much as science. The sailing ship was the gateway to mysterious foreign lands, and only the right combination of knowledge, skill and intuition would open up that path, with the ocean itself continually prowling at your shoulder, ready to humble you if you got too arrogant.

All of that responsibility was funnelled through the captain. Back on land, there were governments and families, networks to provide food, backup and medicine, experts and libraries, and places to shelter from a storm. There were resources, knowledge and people you could depend on. But out in the open ocean, there was nothing. As the captain stood at the helm he would have felt the giant expanse of nature around him and at the same time seen the tiny capsule of humanity under his control. For the time they were at sea, the captain was the absolute ruler of this floating kingdom, ultimately responsible for food supplies, cargo, the money needed to buy foreign goods, crew management and discipline, navigation, the state of the ship and, above all, getting ship, cargo and crew safely from one side of the world to the other. If you got the calculations wrong or misread the wind or the currents, the vast forces of nature could break your masts and starve your crew – or you could simply get lost and never be found. It was a staggeringly isolated position to be in. But there was also huge pride in that self-reliance and independence, in the knowledge and confidence that you could be master of this situation, that you did have the skill to read and cooperate with nature, to corral the forces that shoved and strained your ship, that you could face the powerful rawness of it all and still return to civilization. Once the ship left port, it was locked into a deal with the planet: ride the nuances of this giant uncontrollable system with skill, and it would carry you home. Make too many mistakes – or just one serious one – and you were toast. You had to *earn* a successful voyage the hard way, every time. Captains' logbooks are full of minute detail about wind, rain, currents and temperature, a meticulous collection of clues that could be used to deduce the next best move. But the measurements were only ever a part of it, and the feel of the ship and the surroundings would have been just as important. It's no wonder that sailing was seen as a moral choice as well as a rational one. But this great synthesis of human and nature was about to be jettisoned by economics, technology and the demands of convenience.

Even before *Cutty Sark*'s first voyage, the transition to a new era was well under way. The Suez Canal opened the same week that *Cutty Sark* was launched, connecting the Mediterranean directly with the Red Sea and shaving 5,600 miles off the voyage from China to London. It was too difficult and expensive to tow a sailing ship through this short cut, so the canal conferred a huge advantage on coal-fuelled steamships. Engine-propelled ships had been around since the early 1800s, although their progress from wooden paddle-steamer to a technology that could compete with the fastest sailing ships for cargo had taken decades. But now their moment had come. *Cutty Sark* managed only eight seasons of tea runs before the steamships took over this lucrative and high-profile trade which was dominated by the need for speed. She was banished to the wool trade (where speed was also important) and then to a more mundane existence as a general cargo ship. But this was not a like-for-like replacement, and the transition was not without controversy.

The most fundamental aspect of this upheaval wasn't the shift from wood to metal, or from free wind to expensive coal, or from the irregularity of the weather to timetabled reliability, although all of those were important. It was the change from voyaging *with* nature to voyaging *despite* nature. For a steamship is just a mechanism. Apart from the need to keep shovelling coal into its boilers, you could pretty much switch a steamship on, point it in any direction, and walk away while your ship moved itself around. The centuries of collaboration between humans and nature were over. The crews of sailing ships hated steamships; the majority would have chosen balancing on wet rigging in a storm over shovelling coal any day. The romanticism of sailing is not a modern-day phenomenon; the crews of the nineteenth century knew and felt it, and many of them retired rather than give it up. Most of the small number of crew members required on a steamship were mere servants of the mechanism, shut away inside the bowels of a steel prison, feeding the blazing fires of the 'devil boats', completely insulated from the ocean around them. The ability to travel at a fixed speed in a predetermined direction may well have marked a human triumph over nature, but it also marked a comprehensive defeat for any visceral human connection with the nature of the ocean. Of course, currents and waves would still affect the progress of a ship, but they mattered far less. Commentators at the time expended many newspaper column inches on whether it was morally right to defy God in this way, to have the temerity to go wherever you wanted, and to turn a ship into a mere container that happened to move.

There were also more prosaic considerations that kept sailing ships in the game for a long time. The first was flexibility. There was weather everywhere, and so while you might have to wait for the right season, there was no limit to where a sailing ship could go. There would always be wind to help you move, eventually. Sailing ships also cost nothing more than the crew's wages and food. But a steamship required coal, which was expensive. Even after steamships had taken over most of the cargo routes, sailing ships were still used to transport coal to many ports, so that the steamships could refuel. This is mindblowing to consider in a modern world facing an existential crisis because of our fossil-fuel use: in the late 1800s, a free, clean source of energy was used as an enabler for expensive, dirty energy. Without sailing ships to transport the coal, steamships would have been limited to the small number of ports that had direct access to their fuel. Now, we take pervasive supply chains of fossil fuels for granted, but those chains had to be built from scratch and the sailing ships kick-started that process.

Steamships were critical in the development of the modern world, but they were defiant by design. Humans were no longer taken by the ocean; now they took themselves. Unlike all other ocean voyagers, we became independent of the ocean itself, voyaging through and over it, but caring little about what was going on around us. No engine of this scale could be reduced purely to wallpaper – the ocean still had its influence, even on the metallic voyaging machines. But humans no longer needed to connect to the ocean mentally in the same way, and so the habit was lost. Understanding the intricacies of the blue machine is no longer a matter of raw survival, now that radio, radar, GPS, satellite phones, weather forecasts, current forecasts, AIS and distress signals are taken for granted.[12]

So humanity used its technological achievements as a scythe, slicing through centuries of accumulated connection and knowledge to detach itself and its voyagers from the raw reality of the ocean. The modern world demanded a certain type of efficiency, judged by the ability to deliver precise scheduling, year-round operation, convenience, speed, and profit for investors. By those measures, the loss of human connection to the blue machine was irrelevant, and so was the associated atrophy of culture and identity. But humans are still human. The balance sheet may neglect emotional riches, but it can't eliminate them. And this loss of connection does not have to be permanent.

Voyaging across the ocean has the potential to be far more meaningful than merely getting from A to B across the sea surface, even (perhaps especially) in the modern world. Recognizing this allowed the greatest ocean voyaging culture of them all to save itself from the brink of extinction, to rebuild its identity and to rethink the future. It was a difficult path, carved out by persistent belief in an idea despite continual obstacles and setbacks. Great teachers, leaders, volunteers and community members all rose to the challenge of navigating both their past and their present, until they found their way into the future on a voyaging canoe called *Hōkūleʻa*.[13]

---

[12] AIS stands for 'Automatic Identification System' and almost all ships are required to carry it (although enforcement is tricky and there is still a minority who want to remain invisible). If you look online, you'll find links to AIS data and maps showing the position of every ship in the world, right now.

[13] I will only give a brief overview here, but the full story of *Hōkūleʻa* and the navigator Nainoa Thompson is told in Sam Low's excellent book *Hawaiki Rising*, which I highly recommend.

Revival

Polynesia is a huge triangular expanse in the Pacific that stretches from the Hawaiian Islands in the north 4,500 miles south-west to New Zealand and then all the way across to Rapa Nui (Easter Island) in the south-east. It covers 11 million square miles of ocean, an area similar in size to the entire continent of Africa. Within it there are more than 1,000 islands,[14] most of which are very small. Western explorers from Captain Cook onwards mapped these specks of land, shipped in missionaries and colonized many of them, seeing them as ideal military and diplomatic outposts. By the early 1970s, much of Hawai'i's native culture had been either deliberately suppressed or simply not passed on to the next generation, as the arrival of gasoline, tourists, cargo ships and American administration systems generated a giant wave of change without much regard for what had been there before. Many of the native population struggled to find their place in this imported system.

But a small group was starting to ask questions about how the islands' original inhabitants had arrived. The prevailing academic view was that they must have drifted there accidentally, but the currents and winds made that unlikely – the closest mainland is nearly 2,500 miles away – and anyway, plenty of other Pacific islands were inhabited; it seemed unlikely that people had just fortuitously drifted to all of them. And so it was suggested that perhaps the Hawaiians and other Polynesians had had the skill and knowledge to navigate between the islands and across this vast expanse of ocean, making deliberate and repeated trips. It was known that there had been ancient voyaging canoes, with double hulls connected by a platform and claw sails (which are triangular and held in place by two spars).

---

[14] The exact number depends on how large a protruding piece of rock or coral has to be in order to be counted as an island.

And so the plan grew: they would build a replica of one of those voyaging canoes, and see whether it could be sailed between the islands without modern navigation techniques. In a corner of the Dillingham Corporation shipyard in Honolulu, the canoe *Hōkūleʻa* began to take shape.

Building a voyaging canoe that could sail safely was one thing, but the navigation challenge was an order of magnitude larger. The test would be a voyage from Hawaiʻi to Tahiti, a journey of more than 2,600 miles. This was no task for an amateur: finding tiny islands in this huge ocean permitted only a small margin of error. In the ancient Polynesian world, navigators were revered masters of their art, holders of secret and sacred knowledge which was vital for their communities. But the last of Polynesia's navigators had died, and their knowledge was lost. The chain appeared to have been broken – until the Hawaiians found Mau Piailug, one of the last of the Micronesian navigators.[15] Mau agreed to break the tradition of secrecy and help the Hawaiians navigate *Hōkūleʻa*, because he could see that the old ways were dying, and if he could help save their culture, perhaps he could also save his own.

Preparing for the first long voyage was a steep learning curve with many setbacks, but the local Hawaiian community was drawn to the project, and more and more people began to see its importance. This wasn't only about proving the skills of their ancestors; it was also a voyage of rediscovery of who they really were. On 1 May 1976, *Hōkūleʻa* set out from Hawaiʻi. Mau was the navigator, Kawika Kapahulehua (uncle of my mentor Kimokeo) was the captain, and there were thirteen other crew members, all on a canoe only 18.7 metres long and 4.7 metres

---

[15] Micronesia is a region of the western Pacific, filling part of the gap between Polynesia and the continent of Asia. Mau had received *pwo* in 1951, the formal recognition that he was not just a navigator, but a master of all navigation arts. He was the last person to be initiated on his island of Satawal.

wide. They sailed wherever Mau told them to go, through wind and swell, across the calm waters of the doldrums, across ocean and more ocean. Mau almost never slept, constantly watching the sky, the stars, the swells and every other detail of their surroundings, feeling the movement of the canoe. The ocean itself provided enough information to orientate the canoe, even when the sky filled with clouds, but noticing and correctly interpreting every tiny detail was essential. The crew had no compass, no GPS, and no external updates on their position (although a safety boat was following them). After thirty-one days, they finally sighted the coral atoll of Mataiva. From here it was only a short sail to Tahiti, where they found around 17,000 Tahitians waiting for them. This was an extraordinary moment for all of Polynesia, proving beyond doubt that the ancient Polynesians were incredibly skilled seafarers, and that they could and should be proud of their heritage as ocean people, whatever the modern world told them. The celebration was a turning point in an entire culture's view of itself. Mau returned to his home island, and *Hōkūle'a* returned to Hawai'i to be welcomed with great acclaim.

Proving that this feat of navigation was possible was one thing. Making sure there was something more tangible to pass on to the next generation was an even bigger step. The man who took responsibility for passing on that baton was Nainoa Thompson, a Hawaiian who had been educated in western schools and who was on board *Hōkūle'a* for the return trip to Hawai'i. Over the years following that first voyage he dedicated himself to studying the stars, using the local planetarium to try to figure out how to navigate in the way that Mau had done. He created his own star compass, and tried and tested methods for using the stars to calculate the position of a travelling canoe. He watched the ocean, and looked for what it could tell him. And then, in 1978, Mau returned to Hawai'i to teach Nainoa. In the meantime, the Hawaiians had learned a lot about teamwork, and about the attitude needed by the entire crew in order to succeed. They now knew that technical skills were not enough, and that the commitment, humility and harmony of the whole crew were also essential. Together, they practised and learned and prepared. On 15 March 1980, *Hōkūle'a* left Hawai'i again, this time with Nainoa navigating and Mau on board but not saying anything. The whole crew knew they were carrying the expectations of a whole culture on their shoulders. Had they learned the lessons well enough? Could modern Hawaiians work with the ocean in the way their ancestors had? The ocean and the clues it offered had not changed, but could they read and use them once again?

On the thirtieth day of the voyage, they saw a seabird that lived on land and fed at sea. An island must be near. And on the following day they watched and watched as the waves rolled past until Mau stood up and said: 'The island is right there.' And it was. *Hōkūle'a* had reached Tahiti for a second time, now with a modern Hawaiian finding the way. Mau's student had passed the test with flying colours, and the Hawaiians were once again navigators.

The early voyages of *Hōkūle'a* were an important part of a huge revival in Hawaiian culture. The Polynesians were ocean people, descendants of the greatest ever navigators, and now they knew it. During the 1980s and 1990s, more voyaging canoes were built and navigation schools were set up. Not only the Hawaiians but all other Polynesians saw their past, present and future connection to the ocean, and celebrated it with enthusiasm. The values of the navigators were also shared: teamwork, humility, observation, hard work and the importance of their canoe family. Islands all over the Pacific revived and renewed their culture. Now they saw the ocean differently, as the connection between their islands, and the connection between their people. In 2007 Mau carried out the first *pwo* ceremony to take place in fifty-six years, for eleven men from Satawal and five

Hawaiians (including Nainoa), creating a new generation of master navigators.

In February 2020, at the American Geophysical Union's Ocean Sciences meeting in San Diego, Nainoa emphasized the winning combination: culture and science should be partners. Speaking of the education of the next generation, he said: 'These children know how to voyage and go anywhere in the world, but they also know how to come home because *they know who they are*.' Having a perspective on the world and your place in it allows you to use science and technology to match your values.

From 2013 until 2019, *Hōkūle'a* undertook the Malama Honua Worldwide Voyage alongside her sister canoe *Hikianalia*, sailing more than 150,000 nautical miles around the world, visiting 150 ports in eighteen nations, with a total of 245 crew members rotating throughout the trip, using both traditional methods and modern aids. During the voyage, the crews talked to over a hundred thousand people, sharing their connection with the ocean and building new connections with the people they met. These Polynesians chose to see their future relationship with the ocean differently, and they are putting in the work to make that future happen. They are voyagers, and they understand that the ocean itself is a critical part of every voyage. We in the western world may not be very aware of our own links to the ocean, but the lesson of *Hōkūle'a* is that we have a choice. Enjoying the comforts of modernity does not mean giving up on our connection to the world around us. This is not about the ocean: it's about us and who we choose to be.

We are all citizens of the Earth: an ocean world. Whether we choose to acknowledge the blue machine or not, it dominates the planet, regulating how energy and atoms flow around the globe, and setting the scene for everything else. This great liquid engine is majestic and intricate, dynamic and interconnected, with a vast array of life rippling through its swirling innards. It's far larger than us, and the great rules of ocean physics do not bend to human will. We can pretend to live our lives in spite of the ocean, or we can choose to understand and work with it, and thereby benefit from the natural processes that are in any case out of our direct control. They bring richness and variety, as well as surprises and a degree of unpredictability. That is the beauty of the blue machine.

We are well past the point where we can carry on in ignorance, acting as if it doesn't matter whether we acknowledge the ocean or not. The blue machine may be gigantic, but it is finely balanced, and our human civilization has more than enough influence to disrupt its processes. Our society has a complicated relationship with the ocean, encompassing both great love and great abuse. We can see now what this engine is and what it does, and we can also see and measure the damage we are doing to it. We are already interfering with and polluting it, and the engine is groaning in response, but it is not too late to change our ways. We can't walk away from our relationship with the ocean without walking away from planet Earth. And we can't do that. So it is time to consider the future of that relationship, and the choices that we face as modern citizens of a blue planet.

PART THREE

THE BLUE MACHINE AND US

7

Future

You can't protect what you don't understand.
And you won't, if you don't care.

LACY VEACH, NASA ASTRONAUT

THE BULK OF *The Blue Machine* deliberately sidesteps the damage that we're currently inflicting on the ocean. So many discussions about global change focus on pointing accusing fingers at broken things, without showing the damage in context. My aim has been to draw the outlines of the Earth's wonderful ocean engine, to show how it works, and to share how it all fits together and why it matters. Seeing the full physical and biological complexity of the ocean changes our perspective on our planet and our lives, and that, by itself, is a gift. Our global ocean is a fabulous system, and contains so much to appreciate and rejoice in, but what happens next?

I wish so much that I could have written an ocean book that ended with pure celebration, with the untainted elation of great stories shared, with nothing but a positive, exciting ocean future to look forward to. But with a deeper knowledge of the ocean comes the responsibility to be good citizens of our ocean planet. The benefit of hindsight tells us that, for the most part, we have not been good citizens over the past two hundred years. Perhaps for much of that time we had the excuses of ignorance, or the lack of a global perspective, or a lack of history from which to learn. If they ever were valid excuses, they certainly aren't any more. We can be better citizens of this ocean planet, if we choose to be, but that requires some significant changes.

While doing the research for almost every story in this book, I found that the latest scientific research papers on each topic started by discussing how those systems are changing. The giant ocean engine will keep turning, but its delicate equilibrium and the ways life is woven through it are not fixed. It must turn, but it doesn't have to turn like *this* in every detail. And yet *this* is a system of huge richness, and physical oceanography and evolution would take a long time to replace that bounty if we lost it. We don't understand everything about the global ocean, but we certainly do understand enough to know how valuable it is and the most obvious ways in which to protect it. The primary reason for setting out the damage we have inflicted on the blue machine is not to shock. It's to lift us out of helplessness. As we acquire knowledge, we also acquire the grounds for optimism.

The most fundamental problem

We have nudged the control levers of this planet, by burning fossil fuels and changing land use. The consequence is that the Earth now has an energy imbalance, because more energy is coming in than is going out. We can measure this imbalance – on average, it's about 0.3 per cent of the total energy budget, equivalent to a continuous flow of around 500 terawatts, half a million billion joules every second. This is fundamentally what climate change is: the slow accumulation of extra energy in the Earth's system. And where is it going? The answer is clear and indubitable: into the ocean. More than 90 per cent of all the additional energy accumulating on Earth because of human changes to the climate system has ended up in the ocean as heat. Tracking the total amount of heat in the ocean is one way to create a planetary thermometer, and it's showing a steady rise.[1]

But this extra heat isn't just sitting in storage, parked on a shelf in the back of the ocean equivalent of a cupboard. The extra energy enters the ocean at the top as heat, so the surface waters – the warm mixed layer – are heating up faster than the layers further down. We have seen that a critical feature of the ocean is the set of systems that bring nutrients up from the cold depths towards the sunlight, so that phytoplankton can turn them into the materials of life. But a warmer surface layer makes it much harder for deep nutrients held in cold water to be mixed upwards into the sunlight, because the layering is stronger and harder to overcome. This reduces the raw material for life up in the sunlight, so the whole ecosystem is put under stress. Stronger stratification means that there's less exchange within the ocean engine of everything: heat, gases, nutrients and more, starving the ocean's internal interactions of material. The addition of extra heat at the surface is reinforcing the layered structure and therefore acting as a brake on the vertical turning over of the blue machine. A warmer ocean can also feed more energy into the atmosphere, changing weather patterns and making storms more intense. The impact can be particularly severe in the tropics, where hurricanes and typhoons often hit poor communities that lack resilient infrastructure.

---

[1] Even though the amount of extra energy is reasonably well known, it's hard to translate that into a specific temperature increase in degrees, and that's because this extra heat isn't evenly distributed throughout the ocean. Two-thirds of the extra energy is in the top 700 m, and a third is below that. But the pattern is quite complicated, so scientists tend to add up all the extra energy and talk in terms of the total ocean heat content, rather than ocean temperature. This number still acts as a thermometer for the planet.

We speak lovingly of oxygen as the facilitator of life,[2] and we are well aware that even a relatively small drop in the air's oxygen level quickly causes humans to pass out. But oxygen levels in the open ocean have dropped by 2 per cent in the past 50–100 years. The drivers of that change are complex, and there's a lot of variation across the globe. A small portion of the loss is due to warming water, and the rest is caused by increasing ocean stratification, changing water-flow patterns, and the shifting of life that generates oxygen through photosynthesis or takes it up through respiration. It's a tricky thing to pin down, because oxygen levels at depth can vary a lot naturally. But we can already see the changes, and it looks as though a deoxygenated ocean in the mid-depths will be one of the major consequences of climate change, with potentially severe effects on deep ocean ecosystems. After all, it doesn't matter how much food there is to eat if you can't actually breathe while you're eating it. In many ways, the layers of the ocean have a Goldilocks quality to them: their existence provides a structure to the ocean environment, but if they're too strong the whole machine shuts down because it can't move.

This is all bad enough, but the extra heat comes with a nasty sidekick. If the water is warmer, less carbon dioxide will cross into the ocean. So in a warmer world, with a warmer ocean surface, more CO_2 will be left in the atmosphere. That will make everything heat up faster and will therefore warm the ocean further and leave it with even less capacity to take in our extra carbon dioxide. The ocean has been softening the impact of our carbon dioxide emissions, conveniently for us, but it's not clear

---

[2] At least life like us, our pets and a lot of charismatic wildlife. There are plenty of bacteria and a few fungi which are much happier without any oxygen at all, because it's so reactive that it's toxic to them.

that it's going to continue doing so at the same rate in a warming world.

The additional carbon dioxide in the ocean has another consequence, less well known than global warming but only marginally less ugly. It's known as ocean acidification, and it could add significant stresses to marine ecosystems in the future, by slowing the assembly of ocean life's most useful solid construction material: calcium carbonate. This, as we saw in chapter 5, is the stuff that the shells of oysters, snails and limpets are made from, as well as the solid architecture of coral reefs, and also the hard cases of the coccolithophores – the tiny rotund phytoplankton that make up the bulk of the White Cliffs of Dover.

Carbon dioxide that finds its way into the ocean joins a chemical dance, making temporary shifting alliances with hydrogen and oxygen from the water to form carbonate and bicarbonate. Each of these new clusters has an electric charge, and they are unmade and remade in a continuous waltz. The ocean has plenty of them already, but adding more shifts the proportions of the rest. The consequence is to change a fundamental characteristic of ocean water: its pH.[3] Why does this matter? In part, because construction of calcium carbonate requires an alkaline environment, and the lower the pH, the more difficult that construction process becomes. Around 250 years ago, the average pH of the surface ocean was 8.25. By 2020, it had fallen to 8.1, and it's continuing to drop. From 8.25 to 8.1 may sound like a small change, but because of the way pH units work, it makes a significant difference to the chemistry. Each individual carbonate-building creature may find

---

[3] You may remember pH from school as the scale that tells you how acid or alkaline a liquid is. Pure water has a pH of 7. Lower numbers indicate an acid solution and higher numbers an alkaline one. The ocean has been an alkaline environment for at least the last billion years.

life just a little bit harder, and that stress will trickle down to whatever feeds on the carbonate architects and therefore the rest of the ecosystem (zooplankton, shellfish fishermen, fish, walruses and plenty more).

The additional heat energy in the Earth system has other consequences that directly affect the ocean. Currently, about 2.1 per cent of all the Earth's water is locked up as ice on land, mostly piled up on top of Antarctica and Greenland. As the polar regions warm, ice is melting faster than it's being formed, and there are also changes to the rain and snow on land that eventually drain downward into the ocean, contributing extra liquid water. This extra water takes up extra space, which is one of the main causes of sea-level rise. Another is that water expands as it warms, and so a warmer ocean takes up more space. This thermal expansion is responsible for about one-third of current sea-level rise. Also, the additional water is fresh rather than salty, and so it can change the structure of the ocean engine – the thicknesses and density of the layers (and therefore whether they sink or float). The full consequences aren't clear yet, but it seems very likely that dumping extra fresh water at the ocean surface will change the shape of the ocean engine in those regions.

So two hundred years of humans yanking carbon out of fossil-fuel deep storage and pumping it into the atmosphere comes with some pretty fundamental consequences for the blue machine: changing how easy it is for the ocean engine to turn, and making it harder for life to blossom.

(Some of) the other problems

In chapter 5, we saw the huge range of size scales of life in the ocean: 1 gigatonne of wet weight of life for every size class, all the way down to tiny phytoplankton only a micron across, one-thousandth of a millimetre.[4] It's a stunningly consistent pattern, showing us that healthy ecosystems are immense interlinked networks that need both their tiniest and their most gigantic citizens. And this is the reality of the global ocean, or it was, in 1850. But when you look today, it's no longer true at the largest size scales. In the largest size class, up where the big whales are, nearly 90 per cent of that biomass is missing. If you lump together all the size classes larger than 10 grams, about 60 per cent is missing. It's not exactly lost, because we know where it went. Humans hauled it out of the ocean to eat it, or turn it into fertilizer, or to sell it for body parts (shark fins, shells, swim bladders). If it was big enough for us to see and eat, it was considered fair game. Even though the total mass of humans on Earth is only around 0.4 gigatonnes, we are responsible for about 2.7 gigatonnes of life being missing from the ocean.

With our perspective on the ocean as an engine through which energy flows, we can see that the consequences here are not just fewer fish, but include cutting off a vital source of stored energy in the ocean's living systems and diverting it to the land. Once that energy is out of the ocean, it is unavailable to bigger predators or smaller scavengers, and can't be digested by marine microbes in the depths, and so it has been lost to the living energy flows of the ocean. Removing the fish also diverts nutrients out of the ocean system and into humans, nutrients that would otherwise have been recycled in the ocean to keep ecosystems going.

It's mostly the big animals that we have extracted, so historically there have been advantages to being a small zooplankton that humans can't easily see and therefore can't pillage. This

[4] Just to remind you, if you add up everything that weighs between 0.01 gram and 0.1 gram, you get the same total weight as everything between 0.001 and 0.01 grams, and everything between 0.1 gram and 1 gram, and so on up and down the size scale.

unintentional amnesty may be coming to an end though, thanks to a growing industry interested in harvesting zooplankton to make fish food for pet fish, for aquaculture, and as a source of omega-3 fatty acids that can be used in dietary supplements. Not content with ravaging the biggest parts of the ocean ecosystem, humans are now starting on the smaller stuff. And of course, the further down the food chain we go, the more we're cutting away not just at the citizens of the ecosystem, but at its foundations. Overfishing is a major problem, and it's not one many are really taking seriously. Fish don't care about national borders, and climate change means that fish populations are migrating to stay within water that matches their needs. So the problem moves around and it's easy to claim that it's always someone else's fault.

And then there are the specific pollutants, like ocean plastic. Plastic pollution in the ocean has been known about since the 1970s, but has risen much further up the public agenda in the past few years. Our problems with plastic are a direct consequence of our failure to realize that there is no such place as 'away'. We've put a huge amount of effort into making plastic, and very little effort (by comparison) into working out how to create new useful things out of discarded plastic. The first and biggest step is to stop using plastic where we don't really need it, and the second is to worry about what happens when we do actually need it. Campaigns to reduce plastic packaging and single-use plastics of all types, and to improve recycling, are starting to gain traction.[5] One of the major advantages of plastic as a material is its durability; but just as with CFCs, that engineering advantage in the short

[5] If you live somewhere with drinkable tap water, you can make a big difference by avoiding bottled water. The alternative is very easy: carry one bottle with you and refill it at the tap, or ask at a café. Of all the unnecessary plastic use in society, this is at the top of the pile.

term is what causes problems in the natural environment in the long term. We don't know exactly how long all the different types of plastic last in the ocean, but we do know it's a long time.[6] The problem with the plastics isn't really that they're floating about in the ocean being ugly (although that is definitely an issue). It's the specific effects that they have on ecosystems, and those depend on the plastic size.[7] The largest pieces (and especially discarded fishing gear) aggregate and may choke or entangle bigger animals. Sunlight and wave action will break the larger floating pieces down into smaller fragments, which may be ingested by zooplankton, seabirds and fish that think they're food, or that may become coated in algae until they sink. These fragments may also leach toxic substances into the water, or act to accumulate toxins. It's just not feasible to filter those micro-plastics out of the ocean, because in doing so you would also filter out a substantial fraction of the phytoplankton and zooplankton which are the foundation of ocean ecosystems. The solution is to stop putting the plastic in. We do need plastic for a small number of applications because there is no existing alternative, particularly medical and scientific uses, but we can use as little as possible and make the system circular for everything we do use.

[6] And 'biodegradable' plastics do not generally biodegrade in the way that the name leads people to believe. If you put them on your compost heap, they'll still be there years later. Some may require an industrial composter, with very high temperatures for long periods, and carefully controlled humidity, pH and microbes. And some just make unverified claims about their suitability for composting. Also, plastics made from a plant base, like bamboo fabric, is just as durable as a 'normal' plastic, so they'll hang around and pollute the environment in the same way as a fossil-fuel-based plastic.

[7] Incidentally, we can add up the amount of plastic that has probably ended up in the ocean, and we can only find about 1% of it. We don't know where the rest has gone, which is possibly more worrying than knowing where it is.

Those are the direct effects of humans on ocean ecosystems. But there are also indirect effects. In the record of whale stress over the last 150 years, held in preserved earwax, Richard Sabin showed me the peaks and troughs associated with historical whaling, and then he showed me what has been happening since whaling stopped in the 1970s.[8] You would think that the whales would be able to relax and their stress levels would go down. But their stress has been rising again, with a peak around the year 2000 which was nearly as high as the highest peaks in the whaling years. The biggest culprit appears to be rising ocean temperatures, which change the availability and location of prey, competition between whale species and the location of the whales' preferred habitats. But that's far from the only stress that has been increasing: there are more ships making noise, more pollutants affecting the health of ocean ecosystems, more recreational fishing, reduced polar sea ice and more disease. All of these are known as 'sub-lethal stressors' – they may not actually kill a whale, but they'll make life harder for it. One of them is bad enough, but when they are all piled on at once, it's far worse. Humans may not have noticed that they were treating the ocean as a dumping ground, as a non-place, as 'away', but life in the ocean certainly did notice.

Richard lit up as he told me why he's so excited about whales: 'They don't have opposable thumbs, but they have incredibly complex brains, and wonderful abilities for cognition and remembering and recognition, and language and everything else. We want to show people the similarities between us and an organism that seems weird and bizarre and that lives in an environment that's so alien to us, the ocean.' And of course, there was a 'but' coming: 'But they're reliant on the ocean being a predictable place, and they've adapted to it over time. They've

[8] It still hasn't stopped completely. But there were major international agreements to stop almost all of it.

learned about the ocean currents, the massing of prey species in certain areas at certain times of the year and so on. And we're changing all that, faster than they can adapt. I don't think that we can be ignorant about that any longer.' The issue of predictability is an important one. As we saw with the penguins visiting the ocean fronts to feed, and the tuna visiting warm-core rings, animals use the features of the ocean engine efficiently because they can rely on them. Of course, many of them *could* swim elsewhere to forage, but it takes trial and error to find out where to go, and that costs precious time and energy. Survival in the ocean is tough enough already. The great richness of ocean life is interwoven with the physical nature of the blue machine, and if the engine changes, life must adapt or die. Much of it will never find alternative locations or opportunities, and will therefore struggle to survive.

Even when species can move, they can't take their whole ecosystem with them. Bluefin tuna have recently been found up near Greenland, far north of their normal hunting areas, probably because the water there is warmer than normal and their prey have also moved north. But the previous occupants have not moved out en masse, and so now ecosystems in that area are adjusting to this new visitor. It's not clear what the future ecosystem structure will be, but there will be winners and losers. Food webs will change. Areas that were previously isolated may become connected to others, and areas that were previously connected may become isolated. Uncertainty is guaranteed.

The big picture

All of this is just a very quick overview,[9] covering only the biggest and most fundamental impacts. It's not just that a warmer

[9] A very short list of some of the problems I have not discussed here: the gigantic loss of coral reefs to bleaching, the release of toxic chemicals into ocean

ocean stresses fish, or even that a warmer ocean will shift fish populations towards the poles. It's that the extra heat changes the engine itself: its patterns, its speed of operation and the location of the physical features inside it. The physical ocean patterns will affect the biological ocean patterns, because they're all woven together. And the biology can also affect the physics, because phytoplankton change how light travels through the ocean, and therefore the depth at which energy from the sun is converted into heat.

If we continue to treat the ocean as an absence rather than a presence, and therefore an acceptable dumping ground for all sorts of stuff that we don't want to think about, none of this will get any better. Focusing on isolated problems also won't result in real improvements, because the fundamental problem is a systemic one. But once we think of the ocean as a single integrated, intricate system, interwoven with our own lives, mediator of great flows of energy and materials around the planet, we have the perspective to really see what we're doing to the blue machine.

*

Kimokeo is driving his truck south along the Maui coast to Makena, and four of us are riding with him. It's 5.30 a.m. and

---

ecosystems, bioaccumulation of toxic additions like mercury, invasive species mucking up ecosystems (look up the lionfish in the Caribbean for an example), massive bycatch by the fishing industry and also illegal fishing, marine mammals being hit by ship propellers, the massive ice loss in the polar regions, dumping of human waste (tyres, trash, the contents of cargo containers), immense fishing pressure on the biggest species (sharks for their fins, trophy fish, endangered tuna), fish farming (which is often very intensive and sheds nutrients and disease into the local environment), ocean infrastructure (oil wells, beach 'protection', draining of coastal areas) . . . the list goes on and on.

we're on the way to borrow two canoes from the Makena Club so that we can paddle around the submerged volcanic crater of Molokini, a few miles offshore. In spite of the early start, Kimokeo is talking about everything, all at once. 'The important things are wind, waves, tide and current,' he says. 'Those four things affect everyone in the world.' I ask him about whether they're changing with time, and he frowns. 'It's hot. We used to only get one day every year above 92 degrees [Fahrenheit] on Maui, but now it's eight to ten days every year. That's climate change. When it's hot, the offshore winds don't blow, so yes, the wind is changing.' Then he gets distracted by a tree that's showering yellow seed pods at the side of the road and instructs me to get out, find one that hasn't been trampled on, and try eating it. He tells me about the importance of this tree for feeding bees (I later find out that it's kiawe, or *prosopis pallida*), and about a project to grow more of them to encourage bees somewhere up the mountain. A bit later, after an explanation of the celestial navigation school he's setting up and a primer on the best ways to build sustainable houses, we get back to the subject of nature. 'It's about the connection between humans, the land, the ocean and the heavens. And that connection isn't just given to one culture. That would be . . .' he pauses, looking for the right word, 'a small idea. Everyone has the gift of connection to these things equally. They have wind to push the sails of boats and sun to provide light and to dry fruit, and currents to move canoes. Everyone can take advantage of what nature has to offer, because we're all connected.' Then we're at the beach with the canoes, and we take off across the ocean to admire the huge fizz of white water surrounding the crater, where the swell is thumping into the volcanic rock piercing the ocean surface. If anyone other than Kimokeo had been steering, I'd have been very nervous, but I trust him and the canoe is safe.

Three hours later, we're back in Kīhei and we've arrived at

Nalu's Grill for breakfast. Walking in here with Kimokeo is like walking in with a celebrity. Everyone comes up to say hello – the restaurant owner, the waiting staff, three sets of passing paddlers, and two guys who seem to be on their way to build a house. He knows everyone, is genuinely pleased to see them all, and spends breakfast giggling and cracking jokes, introducing everyone to everyone else, and singing along to the background music at the same time as telling me about more of his and Maui's projects for making their island a better place, and a better contributor to the world. Large groups of people here wove huge *lei* (garlands made from leaves) which Kimokeo then personally delivered to the communities hurt by the Sandy Hook and Christchurch mass shootings to show support. They run an annual Paddle for Life, a fundraiser for cancer survivors. There are plans to get children into canoes, and a project about coconut trees to encourage even more bees. The list seems endless, fuelled by enthusiasm and considerable commitments of time and money from the local community. When the builders leave, delighted to have had time to talk to him, one of them slaps Kimokeo on the shoulder. 'You're doing good things,' he says, 'keep doing them.'

Even so, there are hints of the limits facing a small community. I ask Kimokeo about the last time they made a *lei* for survivors of a disaster, and he says they don't do it any more. 'There are too many shootings now. We can't keep up.'

He drives us back to where we're staying. On the way, there's a beautiful stretch of road right next to the beach. The beaches here are narrow – just a few metres wide – but they're stunning examples of a tropical paradise. So what Kimokeo says as we drive past this one comes as a shock. When he came to Maui in 1972, he tells me, these were great wide beaches, stretching outwards to form a broad belt of golden sand before touching the ocean. He says I should look at old photos from the 1970s to see the difference. The beaches have been lost: 'This is climate change.' Afterwards, I look this up. The US Geological Survey's National Assessment of Shoreline Change for 2012 quotes an average long-term shoreline loss on Maui of 17 centimetres of beach per year. This is an average for the whole island, but over the fifty years since 1972 it represents the loss of a staggering 8 metres of beach. The science is complex, since beaches are always temporary and sediment moves about all the time. Sand availability, the replacement of sand dunes by sea walls, exposure to waves and general geology all affect whether any individual beach is growing or shrinking. But recent scientific papers agree that Maui's beaches are eroding faster than Oʻahu's or Kauaʻi's, and it seems likely that higher relative sea-level rise on Maui might explain the difference. They also all agree that far more shoreline loss is coming in the future. I ask Kimokeo whether Maui can do anything about it. He talks about rebuilding and rerouting roads, about the houses and land that are being lost to the ocean, but it's clear that the wide sandy beaches he remembers have probably been lost from Maui for ever. Kimokeo is vocal about the seriousness of this change, but he doesn't waste any time looking backwards. His mental and physical effort is spent driving forward the many projects that will change the future.

The most powerful tools: perspective, knowledge and humility

Being human has never been easy; but no previous generation has faced such an avalanche of information about the intricacy and richness of the world around us at the same time as an equally overwhelming avalanche of information about the damage our massive societal systems are doing to it. But information alone is not going to save us. Of course, it's important, but we need to ask why we should care. Because if we don't care, it doesn't matter what we know – we won't act on it.

We are defined by our connections of all sorts – our relationships with our families and friends, the interactions we have in our communities, our connection to the physical world around us via what we eat and what we build, and our intellectual connection to the rest of humanity as ideas circulate. The threat of the vacuum of outer space is not that it's harsh (although there is a lot of damaging radiation out there) but that it represents absence. There may be trickles of energy whooshing outward from the stars, but there are almost no atoms for that energy to act on. There is nothing there to connect with. So our connections are all here on Earth. We are wonderful living components of the richest and most beautiful environment any of us will ever see. This is the beauty of being human – to be able to appreciate *participating* in our surroundings and our communities.

I have written previously about each of us having three life-support systems – our own body, our planet and the infrastructure of our civilization.[10] We need all of them to function for us to stay alive, and we need all of them to be healthy. But the next stage is that we need all of them to be integrated with each other, because there is no choice. The atoms *have* to go round and round, and the limited amount of energy that is available to our planet *has* to flow through. More than anything else, this requires long-term thinking. We have to operate within that framework.

For me, the biggest reasons for caring about the ocean are twofold. The first is a deep visceral appreciation of ocean life and its ingenuity, beauty and variety. Knowing that it exists, seeing and exploring it and connecting to it, make our lives richer and better. This is a deeply human thing, and one of the great joys of life. The second reason is pragmatic. For the vast majority of us, Earth is the only place we're ever going to be. Perhaps the

[10] In my first book, *Storm in a Teacup*.

most fundamental driver of human behaviour is the need to survive, and if we want to survive, we need Spaceship Earth to be in full working order. As we have seen, the ocean engine, both its physical and its biological systems, is a major part of that life-support system. So if we want to survive and thrive as a species, being good stewards of the ocean is essential.

Once we care, how should we act?

Nainoa and Kimokeo and the other teachers of their culture will repeat one phrase more than any other: 'He moku he wa'a, he wa'a he moku.' It means: 'A canoe is an island, and an island is a canoe'. For an inhabitant of a small island in the middle of the vast Pacific Ocean, this makes perfect sense. If you want to thrive on your island, the most effective way to do it is to cooperate, and to make the effort to be nice rather than burning bridges (it's a small island – you're going to be bumping into each other a lot in the future). If the teamwork falls apart, no one is going to benefit and everyone is going to suffer. It's the same in a canoe. If you paddle across to the next island, everyone in your canoe must also be cooperative, happy and make a strong team. Once you're in the canoe, there is nowhere else to go, and you cannot run away from your problems, so you have to face them honestly and respectfully. You also need everyone to make their contribution for the canoe to arrive safely.

Some years ago, I happened to be reading the astronaut Chris Hadfield's book *An Astronaut's Guide to Life on Earth* at the same time as Sam Low's book *Hawaiki Rising*, about *Hōkūle'a*. It was very striking that the description of cooperation, openness and teamwork required in a confined space was almost exactly the same in the voyaging canoe *Hōkūle'a* and in the International Space Station. Technology may change, but humans stay the same. Cooperation and openness are the route to the best problem-solving, if you are in a small vessel in a harsh environment. Even the most modern technological advances have had to build that into their systems, because in the most challenging situations, you need a team that works.

And of course, once you start thinking about canoes and islands, it's very obvious that the Earth is our canoe, a fragile oasis voyaging through the emptiness of space with a bunch of people on board. This is the perspective that astronauts talk about when they return to Earth, the consequence of the 'overview effect': the experience of looking down on Earth from outer space and seeing all its beauty in context. Rusty Schweickart, the Apollo 9 astronaut, nailed this when he said 'We're not passengers on Spaceship Earth. We're the crew.' So here we all are, crew members of our biggest life-support system. We're a team, whether we like it or not. What happens to the spaceship affects all of us.

So that's the Earth. But what about its blue ocean, our planet's defining feature? I think we should look at those early dramatic photographs of Earth more often: 'Earthrise' (taken during Apollo 8 in 1968) and 'Blue Marble' (taken during Apollo 17 in 1972). They show us our place in the universe, and add the bigger picture to our identity. We should look at spherical globes instead of flat maps as often as possible. We should turn the globe around sometimes so that all we can see is the Pacific Ocean. And when we look at the blue in all of those representations, we need to see in our mind's eye the blue engine that really is the beating heart of planet Earth. Its restless nature, its intricate interior, and the web of life that permeates all of it; this is what defines our world. But once we have this perspective, what should we do with it?

Values

There's often a perception that 'following the science' should be the primary way to get things done. I'm a scientist and I'm telling you that it's not, for the simple reason that science does not lead. Where leadership comes from is a clear statement of values, followed by using the best available information about how to act according to those values. We have to decide for ourselves and for our communities what we care about. How do we prioritize social justice, economic gains, individual freedom, community values, and the rest? The answer might be different in different situations, and they are often hard questions. They have nuance, and they may depend on things that depend on other things. But we should talk explicitly about values, because that's what tells us what to prioritize and in which direction we want our society to head. Once you've made a decision about your values, then science has a contribution to make, because it gives us all our best collective understanding of how to get there.

Discussions about scientific advice are often confused because there is no explicit discussion of values, often because people assume that everyone has the same value system that they do. That's rarely the case, and so the discussion of values needs to be explicit, just to make sure.

Of course, science has a critical role to play in guiding the right questions to ask, and the topics to which values need to be applied. But it doesn't tell you 'what to do'.

So why does all this matter for the ocean? Because perspective gives you a context for your value system. Once you see the ocean as a vast, dynamic engine that is the heart of our planetary life-support system, following your values has to include that larger context. Knowing about the ocean engine doesn't tell you what to do, but it sets the context for you to think about what you want.

For humans, dealing with problems in the ocean is harder than dealing with problems on land, partly because the ocean is less accessible in general, but mostly because ocean interactions are far trickier to see. They involve tiny mobile phytoplankton instead of huge fixed trees, and ecosystems that move around with the currents, constantly shifting and adapting with the weather and the seasons, and relatively small shifts in the temperature and salinity of water that have huge consequences for the turning of the ocean engine. If someone builds a cement factory in an old woodland, there's at least a decent chance that most of the consequences will be visible: the extra traffic, the extra jobs created, the extra pollution, the disappearance of wildlife and the money brought into the local area. A local person at least has a chance of seeing the outline of the trade-offs involved. But if someone dredges the sea floor to harvest fish, no one can really see the state of what's left behind, how long the damage remains and which other species are affected, although they might be able to find out what the fish sold for and how many people were employed on the fishing vessel. So we have to be very careful about what we do in the ocean, because it's easy to be blind, either deliberately or accidentally, to the full picture. Are we doing things because they make us feel better about ourselves, or because they're actually beneficial to the natural world in the long term? Or just because there's an opportunity for a company to make money in the short term?

There is a particular conflict here which is often poorly resolved. It's not uncommon to hear two aims for ocean projects stated simultaneously: first, a return to a 'pristine' natural environment,[11] and second, the creation of a state where the ocean does specific things for us – stores carbon, prevents coastal

---

[11] 'Pristine' is not necessarily a useful word, because it's poorly defined. Natural systems change over time, and humans have been around for a long time too. Which version of the past do you want? Is it even possible to shift a chunk of the world to some time in the past when everything in its surroundings is firmly in the modern day? And what if, left to itself, those places become new, healthy ecosystems that fit in better with the rest of the modern world, but are unlike anything seen before? Is that 'pristine'?

erosion, provides food and provides a source of energy. It's certainly the case that having healthier ocean ecosystems can be beneficial for lots of reasons, but the conflict here is between letting nature get on with it, over time, and thinking that humans can manage natural systems better than nature itself. That second category slides from prudent management of natural systems all the way to full-blown geoengineering. We can't claim to be recreating 'pristine' environments if we are also engineering them to do stuff for us. In the end, there's likely to be a balance between the two. But we can't have everything, and we need to be explicit about the trade-offs.

Pretending that things are simple, or that there are silver bullet solutions, isn't going to help.[12] 'Managing' the ocean implies taking on responsibility for a complex system that we don't fully understand. They say that the road to hell is paved with good intentions, and often, so is the road to environmental disaster. So the task before us is to have the humility to work with our planetary systems, rather than fighting them. Nobody owns the global ocean, and only hubris would lead us to think that we can drive the blue engine like a motor car. And anyway, it's often tricky to calculate the long-term outcome of a deliberate change. For example, people are starting to propose schemes involving intensive technological manipulations of ocean chemistry or biology to remove carbon dioxide from the atmosphere, but it's extremely hard to track and verify how well they will work in practice in the complicated ocean engine. Are they really going to have the advertised effect in the long term? Are there any other downsides for other parts of the engine? We should have

---

[12] Possibly the one exception is stopping the burning of fossil fuels. It's a single stated aim that, if realized (well, when realized, because the consequences of not doing it are so bad that we don't have a choice), would remove a huge amount of pressure on the Earth system. But this would be a choice to stop doing something, not to start new things in the system.

a reasonable degree of certainty on these points before we commit ourselves to these actions.

And then we need to consider who will be making and influencing these decisions. Indigenous citizens of ocean regions, especially in the Pacific and Arctic, have long-established and deep connections to their natural environment. They're not the ones who have inflicted damage on our global ocean, but they certainly do experience the consequences. The concept of connectedness is deeply embedded in their entire way of life, and they have fought for years against the separatist and extractive practices that arrived with visitors from the western nations. These communities vary hugely in their lifestyles and their opinions, but all of them have historically been excluded from any decisions about ocean management. Their voices are finally starting to be heard, but often they are not yet considered full partners. While westerners focused on measurement, calculation, ownership and power play, these communities often start discussions by considering their values. We would do well to learn from that. The middle of a crisis is a hard place to rebuild relationships and collaborative discussions, but we must try. In the Arctic in particular, many in these communities are unhappy that climate change is being seen as the world's latest excuse for bossing them about without listening to their point of view. We in the western world have severed many of our ties not only with the natural world, but also with many of our fellow humans. But we are all in this canoe together, connected whether we like it or not, so rebuilding relationships of all kinds has to be a priority.

Thinking about values also means thinking about being human. It's easier to talk about economic value, because it comes down to numbers, and the nature of mathematics means that we always know whether one number is bigger than another one.[13]

[13] Early economists – in fact, the inventor of GDP, Simon Kuznets – were very clear in saying that this is the great failing of a system that elevates the

But really talking about values – how we think about respect, equality, fairness, opportunity, freedom, community – is much more difficult because it involves introspection and a lot of time, and it involves asking hard questions which don't have simple answers, and it requires moving away from the mindset that there's 'right' and 'wrong' and anyone on the 'wrong' side is either an idiot or a villain.

Adrian Glover from the Natural History Museum thinks a lot about how decisions about the natural world are made, because the area of the planet that he studies – the extensive deep seabed plains covered in polymetallic nodules – are being eyed by mining corporations who see vacuuming these sea potatoes out of the depths as an easy win. Plenty of deep-sea biologists feel very strongly that it would be wise to understand what this environment really is, and to explore its sedate processes and nuances, before we march in and plunder it. But since most people have never seen the deep ocean, and anyway, it sounds dark and cold and very far away, it's hard to have a conversation about what we should do and who should decide. 'I think what we want', he says, 'is for the debate to be better informed and people to be better engaged and better connected to that decision than they have been for past industrial disasters.' He also makes the point that it's more and more common for the case for conservation to be made only in terms of direct economic benefits to humans – the possible discovery of new medicines or new food sources, or as a convenient place to dump carbon. This bothers him. 'Historically we've been just going on about how it could be really important for this and for that. But we never say that it's just a wilderness environment which is filled with biodiversity, and that's enough. People will visit a wilderness region or National

---

numbers above all else. Kuznets himself told the US Congress in 1937: 'The welfare of a nation can scarcely be inferred from a measurement of national income.'

Park, and they just want to read the information about the inter-esting birds and the weird worms that live there and be engaged with their natural world. And it doesn't require an economic justification. It's innate.'

We don't have long to work all of this out. Ocean systems, both physical and biological, are changing *now*. The best immediate action is to stop causing the damage. But longer-term interven-tions to repair the damage that's already been done are going to be complicated and we can't run away from that complication. Robust science coupled with clearly stated values, genuine col-laboration and a planetary perspective on the blue machine will give us the foundation to choose a healthier future ocean.

Humans and the ocean

We built a culture based on ignoring the realities of living on a finite planet, a culture based on building and expanding and consuming and creating without having to think too much about the planetary life-support system that was keeping us alive. At the same time, we lost much of our relationship with the natural world, and deprived ourselves of the delight and the wonder of being part of something far bigger. But even while we were treating the ocean as an absence rather than a presence, it was still there, making this planet habitable, shaping our history and our culture, and setting the patterns for the ocean life that we did see and interact with. As the voyaging canoe *Hōkūleʻa* showed, it's perfectly possible for us to reconnect if we choose to. The ocean is still there, it is still shaping our physical reality, and it is still a stunningly beautiful place. We have not lost it. But I think that we in the western world do need to rediscover it.

We are not helpless in the face of our past actions. Knowledge and understanding give us agency. It is very easy to feel depressed about the problems we have inflicted on the ocean (and therefore indirectly on ourselves), but we can't let that become apathy. We need action, and action takes enthusiasm, energy and determination. If you are dubious about whether it's possible, just think about how much our society has changed even in the past fifty years, and you will see that we have no problem with achieving change. We just have to want it. So . . . what do we want?

I know what I want. I want a future where the ocean continues to be a healthy heart for our planetary life-support system. And this is not a hair-shirt exercise. There are *better* systems out there, better for us as humans as well as for the planet. We can kick our fossil-fuel addiction and move on. This is all about going towards that better world.

So what do we actually do in practice? The first and biggest task is to learn about and articulate our status as citizens of an ocean planet. We need to talk about the blue machine, to share and develop an ocean perspective, and to build it into our world-view. And then we need to act. The problems the ocean faces are caused by human *systems*, not by a list of isolated bad behaviours. It's only by considering the big picture that we can rebuild and improve the problematic aspects of our civilization so that they coexist with our huge planetary life-support systems. We are all crew, and we can all contribute. Much of that contribution may take the form of voting, writing to politicians, making our preferences clear through what we buy, and making sure that the small, local decisions are constructive – but those all count. If we want change, we can make it happen.

Appreciating the blue machine for what it really is changes us. We are tiny specks of life wandering around the edge of this vast expanse of blue, and imagining ourselves on our blue planet can bring us both humility and comfort. Our influence is already tweaking this giant engine, but the future can be different. More than anything else, that future rests on our human relationship with the global ocean, which comes down to a question of identity. Who do we choose to be each morning? Do we see Earth's defining feature as a critical part of our lives, and choose to see ourselves as citizens of this blue planet? Or do we choose to turn away?

*

The darkness of the universe and the bright pinpricks of the distant stars are fading from view, as pink and purple seep into the sky. Today is a new ocean day, and we sit with our paddles, waiting for the sun to rise. The universe is still out there, but our focus is here, in our canoe. We are all here, citizens of Earth, surrounded by ocean. We have to choose how and how much to connect with the blue machine, because the one thing we can't do is ignore it. The ocean can embrace us, and the ocean can break us. We can work with it, or against it.

A single sunbeam glints across the front of the canoe. Raise your paddle, because the time is now.

Imua!

Acknowledgements

The ocean has been a source of fascination for me for years, ever since I accidentally fell into the world of ocean science as a postdoc at the Scripps Institution of Oceanography. I owe Grant Deane an enormous debt of gratitude for giving me the opportunity to work with him, and for opening the door to the wonders and complexities of the blue of our blue planet. Navigating those complexities to share stories that show how and why the ocean works, without short-changing either the grandeur or the science, has been a challenging task, and I could not have done it without a considerable amount of help and support. I am extremely grateful to everyone who generously shared their time, their expertise and their enthusiasm as I wrote this book.

The only way to study the ocean is collaboratively, and I have learned so much from the worldwide community of ocean scientists. I owe special thanks to David Thornalley, Chris Brierley, Arnaud Czaja, Adrian Glover and Adrian Callaghan, among many others, for their help and support with the London Ocean Group, which we created together and which continues to help us learn from each other. The huge privilege of seeing the ocean at its most raw while living on research ships in the Atlantic, Pacific, Southern and Arctic Oceans, and the Baltic and North Seas, was enriched still further by the amazing teams of people on board, who were committed to both the best science and the best companionship. They are too numerous to name, but they know who they are.

Kimokeo Kapahulehua has been extremely supportive throughout the writing of this book, taking time to teach me about Hawaiian culture and the importance of the canoe, putting me in contact with others who could help, and sharing his world-view. I am very grateful for his trust in me, and especially for his permission to write about some of the time we spent together, both on land and on the water. I met the outrigger canoe paddling community in London, at my club OCUK, and I owe particular thanks to Cameron Taylor and Siobhan Thomas for introducing so many of us not just to the sport of outrigger paddling, but also to its culture.

I am honoured that Nainoa Thompson gave his permission for me to include the section about *Hōkūle'a*. His tireless work to share the stories and the lessons of the revival of traditional Polynesian navigation techniques, especially with the younger generation, is an inspiration. The Polynesian Voyaging Society does extraordinary work, and I recommend their website for information about the current location and voyages of *Hōkūle'a* and the other Polynesian voyaging canoes. Myrna Ahee and Sonia Swenson Rogers offered invaluable help with my requests.

One of the best things about writing a book is the opportunity to go directly to the places and projects that matter and the people who can explain why. It was a privilege to talk to every one of my interviewees: Keith Olson and Laurence Sombardier at NELHA, Deb Kelley at the University of Washington, Steph Henson and her team at the National Oceanography Centre, Southampton, David Johns at the Marine Biological Association in Plymouth, Adrian Glover and Richard Sabin at the Natural History Museum in London, and Tim van Berkel of the Cornish Seaweed Company. I'm also grateful to Hannah Stockton, curator at Royal Museums Greenwich, for taking the time to talk to me about *Cutty Sark,* and to Karen Eden-Tuxford for helping arrange that conversation.

Helen Farr, Ruth Geen, Tina van der Flierdt, Leanna Heffner and Yogini Raste all helped me develop ideas and fact-check certain points. David Ho, Smith Mordak, Ian Brooks and Laurent Bopp heroically read the whole manuscript and offered invaluable comments and corrections. Steph Henson, David Johns, Mario Hoppmann and Jaime Palter all read and provided feedback and reassurance on individual chapters. Any errors that remain are my responsibility alone, of course.

Will Francis, my literary agent at Janklow & Nesbit, has been a warm and wise guiding voice on this project, offering much-valued encouragement and friendship, especially during the development of the initial idea, and in the inevitable patches in the middle where I began to doubt myself. Susanna Wadeson at Transworld believed in the idea from the start and provided constructive and positive support throughout. The manuscript was improved significantly by careful and detailed comments and corrections from Gillian Somerscales in the UK and John Glusman and Helen Thomaides from W. W. Norton in the US. I'm extremely grateful for their eagle-eyed attention to detail, and their commitment to making sure that the book met their high standards.

Throughout the writing process, I have been incredibly grateful for general support and encouragement from many friends, particularly my cohort from Churchill College, Cambridge, and Jem Stansfield, Trent and Melinda Burton, Robin Ince, Smith Mordak, Gaia Vince, Mario Hoppmann, Ian Brooks, Tom Wells, Adam Rutherford and Alice Roberts, among many others. Jack Wormald was a valued companion for some of the early trips and adventures mentioned in the book.

My family has always provided a secure and loving foundation for my adventures, particularly my mum, Susan Czerski, and my cousins, Carl, James and Nicki, and their families. My dad, Jan, died while this book was being written, but his influence lives on and I'm incredibly sad that he never got to see the finished product. My sister, Irena, is always my biggest fan, whether I deserve it or not, and I'm in awe of her constant love and support. My partner, David, is the best, always encouraging, always there, an awesome teammate in life and, as a bonus, incredibly helpful with some of the oceanographic details and sources. The opportunity to fall in love with him is one of the best gifts the ocean has ever given me.

References

Introduction

University Corporation for Atmospheric Research, Center for Science Education, *The Water Cycle*, https://scied.ucar.edu/learning-zone/how-weather-works/water-cycle.

Ernest Henry Shackleton, *South: the Story of Shackleton's Last Expedition, 1914–1917* (London: William Heinemann, 1919), ch. 9, 'The boat journey'.

G. A. Belcher, C. Tarling, A. Manno, A. Atkinson, P. Ward, G. Skaret, S. Fielding, S. A. Henson and R. Sanders, 'The potential role of Antarctic krill faecal pellets in efficient carbon export at the marginal ice zone of the South Orkney Islands in spring', *Polar Biology* 40 (2017), pp. 2001–13, https://doi.org/10.1007/s00300-017-2118-z.

Chapter 1: The Nature of the Sea

Adele K. Morrison, Thomas L. Frölicher and Jorge L. Sarmiento, 'Upwelling in the Southern Ocean', *Physics Today* 68 (2015), https://doi.org/10.1063/PT.3.2654.

NELHA

Guy Toyama, 'Deep ocean water as a catalyst for economic development at NELHA (Natural Energy Laboratory of Hawai'i Authority)', *Deep Ocean Water Research* 11 (2010), pp. 21–3.

J. C. War, 'Land based temperate species mariculture in warm tropical Hawai'i', *Oceans'11 MTS/IEEE KONA* (2011), pp. 1–8, https://doi.org/10.23919/OCEANS.2011.6107220.

Rod Fujita, Alexander C. Markham, Julio E. Diaz Diaz, Julia Rosa Martinez Garcia, Courtney Scarborough, Patrick Greenfield, Peter Black and Stacy E. Aguilera, 'Revisiting ocean thermal energy conversion', *Marine Policy* 36 (2012), pp. 463–5.

Energy balance

Andrew C. Kren, Peter Pilewskie and Odele Coddington, 'Where does Earth's atmosphere get its energy?', *Journal of Space Weather and Space Climate* 7 (2017), p. A10.

World Nuclear Association, *Nuclear Power in the World Today*, Oct. 2022, https://www.world-nuclear.org/information-library/current-and-future-generation/nuclear-power-in-the-world-today.aspx.

M. Wild, D. Folini, M. Z. Hakuba, Christoph Schär, Sonia I. Seneviratne, Seiji Kato, David Rutan, Christof Ammann, Eric F. Wood and Gert König-Langlo, 'The energy balance over land and oceans: an assessment based on direct observations and CMIP5 climate models', *Climate Dynamics* 44 (2015), pp. 3393–429, https://doi.org/10.1007/s00382-014-2430-z.

E. W. Wong and P. J. Minnett, 'The response of the ocean thermal skin layer to variations in incident infra-red radiation', *Journal of Geophysical Research: Oceans* 123 (2018), pp. 2475–93, https://doi.org/10.1002/2017JC013351.

Zhongping Lee, Chuanmin Hu, Shaoling Shang, Keping Du, Marlon Lewis, Robert Arnone and Robert Brewin, 'Penetration of UV-visible solar radiation in the global oceans: insights from ocean color remote sensing', *Journal of Geophysical Research: Oceans* 118 (2013), pp. 4241–55.

Greenland shark

Julius Nielsen, Rasmus B. Hedeholm, Jan Heinemeier, Peter G. Bushnell, Jørgen S. Christiansen, Jesper Olsen, Christopher Bronk Ramsey, Richard W. Brill, Malene Simon, Kirstine F. Steffenson and John F. Steffenson, 'Eye lens radiocarbon reveals centuries of longevity in the Greenland shark (*Somniosus microcephalus*)', *Science* 353 (2016), pp. 702–4.

M. A. MacNeil, B. C. McMeans, N. E. Hussey, P. Vecsei, J. Svavarsson, K. M. Kovacs, C. Lydersen, M. A. Treble, G. B. Skomal, A. Ramsey and A. T. Fisk, 'Biology of the Greenland shark *Somniosus microcephalus*', *Journal of Fish Biology* 80 (2012), pp. 991–1018.

Julius Nielsen, Rasmus B. Hedeholm, Malene Simon and John F. Steffensen, 'Distribution and feeding ecology of the Greenland shark (*Somniosus microcephalus*) in Greenland waters', *Polar Biology* 37 (2014), pp. 37–46, https://doi.org/10.1007/s00300-013-1408-3.

Julius Nielsen, Jørgen Schou Christiansen, Peter Grønkjær, Peter Bushnell, John Fleng Steffensen, Helene Overgaard Kiilerich, Kim Præbel and Rasmus Hedeholm, 'Greenland shark (*Somniosus microcephalus*) stomach contents and stable isotope values reveal an ontogenetic dietary shift', *Frontiers in Marine Science* 6 (2019), p. 125.

David Costantini, Shona Smith, Shaun S. Killen, Julius Nielsen and John F. Steffensen, 'The Greenland shark: a new challenge for the oxidative stress theory of ageing?', *Comparative Biochemistry and Physiology Part A: Molecular & Integrative Physiology* 203 (2017), pp. 227–32.

Frank E. Muller-Karger, Joseph P. Smith, Sandra Werner, Robert Chen, Mitchell Roffer, Yanyun Liu, Barbara Muhling, David Lindo-Atichati, John Lamkin, Sergio Cerdiero-Estrada and David B. Enfield, 'Natural variability of surface oceanographic conditions in the offshore Gulf of Mexico', *Progress in Oceanography* 134 (2015), pp. 54–76.

Paula Pérez-Brunius, Heather Furey, Amy Bower, Peter Hamilton, Julio Candela, Paula García-Carrillo and Robert Leben, 'Dominant circulation patterns of the deep Gulf of Mexico', *Journal of Physical Oceanography* 48 (2018), pp. 511–29.

David Rivas, Antoine Badan and José Ochoa, 'The ventilation of the deep Gulf of Mexico', *Journal of Physical Oceanography* 35 (2005), pp. 1763–81.

Ahmed Ibrahim, Are Olsen, Siv Lauvset and Francisco Rey, 'Seasonal variations of the surface nutrients and hydrography in the Norwegian Sea', *International Journal of Environmental Science and Development* 5 (2014), pp. 496–505.

Brynn M. Devine, Laura J. Wheeland and Jonathan A. D. Fisher, 'First estimates of Greenland shark (*Somniosus microcephalus*) local abundances in Arctic waters', *Scientific Reports* 8 (2018), pp. 1–10.

Layers

Andrea Wulf, *The Invention of Nature: The Adventures of Alexander von Humboldt, the Lost Hero of Science* (Hachette, 2015).

Bin Wang, Renguang Wu and Roger Lukas, 'Annual adjustment of the thermocline in the tropical Pacific Ocean', *Journal of Climate* 13 (2000), pp. 596–616.

Pierrick Penven, Vincent Echevin, J. Pasapera, François Colas and Jorge Tam, 'Average circulation, seasonal cycle, and mesoscale dynamics of the Peru Current System: a modeling approach', *Journal of Geophysical Research: Oceans* 110 (2005), C10021.

Damián Oyarzún and Chris M. Brierley, 'The future of coastal upwelling in the Humboldt Current from model projections', *Climate Dynamics* 52 (2019), pp. 599–615.

Alice Pietri, Pierre Testor, Vincent Echevin, Alexis Chaigneau, Laurent Mortier, Gérard Eldin and Carmen Grados, 'Finescale vertical structure of the upwelling system off southern Peru as observed from glider data', *Journal of Physical Oceanography* 43 (2013), pp. 631–46.

Villy Christensen, Santiago de la Puente, Juan Carlos Sueiro, Jeroen Steenbeek and Patricia Majluf, 'Valuing seafood: the Peruvian fisheries sector', *Marine Policy* 44 (2014), pp. 302–11.

AFP, 'Peru mines new gold in guano', *Independent*, 9 Oct. 2010, https://www.independent.co.uk/climate-change/news/peru-mines-new-gold-in-guano-2102574.html.

Vivian Montecino and Carina B. Lange, 'The Humboldt Current system: ecosystem components and processes, fisheries, and sediment studies', *Progress in Oceanography* 83 (2009), pp. 65–79.

Alexander von Humboldt, *Cosmos: A Sketch of the Physical Description of the Universe*, vol. 1 (Harper, 1858), https://www.gutenberg.org/cache/epub/14565/pg14565-images.html.

Timothy Rooks, *How Humboldt Put South America on the Map*, Deutsche Welle, 7 Dec. 2019, https://www.dw.com/en/how-scientist-alexander-von-humboldt-put-spanish-south-america-on-the-global-map/a-46693502.

P. Espinoza and Arnaud Bertrand, 'Ontogenetic and spatiotemporal variability in anchoveta *Engraulis ringens* diet off Peru', *Journal of Fish Biology* 84 (2014), pp. 422–35.

David M. Checkley Jr, Rebecca G. Asch and Ryan R. Rykaczewski, 'Climate, anchovy, and sardine', *Annual Review of Marine Science* 9 (2017), pp. 469–93.

Manuel Barange, Janet Coetzee, Akinori Takasuka, Kevin Hill, Mariano Gutierrez, Yoshioki Oozeki, Carl van der Lingen and Vera Agostini, 'Habitat expansion and contraction in anchovy and sardine populations', *Progress in Oceanography* 83 (2009), pp. 251–60.

UN Food and Agriculture Organization, *The State of World Fisheries and Aquaculture 2018: Meeting the Sustainable Development Goals* (Rome, 2018), licence: CC BY-NC-SA 3.0 IGO.

Kristin Wintersteen, *Protein from the sea: the global rise of fishmeal and the industrialization of southeast Pacific fisheries, 1918–1973*, desiguALdades.net, working paper series no. 26 (Berlin: desiguALdades.net Research Network on Interdependent Inequalities in Latin America, 2012).

C. J. Shepherd and A. J. Jackson, 'Global fishmeal and fish-oil supply: inputs, outputs and markets', *Journal of Fish Biology* 83 (2013), pp. 1046–66.

Andrew C. Godley and Bridget Williams, *The Chicken, the Factory Farm and the Supermarket: The Emergence of the Modern Poultry Industry in Britain* (Reading: University of Reading Department of Economics, 2007).

Hannah Ritchie, Pablo Rosado and Max Roser, 'Meat and dairy production', *Our World in Data*, Aug. 2017, last revised Nov. 2019, https://ourworldindata.org/meat-production.

Tim Cashion, Frédéric le Manach, Dirk Zeller and Daniel Pauly, 'Most fish destined for fishmeal production are food-grade fish', *Fish and Fisheries* 18 (2017), pp. 837–44.

Sloths and ocean effects on precipitation

D. P. Gilmore, C. Peres da Costa and D. P. F. Duarte, 'Sloth biology: an update on their physiological ecology, behavior and role as vectors of arthropods and arboviruses', *Brazilian Journal of Medical and Biological Research* 34 (2001), pp. 9–25.

Jonathan N. Pauli, Jorge E. Mendoza, Shawn A. Steffan, Cayelan C. Carey, Paul J. Weimer and M. Zachariah Peery, 'A syndrome of mutualism reinforces the lifestyle of a sloth', *Proceedings of the Royal*

Society B: Biological Sciences 281 (2014), 20133006, https://doi.org/10.1098/rspb.2013.3006.

Mohammed Faizal and M. Rafiuddin Ahmed, 'On the ocean heat budget and ocean thermal energy conversion', *International Journal of Energy Research* 35 (2011), pp. 1119–44.

Andrew C. Kren, Peter Pilewskie and Odele Coddington, 'Where does Earth's atmosphere get its energy?', *Journal of Space Weather and Space Climate* 7 (2017), pp. A10.

Kevin E. Trenberth, John T. Fasullo and Jeffrey Kiehl, 'Earth's global energy budget', *Bulletin of the American Meteorological Society* 90 (2009), pp. 311–24.

Peter R. Waylen, César N. Caviedes and Marvin E. Quesada, 'Interannual variability of monthly precipitation in Costa Rica', *Journal of Climate* 9 (1996), pp. 2606–13.

Xin-Yue Wang, Xichen Li, Jiang Zhu and Clemente A. S. Tanajura, 'The strengthening of the Amazonian precipitation in the wet season driven by tropical sea surface temperature forcing', *Environmental Research Letters* 13 (2018), https://doi.org/ 10.1088/1748-9326/aadbb9.

John Abraham, 'Warming oceans are changing the world's rainfall', *Guardian*, 12 Sep. 2018, https://www.theguardian.com/environment/climate-consensus-97-per-cent/2018/sep/12/warming-oceans-are-changing-the-worlds-rainfall.

Guano

William Mitchell Mathew, 'Peru and the British guano market, 1840–1870', *Economic History Review* 23 (1970), pp. 112–28.

Ewald Schnug, Frank Jacobs and Kirsten Stöven, 'Guano: the white gold of the seabirds', *Seabirds* (2018), pp. 81–100. https://doi.org/ 10.5772/intechopen.79501

Salt

Mark Kurlansky, *Salt* (Random House, 2011).

James E. Breck, 'Body composition in fishes: body size matters', *Aquaculture* 433 (2014), pp. 40–9, https://doi.org/10.1016/j.aquaculture.2014.05.049.

John M. Wright and Angela Colling, *Sea Water: Its Composition, Properties and Behaviour, Prepared by an Open University Course Team* (Elsevier, 2013).

William W. Hay, Areg Migdisov, Alexander N. Balukhovsky, Christopher N. Wold, Sascha Flögel and Emanuel Söding, 'Evaporites and the salinity of the ocean during the Phanerozoic: implications for climate, ocean circulation and life', *Palaeogeography, Palaeoclimatology, Palaeoecology* 240 (2006), pp. 3–46.

Nadya Vinogradova, Tong Lee, Jacqueline Boutin, Kyla Drushka, Severine Fournier, Roberto Sabia, Detlef Stammer et al., 'Satellite salinity observing system: Recent discoveries and the way forward', *Frontiers in Marine Science* (2019), p. 243.

Robert Boyle, *The Saltness of the Sea* (1674), https://digital.nmla.metoffice.gov.uk/IO_845a61a4-3988-4e0b-98ee-ed216651b3e5/.

HMS Challenger

Heidi M. Dierssen, A. E. Theberge and Y. Wang, 'Bathymetry: history of sea floor mapping', *Encyclopedia of Natural Resources* 2 (2014), pp. 564–8.

J. J. Middelburg, K. Soetaert and M. Hagens, 'Understanding alkalinity to quantify ocean buffering', *Eos,* 29 July 2020, https://eos.org/editors-vox/understanding-alkalinity-to-quantify-ocean-buffering.

Thomas Henry Tizard, *Narrative of the Cruise of HMS Challenger: With a General Account of the Scientific Results of the Expedition*, vol. 2 (HM Stationery Office, 1882).

Joel W. Hedgpeth, 'The voyage of the *Challenger*', *Scientific Monthly* 63 (1946), pp. 194–202.

J. Y. Buchanan, 'On the distribution of salt in the ocean, as indicated by the specific gravity of its waters', *Journal of the Royal Geographical Society of London* 47 (1877), pp. 72–86.

C. Spencer Jones and Paola Cessi, 'Size matters: another reason why the Atlantic is saltier than the Pacific', *Journal of Physical Oceanography* 47 (2017), pp. 2843–59.

Turtles

Rebecca Rash and Harvey B. Lillywhite, 'Drinking behaviors and water balance in marine vertebrates', *Marine Biology* 166 (2019), pp. 1–21.

John Davenport, 'Crying a river: how much salt-laden jelly can a leatherback turtle really eat?', *Journal of Experimental Biology* 220 (2017), pp. 1737–44.

Whales

Bertrand Bouchard, Jean-Yves Barnagaud, Philippe Verborgh, Pauline Gauffier, Sylvie Campagna and Aurélie Célérier, 'A field study of chemical senses in bottlenose dolphins and pilot whales', *Anatomical Record* 305 (2022), pp. 668–79.

Danielle Venton, 'Highlight: a matter of taste—whales have abandoned their ability to taste food', *Genome Biology and Evolution* 6 (2014), pp. 1266.

Bertrand Bouchard, Jean-Yves Barnagaud, Marion Poupard, Hervé Glotin, Pauline Gauffier, Sara Torres Ortiz, Thomas J. Lisney, Sylvie Campagna, Marianne Rasmussen and Aurélie Célérier, 'Behavioural responses of humpback whales to food-related chemical stimuli', *PloS One* 14 (2019), e0212515, https://doi.org/10.1371/journal.pone.0212515.

Dorothee Kremers, Aurélie Célérier, Benoist Schaal, Sylvie Campagna, Marie Trabalon, Martin Böye, Martine Hausberger and Alban Lemasson, 'Sensory perception in cetaceans, part I: current knowledge about dolphin senses as a representative species', *Frontiers in Ecology and Evolution* 4 (2016), art. 49, https://doi.org/10.3389/fevo.2016.00049.

Takushi Kishida, J. G. M. Thewissen, Takashi Hayakawa, Hiroo Imai and Kiyokazu Agata, 'Aquatic adaptation and the evolution of smell and taste in whales', *Zoological Letters* 1 (2015), pp. 1–10.

Kangli Zhu, Xuming Zhou, Shixia Xu, Di Sun, Wenhua Ren, Kaiya Zhou and Guang Yang, 'The loss of taste genes in cetaceans', *BMC Evolutionary Biology* 14 (2014), pp. 1–10.

Rebecca Rash and Harvey B. Lillywhite, 'Drinking behaviors and water balance in marine vertebrates', *Marine Biology* 166 (2019), pp. 1–21.

Robert Kenney, 'How can sea mammals drink saltwater?', *Scientific American*, 30 April 2001, https://www.scientificamerican.com/article/how-can-sea-mammals-drink/.

Martin G. Greenwell, Johanna Sherrill and Leigh A. Clayton, 'Osmoregulation in fish: mechanisms and clinical implications', *Veterinary Clinics: Exotic Animal Practice* 6 (2003), pp. 169–89.

Snails

Celia K. C. Churchill, Diarmaid Ó. Foighil, Ellen E. Strong and Adriaan Gittenberger, 'Females floated first in bubble-rafting snails', *Current Biology* 21 (2011), pp. R802–3.

Patrick A. Rühs, Jotam Bergfreund, Pascal Bertsch, Stefan J. Gstöhl and Peter Fischer, 'Complex fluids in animal survival strategies', *Soft Matter* 17 (2021), pp. 3022–36.

Alan Glenn Beu, *Evolution of Janthina and Recluzia (Mollusca: Gastropoda: Epitoniidae)* (Australian Museum, 2017).

Sea ice

Kerstin Jochumsen, Manuela Köllner, Detlef Quadfasel, Stephen Dye, Bert Rudels and Heðinn Valdimarsson, 'On the origin and propagation of Denmark Strait overflow water anomalies in the Irminger Basin', *Journal of Geophysical Research: Oceans* 120 (2015), pp. 1841–55.

Paul R. Holland and Daniel L. Feltham, 'Frazil dynamics and precipitation in a water column with depth-dependent supercooling', *Journal of Fluid Mechanics* 530 (2005), pp. 101–24.

David W. Rees Jones and Andrew J. Wells, 'Frazil-ice growth rate and dynamics in mixed layers and sub-ice-shelf plumes', *The Cryosphere* 12 (2018), pp. 25–38.

Ellen Kathrine Bludd, 'On a groundbreaking 1893 expedition Nansen froze his ship in Arctic ice for a year – now MOSAiC is following his path', *Journal of the North Atlantic and Arctic*, 2022, https://www.jonaa.org/content/on-a-groundbreaking-1893-expedition-nansen-froze-his-ship-in-arctic-ice-for-a-year-now-mosaic-is-following-his-path.

Mary-Louise Timmermans and John Marshall, 'Understanding Arctic Ocean circulation: a review of ocean dynamics in a changing climate', *Journal of Geophysical Research: Oceans* 125 (2020), e2018JC014378, https://doi.org/10.1029/2018JC014378.

Paul Webb, *Introduction to Oceanography* (Pressbooks, 2021), https://rwu.pressbooks.pub/webboceanography/, ch. 6.3.

Denmark Strait Cataract

World Waterfall Database, *World's Largest Waterfalls*, 2002–22, https://www.worldwaterfalldatabase.com/largest-waterfalls/volume.

V. M. Zhurbas, V. T. Paka, B. Rudels and D. Quadfasel, 'Estimates of entrainment in the Denmark Strait overflow plume from CTD/LADCP data', *Oceanology* 56 (2016), pp. 205–13.

Spin

Niels J. de Winter, Steven Goderis, Stijn J. M. van Malderen, Matthias Sinnesael, Stef Vansteenberge, Christophe Snoeck, Joke Belza, Frank Vanhaecke and Philippe Claeys, 'Subdaily-scale chemical variability in a Torreites sanchezi rudist shell: implications for rudist paleobiology and the Cretaceous day-night cycle', *Paleoceanography and Paleoclimatology* 35 (2020), e2019PA003723, https://doi.org/10.1029/2019PA003723.

Erik Brown, 'Mankind launched the first object into the stratosphere in 1918', *Lessons from History*, 6 Dec. 2020, https://medium.com/lessons-from-history/mankind-launched-the-first-object-into-the-stratosphere-in-1918-39571ad1c092.

David T. Zabecki, 'Paris under the gun', *Historynet*, 24 Jan. 2017, https://www.historynet.com/paris-under-the-gun/?f.

Kenneth Hunkins, 'Ekman drift currents in the Arctic Ocean', *Deep Sea Research and Oceanographic Abstracts* 13 (1966), pp. 607–20.

V. W. Ekman, *On the influence of the earth's rotation on ocean-currents* (1905), https://jscholarship.library.jhu.edu/bitstream/handle/1774.2/33989/31151027498728.pdf.

Chapter 2: The Shape of Seawater

David Eugene Hitzl, Yi-Leng Chen and Hiep van Nguyen, 'Numerical simulations and observations of airflow through the 'Alenuihāhā Channel, Hawai'i', *Monthly Weather Review* 142 (2014), pp. 4696–718.

Walter Munk

Hugh Aldersey-Williams, *The Tide: The Science and Stories Behind the Greatest Force on Earth* (Norton, 2016).

Chris Garrett and Carl Wunsch, 'Walter Heinrich Munk, 19 October 1917–8 February 2019', *Biographical Memoirs of the Fellows of the Royal Society* (2020), pp. 393–424.

Hans Storch and Klaus Hasselmann, *Seventy Years of Exploration in Oceanography* (Springer, 2010).

Guoqiang Liu, Yijun He, Yuanzhi Zhang and Hui Shen, 'Estimation of global wind energy input to the surface waves based on the scatterometer', *IEEE Geoscience and Remote Sensing Letters* 9 (2012), pp. 1017–20.

David K. Lynch, David S. P. Dearborn and James A. Lock, 'Glitter and glints on water', *Applied Optics* 50 (2011), pp. F39–49.

Paul C. Liu, 'Fifty years of wave growth curves', in *Proceedings of 25th International Conference on Coastal Engineering*, Orlando, FL, 2–6 Sept. 1996 (American Society of Civil Engineers, 1997), pp. 457–64.

Owen M. Phillips, 'On the generation of waves by turbulent wind', *Journal of Fluid Mechanics* 2 (1957), pp. 417–45.

Nick Pizzo, Luc Deike and Alex Ayet, 'How does the wind generate waves?', *Physics Today* 74 (2021), https://doi.org/10.1063/PT.3.4880.

William Beebe

William Beebe, *Half Mile Down* (New York: Harcourt, Brace and Company, 1934).

Mohole

Sara E. Pratt, 'Benchmarks: March 1961: Project Mohole undertakes the first deep-ocean drilling', *Earth: the science behind the headlines*, 6 July 2016, https://www.earthmagazine.org/article/benchmarks-march-1961-project-mohole-undertakes-first-deep-ocean-drilling.

Byrd Pinkerton, 'Unexplainable', episode 3, 'How an ill-fated undersea adventure in the 1960s changed the way scientists see the Earth', *Vox*, 17 March 2021, https://www.vox.com/unexplainable/22276597/project-mohole-deep-ocean-drilling-unexplainable-podcast.

John Steinbeck, 'High drama of bold thrust through ocean floor', *Life*, 14 April 1961, https://books.google.fr/books?id=9lEEAAAA MBAJ&lpg=PA110&ots=iFqfSHZdb0&dq=John+Steinbeck+1961+crust+pacific+life&pg=PA110%23v%3Dtwopage&q=&hl=fr#v=onepage&q&f=false.

N. Sönnichsen, 'Distribution of global crude oil production onshore and offshore 2005–2025', *Statista*, 28 Jan. 2021, https://www.statista.com/statistics/624138/distribution-of-crude-oil-production-worldwide-onshore-and-offshore/.

HMS Challenger

J. Murray, *A Summary of the Scientific Results Obtained at the Sounding, Dredging and Trawling Stations of HMS* Challenger, vol. 1 (HM Stationery Office, 1895).

James Hanley, 'How deep is the ocean?', *Significance* 11 (2014), pp. 30–3.

Robert Kunzig, *The Restless Sea* (Norton, 1999).

The Voyage of HMS Challenger, vol. 1 (Johnson, 1885), https://archimer.ifremer.fr/doc/1885/publication-4746.pdf.

Regional Cabled Array

'Shore station', Interactive Oceans, University of Washington, n.d., https://io.ocean.washington.edu/story/Shore_Station; the live camera system is at https://interactiveoceans.washington.edu/instruments/high-definition-video-camera/.

'2015 Axial Seamount eruption', *Ocean Data Labs*, n.d., https://datalab.marine.rutgers.edu/data-nuggets/axial-eruption/.

Marvin D. Lilley, David A. Butterfield, John E. Lupton and Eric J. Olson, 'Magmatic events can produce rapid changes in hydrothermal vent chemistry', *Nature* 422 (2003), pp. 878–81.

George W. Luther, Tim F. Rozan, Martial Taillefert, Donald B. Nuzzio, Carol di Meo, Timothy M. Shank, Richard A. Lutz and S. Craig Cary, 'Chemical speciation drives hydrothermal vent ecology', *Nature* 410 (2001), pp. 813–16.

Polymetallic nodules

Alex Rogers, *The Deep* (Headline Publishing Group, 2019).

Helen Scales, *The Brilliant Abyss* (Atlantic Monthly Press, 2021).

James R. Hein, Andrea Koschinsky and Thomas Kuhn, 'Deep-ocean polymetallic nodules as a resource for critical materials', *Nature Reviews Earth & Environment* 1 (2020), pp. 158–69.

'Mineral resources', *World Ocean Review 2014*, https://worldoceanreview.com/en/wor-3/mineral-resources/manganese-nodules/.

Robin McKie, 'Is deep-sea mining a cure for the climate crisis or a curse?', *Guardian*, 29 Aug. 2021, https://www.theguardian.com/world/2021/aug/29/is-deep-sea-mining-a-cure-for-the-climate-crisis-or-a-curse.

Eels

Estuaries and Wetlands Conservation Programme, Zoological Society of London, *The Thames European Eel Project Report*, Nov. 2018, https://www.zsl.org/sites/default/files/media/2018-12/ZSL%202018%20eel%20report_FINAL.pdf.

Port of London Authority, *History of the Port of London pre 1908*, n.d., https://pla.co.uk/Port-Trade/History-of-the-Port-of-London-pre-1908#18.

Vincent J. T. van Ginneken and Gregory E. Maes, 'The European eel (*Anguilla anguilla*, Linnaeus), its lifecycle, evolution and reproduction: a literature review', *Reviews in Fish Biology and Fisheries* 15 (2005), pp. 367–98.

J. F. López-Olmeda, I. López-García, M. J. Sánchez-Muros, B. Blanco-Vives, R. Aparicio and F. J. Sánchez-Vázquez, 'Daily rhythms of digestive physiology, metabolism and behaviour in the European eel (*Anguilla anguilla*)', *Aquaculture International* 20 (2012), pp. 1085–96.

Arjan P. Palstra and Guido E. E. J. M. van den Thillart, 'Swimming physiology of European silver eels (*Anguilla anguilla* L.): energetic costs and effects on sexual maturation and reproduction', *Fish Physiology and Biochemistry* 36 3 (2010), pp. 297–322.

Caroline M. F. Durif, Howard I. Browman, John B. Phillips, Anne Berit Skiftesvik, L. Asbjørn Vøllestad and Hans H. Stockhausen, 'Magnetic compass orientation in the European eel', *PloS One* 8 (2013), e59212, https://doi.org/10.1371/journal.pone.0059212.

Bernd Pelster, 'Swimbladder function and the spawning migration of the European eel *Anguilla anguilla*', *Frontiers in Physiology* 5 (2015), art. 486, https://doi.org/10.3389/fphys.2014.00486.

Thomas D. Als, Michael M. Hansen, Gregory E. Maes, Martin Castonguay, Lasse Riemann, K. I. M. Aarestrup, Peter Munk, Henrik Sparholt, Reinhold Hanel and Louis Bernatchez, 'All roads lead to home: panmixia of European eel in the Sargasso Sea', *Molecular Ecology* 20 (2011), pp. 1333–46.

Michael J. Miller, Håkan Westerberg, Henrik Sparholt, Klaus Wysujack, Sune R. Sørensen, Lasse Marohn, Magnus W. Jacobsen et al., 'Spawning by the European eel across 2000 km of the Sargasso Sea', *Biology Letters* 15 (2019), 20180835, https://doi.org/10.1098/rsbl.2018.0835.

Robert J. Lennox, Finn Økland, Hiromichi Mitamura, Steven J. Cooke and Eva B. Thorstad, 'European eel *Anguilla anguilla* compromise speed for safety in the early marine spawning migration', *ICES Journal of Marine Science* 75 (2018), pp. 1984–91.

David Righton, Håkan Westerberg, Eric Feunteun, Finn Økland, Patrick Gargan, Elsa Amilhat, Julian Metcalfe et al., 'Empirical observations of the spawning migration of European eels: The long and dangerous road to the Sargasso Sea', *Science Advances* 2 (2016), e1501694, https://doi.org/10.1126/sciadv.1501694.

Alessandro Cresci, 'A comprehensive hypothesis on the migration of European glass eels (*Anguilla anguilla*)', *Biological Reviews* 95 (2020), pp. 1273–86.

Lewis C. Naisbett-Jones, Nathan F. Putman, Jessica F. Stephenson, Sam Ladak and Kyle A. Young, 'A magnetic map leads juvenile European eels to the Gulf Stream', *Current Biology* 27 (2017), pp. 1236–40.

B. R. Quintella, C. S. Mateus, J. L. Costa, I. Domingos and Pedro R. Almeida, 'Critical swimming speed of yellow-and silver-phase European eel (*Anguilla anguilla*, L.)', *Journal of Applied Ichthyology* 26 (2010), pp. 432–5.

V. N. Mikhailov and M. V. Mikhailova, 'Tides and storm surges in the Thames River Estuary', *Water Resources* 39 (2012), pp. 351–65.

Willem Dekker, 'The history of commercial fisheries for European eel commenced only a century ago', *Fisheries Management and Ecology* 26 (2019), pp. 6–19.

Helene Burningham and Jon French, 'Seabed dynamics in a large coastal embayment: 180 years of morphological change in the outer Thames estuary', *Hydrobiologia* 672 (2011), pp. 105–19.

Quanquan Cao, Jie Gu, Dan Wang, Fenfei Liang, Hongye Zhang, Xinru Li and Shaowu Yin, 'Physiological mechanism of osmoregulatory adaptation in anguillid eels', *Fish Physiology and Biochemistry* 44 (2018), pp. 423–33.

I. A. Naismith and B. Knights, 'Migrations of elvers and juvenile European eels, *Anguilla anguilla* L., in the River Thames', *Journal of Fish Biology* 33 (1988), pp. 161–75.

Johs Schmidt, 'IV. The breeding places of the eel', *Philosophical Transactions of the Royal Society of London, Series B, Containing Papers of a Biological Character* 211 (1923), pp. 179–208.

Seaweed

Geoff Maynard, 'Storm Tide Washes Away Resort Beach', *Daily Express*, 23 Jan. 2015, https://www.express.co.uk/news/nature/553709/Storm-tide-swept-away-resort-beach-Porthleven.

Robert D. Kinley, Gonzalo Martinez-Fernandez, Melissa K. Matthews, Rocky de Nys, Marie Magnusson and Nigel W. Tomkins, 'Mitigating the carbon footprint and improving productivity of ruminant livestock agriculture using a red seaweed', *Journal of Cleaner Production* 259 (2020), 120836, https://doi.org/10.1016/j.jclepro.2020.120836.

Kelp highway

Tom D. Dillehay, Carlos Ocampo, José Saavedra, Andre Oliveira Sawakuchi, Rodrigo M. Vega, Mario Pino, Michael B. Collins et al., 'New archaeological evidence for an early human presence at Monte Verde, Chile', *PloS One* 10 (2015), e0141923, https://doi.org/10.1371/journal.pone.0141923.

Loren G. Davis and David B. Madsen, 'The coastal migration theory: formulation and testable hypotheses', *Quaternary Science Reviews* 249 (2020), 106605, https://doi.org/10.1016/j.quascirev.2020.106605.

Tom D. Dillehay, Carlos Ramírez, Mario Pino, Michael B. Collins, Jack Rossen and Jimena Daniela Pino-Navarro, 'Monte Verde: seaweed, food, medicine, and the peopling of South America', *Science* 320 (2008), pp. 784–6.

Todd J. Braje, Tom D. Dillehay, Jon M. Erlandson, Richard G. Klein and Torben C. Rick, 'Finding the first Americans', *Science* 358 (2017), pp. 592–4.

Jon M. Erlandson, Todd J. Braje, Kristina M. Gill and Michael H. Graham, 'Ecology of the kelp highway: did marine resources facilitate human dispersal from northeast Asia to the Americas?', *Journal of Island and Coastal Archaeology* 10 (2015), pp. 392–411.

Michael H. Graham, Brian P. Kinlan and Richard K. Grosberg, 'Post-glacial redistribution and shifts in productivity of giant kelp forests', *Proceedings of the Royal Society B: Biological Sciences* 277 (2010), pp. 399–406.

Robert S. Steneck, Michael H. Graham, Bruce J. Bourque, Debbie Corbett, Jon M. Erlandson, James A. Estes and Mia J. Tegner, 'Kelp forest ecosystems: biodiversity, stability, resilience and future', *Environmental Conservation* 29 (2002), pp. 436–59.

Jon M. Erlandson, Michael H. Graham, Bruce J. Bourque, Debra Corbett, James A. Estes and Robert S. Steneck, 'The kelp highway hypothesis: marine ecology, the coastal migration theory, and the peopling of the Americas', *Journal of Island and Coastal Archaeology* 2 (2007), pp. 161–74.

Thomas Lamy, Craig Koenigs, Sally J. Holbrook, Robert J. Miller, Adrian C. Stier and Daniel C. Reed, 'Foundation species promote community stability by increasing diversity in a giant kelp forest', *Ecology* 101 (2020), e02987, https://doi.org/10.1002/ecy.2987.

Morten Rasmussen, Sarah L. Anzick, Michael R. Waters, Pontus Skoglund, Michael DeGiorgio, Thomas W. Stafford, Simon Rasmussen et al., 'The genome of a Late Pleistocene human from a Clovis burial site in western Montana', *Nature* 506 (2014), pp. 225–9.

Maui fishpond

'Fishpond basics', *Maui Fishpond*, n.d., http://mauifishpond.com/koieie/fishpond-basics/, https://seagrant.soest.hawaii.edu/the-return-of-kuula/.

Paula Möhlenkamp, Charles Kaiaka Beebe, Margaret A. McManus, Angela Hiʻilei Kawelo, Keliʻiahonui Kotubetey, Mirielle Lopez-Guzman, Craig E. Nelson and Rosanna 'Anolani Alegado, 'Kū hou kuapā: cultural restoration improves water budget and water quality dynamics in Heʻeia Fishpond', *Sustainability* 11 (2018), art. 161, https://www.mdpi.com/2071-1050/11/1/161.

Graydon 'Buddy' Keala with James R. Hollyer and Luisa Castro, *LOKO IʻA: A Manual on Hawaiian Fishpond Restoration and Management* (College of Tropical Agriculture and Human Resources, 2017)

Chapter 3: The Anatomy of the Ocean

Barnacles

Ryan M. Pearson, Jason P. van de Merwe and Rod M. Connolly, 'Global oxygen isoscapes for barnacle shells: application for tracing movement in oceans', *Science of the Total Environment* 705 (2020), 135782, https://doi.org/10.1016/j.scitotenv.2019.135782.

Ryan M. Pearson, Jason P. Van de Merwe, Michael K. Gagan and Rod M. Connolly, 'Unique post-telemetry recapture enables development of multi-element isoscapes from barnacle shell for retracing host movement', *Frontiers in Marine Science* 7 (2020), art. 596, https://www.frontiersin.org/articles/10.3389/fmars.2020.00596/full.

John D. Zardus, 'A global synthesis of the correspondence between epizoic barnacles and their sea turtle hosts', *Integrative Organismal Biology* 3 (2021), obab002, https://doi.org/10.1093/iob/obab002.

Ryan M. Pearson, Jason P. van de Merwe, Michael K. Gagan, Colin J. Limpus and Rod M. Connolly, 'Distinguishing between sea turtle foraging areas using stable isotopes from commensal barnacle shells', *Scientific Reports* 9 (2019), pp. 1–11.

Sophie A. Doell, Rod M. Connolly, Colin J. Limpus, Ryan M. Pearson and Jason P. van de Merwe, 'Using growth rates to estimate age of the sea turtle barnacle *Chelonibia testudinaria*', *Marine Biology* 164 (2017), pp. 1–7.

Battle of Actium

Johan Fourdrinoy, Clément Caplier, Yann Devaux, Germain Rousseaux, Areti Gianni et al., 'The naval battle of Actium and the myth of the ship-holder: the effect of bathymetry', 5th MASHCON: International Conference on Ship Manoeuvring in Shallow and Confined Water with non-exclusive focus on manoeuvring in waves, wind and current, Flanders Hydraulics Research, Maritime Technology Division, Ghent University, Ostend, Belgium, May 2019, WWC007 (pp. 104–33), hal-02139218, https://doi.org/10.48550/arXiv.1905.13024.

Annalisa Marzano, 'Fish and fishing in the Roman world', *Journal of Maritime Archaeology* 13 (2018), pp. 437–47.

Julia Calderone, 'Elusive underwater waves hold an uncertain grip on the climate', *Hakai Magazine*, 30 April 2015, https://hakaimagazine.com/news/elusive-underwater-waves-hold-uncertain-grip-climate/.

J. M. Walker, 'Farthest north, dead water and the Ekman spiral', part 2: 'Invisible waves and a new direction in current theory', *Weather* 46 (1991), pp. 158–64.

Johan Fourdrinoy, Julien Dambrine, Madalina Petcu, Morgan Pierre and Germain Rousseaux, 'The dual nature of the dead-water phenomenology: Nansen versus Ekman wave-making drags', *Proceedings of the National Academy of Sciences* 117 (2020), pp. 16770–5.

Thomas Midgley and CFCs

Monika Rhein, Dagmar Kieke and Reiner Steinfeldt, 'Advection of North Atlantic deep water from the Labrador Sea to the southern hemisphere', *Journal of Geophysical Research: Oceans* 120 (2015), pp. 2471–87.

Monika Rhein, Reiner Steinfeldt, Dagmar Kieke, Ilaria Stendardo and Igor Yashayaev, 'Ventilation variability of Labrador Sea Water and its impact on oxygen and anthropogenic carbon: a review', *Philosophical Transactions of the Royal Society A: Mathematical, Physical and Engineering Sciences* 375 (2017), 20160321, https://doi.org/10.1098/rsta.2016.0321.

John Bullister, 'Chlorofluorocarbons as time-dependent tracers in the ocean', *Oceanography* 2, 1989, pp. 12–17.

J. L. Bullister, *Atmospheric Histories (1765–2015) for CFC-11, CFC-12, CFC-113, CCl4, SF6 and N2O (NCEI Accession 0164584)*, NOAA National Centers for Environmental Information, 2017, unpublished dataset, https://doi.org/10.3334/CDIAC/otg.CFC_ATM_Hist_2015.

Bill Kovarik, 'A century of tragedy: how the car and gas industry knew about the health risks of leaded fuel but sold it for 100 years anyway', *The Conversation*, 8 Dec. 2021, https://theconversation.com/a-century-of-tragedy-how-the-car-and-gas-industry-knew-about-the-health-risks-of-leaded-fuel-but-sold-it-for-100-years-anyway-173395.

Joseph A. Williams, 'This 1920s inventor sped up climate change with his chemical creations', *History*, 23 Aug. 2019, https://www.history.com/news/cfcs-leaded-gasoline-inventions-thomas-midgley.

Danny Schleien, 'Meet Thomas Midgley Jr., arguably the most dangerous man of all time', *Climate Conscious*, 12 Oct. 2020, https://medium.com/climate-conscious/meet-thomas-midgley-jr-arguably-the-most-dangerous-man-of-all-time-2ae66b1cc101.

M. Susan Lozier, Feili Li, Sheldon Bacon, F. Bahr, Amy S. Bower, S. A. Cunningham, M. Femke de Jong et al., 'A sea change in our view of overturning in the subpolar North Atlantic', *Science* 363 (2019), pp. 516–21.

Titanic

John Roach, '*Titanic* was found during secret Cold War navy mission', *National Geographic*, 23 Dec. 2018, https://www.nationalgeographic.co.uk/history-and-civilisation/2018/11/titanic-was-found-during-secret-cold-war-navy-mission.

Lonny Lippsett and Marine Conservation, 'The quest to map *Titanic*', *Oceanus* 49 (2012), pp. 26–36.

Kenneth J. Vrana, Paul-Henry Nargeolet, William Sauder, Alexandra Klingelhofer, Rebecca King, Laura Pasch, Sarah J. AcMoody et al., 'Mapping RMS *Titanic* with GIS: implications for forensic investigations', *Marine Technology Society Journal* 46 (2012), pp. 111–28.

'Full *Titanic* site mapped for 1st time', *New York Post*, 8 March 2012, https://nypost.com/2012/03/08/full-titanic-site-mapped-for-1st-time/.

Robert D. Ballard and Will Hively, *The Eternal Darkness: A Personal History of Deep-Sea Exploration*, (Princeton University Press, 2017).

Calanus finmarchicus

U. V. Bathmann, T. T. Noji, Max Voss and R. Peinert, *Copepod fecal pellets: abundance, sedimentation and content at a permanent station in the Norwegian Sea in May/June 1986*, Marine Ecology Progress Series (Inter-Research Science Center, 1987), vol. 38(1), pp. 45–51.

Webjørn Melle, Jeffrey Runge, Erica Head, Stéphane Plourde, Claudia Castellani, Priscilla Licandro, James Pierson et al., 'The North Atlantic Ocean as habitat for *Calanus finmarchicus*: environmental factors and life history traits', *Progress in Oceanography* 129 (2014), pp. 244–84.

Laura A. Bristow, Wiebke Mohr, Soeren Ahmerkamp and Marcel M. M. Kuypers, 'Nutrients that limit growth in the ocean', *Current Biology* 27 (2017), pp. R474–8.

Mark F. Baumgartner, Nadine S. J. Lysiak, Carrie Schuman, Juanita Urban-Rich and Frederick W. Wenzel, *Diel vertical migration behavior of* Calanus finmarchicus *and its influence on right and sei whale occurrence*, Marine Ecology Progress Series no. 423 (Inter-Research Science Center, 2011), pp. 167–84.

David W. Pond and Geraint A. Tarling, 'Phase transitions of wax esters adjust buoyancy in diapausing *Calanoides acutus*', *Limnology and Oceanography* 56 (2011), pp. 1310–18.

Jonathan H. Cohen, Kim S. Last, Jack Waldie and David W. Pond, 'Loss of buoyancy control in the copepod *Calanus finmarchicus*', *Journal of Plankton Research* 41 (2019), pp. 787–90.

Mixing

Kakani Katija, 'Biogenic inputs to ocean mixing', *Journal of Experimental Biology* 215 (2012), pp. 1040–9.

Daisuke Hasegawa, 'Island mass effect', in T. Nagai, H. Saito, K. Suzuki and M. Takahashi, eds, *Kuroshio Current* (AGU, 2019), https://doi.org/10.1002/9781119428428.ch10.

Quirin Schiermeier, 'Oceanography: churn, churn, churn', *Nature* 447 (2007), pp. 522–5.

Jessica C. Garwood, Ruth C. Musgrave and Andrew J. Lucas, 'Life in internal waves', *Oceanography* 33 (2020), pp. 38–49.

Chris Garrett, 'Internal tides and ocean mixing', *Science* 301 (2003), pp. 1858–9.

S. Sarkar and A. Scotti, 'From topographic internal gravity waves to turbulence', *Annual Review of Fluid Mechanics* 49 (2017), pp. 195–220.

Daniel L. Rudnick, Timothy J. Boyd, Russell E. Brainard, Glenn S. Carter, Gary D. Egbert, Michael C. Gregg, Peter E. Holloway et al., 'From tides to mixing along the Hawaiian Ridge', *Science* 301 (2003), pp. 355–7.

Joseph P. Martin and Daniel L. Rudnick, 'Inferences and observations of turbulent dissipation and mixing in the upper ocean at the Hawaiian Ridge', *Journal of Physical Oceanography* 37 (2007), pp. 476–94.

Jody M. Klymak, Robert Pinkel and Luc Rainville, 'Direct breaking of the internal tide near topography: Kaena Ridge, Hawai'i', *Journal of Physical Oceanography* 38 (2008), pp. 380–99.

Luc Rainville, T. M. Shaun Johnston, Glenn S. Carter, Mark A. Merrifield, Robert Pinkel, Peter F. Worcester and Brian D. Dushaw, 'Interference pattern and propagation of the M2 internal tide south of the Hawaiian Ridge', *Journal of Physical Oceanography* 40 (2010), pp. 311–25.

Katharine A. Smith, Greg Rocheleau, Mark A. Merrifield, Sergio Jaramillo and Geno Pawlak, 'Temperature variability caused by internal tides in the coral reef ecosystem of Hanauma Bay, Hawai'i', *Continental Shelf Research* 116 (2016), pp. 1–12.

Jerome Aucan and Mark Merrifield, 'Boundary mixing associated with tidal and near-inertial internal waves', *Journal of Physical Oceanography* 38 (2008), pp. 1238–52.

Walter Munk and Carl Wunsch, 'Abyssal recipes II: energetics of tidal and wind mixing', *Deep Sea Research Part I: Oceanographic Research Papers* 45 (1998), pp. 1977–2010.

Carl Wunsch and Raffaele Ferrari, 'Vertical mixing, energy, and the general circulation of the oceans', *Annual Review of Fluid Mechanics* 36 (2004), pp. 281–314.

Chu-Fang Yang, Wu-Cheng Chi, Hans van Haren, Ching-Ren Lin and Ban-Yuan Kuo, 'Tracking deep-sea internal wave propagation with a differential pressure gauge array', *Scientific Reports* 11 (2021), pp. 1–9.

Narragansett Bay

Todd McLeish, 'More tropical fish arriving in Narragansett Bay earlier', *ecoRI News*, 17 Aug. 2016, https://ecori.org/2016-8-17-more-tropical-fish-arriving-in-narragansett-bay-earlier-1/.

Charles Avenengo, '"Gulf Stream orphans" make their way to the bay', *Newport This Week*, 12 Nov. 2022, https://www.newport

thisweek.com/articles/gulf-stream-orphans-make-their-way-to-the-bay/.

Weifeng Zhang and Dennis J. McGillicuddy Jr, 'Warm spiral streamers over Gulf Stream warm-core rings', *Journal of Physical Oceanography* 50 (2020), pp. 3331–51.

Monsoon

'The Ming voyages' BBC Radio 4, *In Our Time*, 13 Oct. 2011, https://www.bbc.co.uk/programmes/b015p8c2.

Friedrich A. Schott and Julian P. McCreary Jr, 'The monsoon circulation of the Indian Ocean', *Progress in Oceanography* 51 (2001), pp. 1–123.

D. Shankar, P. N. Vinayachandran and A. S. Unnikrishnan, 'The monsoon currents in the north Indian Ocean', *Progress in Oceanography* 52 (2002), pp. 63–120.

Panickal Swapna, R. Krishnan and J. M. Wallace, 'Indian Ocean and monsoon coupled interactions in a warming environment', *Climate Dynamics* 42 (2014), pp. 2439–54.

Ruth Geen, Simona Bordoni, David S. Battisti and Katrina Hui, 'Monsoons, ITCZs, and the concept of the global monsoon', *Reviews of Geophysics* 58 (2020), e2020RG000700, https://doi.org/10.1029/2020RG000700.

Sulochana Gadgil, 'The Indian monsoon and its variability', *Annual Review of Earth and Planetary Sciences* 31 (2003), pp. 429–67.

Edward Dreyer, *Zheng He: China and the Oceans in the Early Ming Dynasty, 1405–1433* (Pearson, 2006).

Poles

Rebecca Woodgate, 'Arctic Ocean circulation: going around at the top of the world', *Nature Education Knowledge Nature Education Knowledge* 4, no. 8 (2013): 8.

Mary-Louise Timmermans and John Marshall, 'Understanding Arctic Ocean circulation: a review of ocean dynamics in a changing climate', *Journal of Geophysical Research: Oceans* 125 (2020), e2018JC014378, https://doi.org/10.1029/2018JC014378.

Eddy C. Carmack, 'The alpha/beta ocean distinction: a perspective on freshwater fluxes, convection, nutrients and productivity in high-latitude seas', *Deep Sea Research Part II: Topical Studies in Oceanography* 54 (2007), pp. 2578–98.

Ceridwen Fraser, Christina Hulbe, Craig Stevens and Huw Griffiths, 'An ocean like no other: the Southern Ocean's ecological richness and significance for global climate', *The Conversation*, 6 Dec. 2020, https://theconversation.com/an-ocean-like-no-other-the-southern-oceans-ecological-richness-and-significance-for-global-climate-151084.

Chapter 4: Messengers

Scattering

Xiaodong Zhang and Lianbo Hu, 'Light scattering by pure water and sea water: recent development', *Journal of Remote Sensing* (2021), art. 9753625, https://doi.org/10.34133/2021/9753625.

Xiaodong Zhang, 'Molecular light scattering by pure sea water', in Alexander A. Kokhanovsky, ed., *Light Scattering Reviews*, vol. 7 (Springer, 2013), pp. 225–43, https://doi.org/10.1007/978-3-642-21907-8_7.

Robin M. Pope and Edward S. Fry, 'Absorption spectrum (380–700 nm) of pure water. II. Integrating cavity measurements', *Applied Optics* 36 (1997), pp. 8710–23.

Morse

Kenneth Silverman, *Lightning Man: The Accursed Life of Samuel F. B. Morse* (Da Capo Press, 2003).

Samuel Finley Breese Morse, *Examination of the telegraphic apparatus and the processes in telegraphy* (US Government Printing Office, 1869).

Tejaswini R. Murgod and S. Meenakshi Sundaram, 'Survey on underwater optical wireless communication: perspectives and challenges', *Indonesian Journal of Electrical Engineering and Computer Science* 13 (2019), pp. 138–46.

P. Nickolaenko, A. V. Shvets and M. Hayakawa, 'Extremely low frequency (ELF) radio wave propagation: a review', *International Journal of Electronics and Applied Research* 3 (2016), pp. 1–91.

Joseph Stromberg, 'Why the US Navy once wanted to turn Wisconsin into the world's largest antenna', *Vox*, 10 April 2015, https://www.vox.com/2015/4/10/8381983/project-sanguine.

Rajat Pandit, 'Navy gets new facility to communicate with nuclear submarines', *Times of India*, 31 July 2014, https://timesofindia.indiatimes.com/india/Navy-gets-new-facility-to-communicate-with-nuclear-submarines-prowling-underwater/articleshow/39371121.cms.

Walter Sullivan, 'How huge antenna can broadcast into the silence of the sea', *New York Times*, 13 Oct. 1981, https://www.nytimes.com/1981/10/13/science/how-huge-antenna-can-broadcast-into-the-silence-of-the-sea.html?pagewanted=all.

'Sending signals to submarines', *New Scientist*, 4 July 1985, https://books.google.co.uk/books?id=NOPpwVvNu44C&pg=PA39&redir_esc=y#v=onepage&q&f=false.

Bob Aldridge, *ELF History: Extremely Low Frequency Communication* (Santa Clara, CA: Pacific Life Research Center, 18 Feb. 2001, http://www.plrc.org/docs/941005B.pdf.

Lloyd Butler, 'Underwater radio communication', *Amateur Radio*, April 1987.

Whale vision

Jeremy A. Goldbogen, John Calambokidis, Donald A. Croll, James T. Harvey, Kelly M. Newton, Erin M. Oleson, Greg Schorr and Robert E. Shadwick, 'Foraging behavior of humpback whales: kinematic and respiratory patterns suggest a high cost for a lunge', *Journal of Experimental Biology* 211 (2008), pp. 3712–19.

A. S. Friedlaender, R. B. Tyson, A. K. Stimpert, A. J. Read and D. P. Nowacek, *Extreme diel variation in the feeding behavior of humpback whales along the western Antarctic Peninsula during autumn*, Marine Ecology Progress Series no. 494 (Inter-Research Science, 2013), pp. 281–9.

Humboldt squid

Lars Schmitz, Ryosuke Motani, Christopher E. Oufiero, Christopher H. Martin, Matthew D. McGee, Ashlee R. Gamarra, Johanna J. Lee and Peter C. Wainwright, 'Allometry indicates giant eyes of giant squid are not exceptional', *BMC Evolutionary Biology* 13 (2013), pp. 1–9.

Gabriela A. Galeazzo, Jeremy D. Mirza, Felipe A. Dorr, Ernani Pinto, Cassius V. Stevani, Karin B. Lohrmann and Anderson G. Oliveira, 'Characterizing the bioluminescence of the Humboldt squid, *Dosidicus gigas* (d'Orbigny, 1835): one of the largest luminescent animals in the world', *Photochemistry and Photobiology* 95 (2019), pp. 1179–85.

Séverine Martini and Steven H. D. Haddock, 'Quantification of bioluminescence from the surface to the deep sea demonstrates its predominance as an ecological trait', *Scientific reports* 7 (2017), pp. 1–11.

Steven H. D. Haddock, Mark A. Moline and James F. Case, 'Bioluminescence in the sea', *Annual Review of Marine Science* 2 (2010), pp. 443–93.

Benjamin P. Burford and Bruce H. Robison, 'Bioluminescent backlighting illuminates the complex visual signals of a social squid in the deep sea', *Proceedings of the National Academy of Sciences* 117 (2020), pp. 8524–31.

Yuichi Oba, Cassius V. Stevani, Anderson G. Oliveira, Aleksandra S. Tsarkova, Tatiana V. Chepurnykh and Ilia V. Yampolsky, 'Selected least studied but not forgotten bioluminescent systems', *Photochemistry and Photobiology* 93 (2017), pp. 405–15.

Thomas L. Williams, Stephen L. Senft, Jingjie Yeo, Francisco J. Martín-Martínez, Alan M. Kuzirian, Camille A. Martin, Christopher W. DiBona et al., 'Dynamic pigmentary and structural coloration within cephalopod chromatophore organs', *Nature Communications* 10 (2019), pp. 1–15.

Justin Marshall, 'Vision and lack of vision in the ocean', *Current Biology* 27 (2017), pp. R494–502.

Fish hearing

M. Clara P. Amorim, 'Diversity of sound production in fish', *Communication in Fishes* 1 (2006), pp. 71–104.

Eric Parmentier, Jean-Paul Lagardère, Jean-Baptiste Braquegnier, Pierre Vandewalle and Michael L. Fine, 'Sound production mechanism in carapid fish: first example with a slow sonic muscle', *Journal of Experimental Biology* 209 (2006), pp. 2952–60.

Sherrylynn Rowe and Jeffrey A. Hutchings, 'The function of sound production by Atlantic cod as inferred from patterns of variation in drumming muscle mass', *Canadian Journal of Zoology* 82 (2004), pp. 1391–8.

Jarle Tryti Nordeide and Ivar Folstad, 'Is cod lekking or a promiscuous group spawner?', *Fish and Fisheries* 1 (2000), pp. 90–3.

Sherrylynn Rowe and Jeffrey A. Hutchings, 'Sound production by Atlantic cod during spawning', *Transactions of the American Fisheries Society* 135 (2006), pp. 529–38.

Bone conduction in humans

Tatjana Tchumatchenko and Tobias Reichenbach, 'A cochlear-bone wave can yield a hearing sensation as well as otoacoustic emission', *Nature Communications* 5 (2014), pp. 1–10.

Bubbles

Grant B. Deane and Helen Czerski, 'A mechanism stimulating sound production from air bubbles released from a nozzle', *Journal of the Acoustical Society of America* 123 (2008), pp. EL126–32.

Haddock

Anthony D. Hawkins and Marta Picciulin, 'The importance of underwater sounds to gadoid fishes', *Journal of the Acoustical Society of America* 146 (2019), pp. 3536–51.

G. Buscaino, M. Picciulin, D. E. Canale, E. Papale, M. Ceraulo, R. Grammauta and S. Mazzola, 'Spatio-temporal distribution and acoustic characterization of haddock (*Melanogrammus aeglefinus*, Gadidae) calls in the Arctic fjord Kongsfjorden (Svalbard Islands)', *Scientific Reports* 10 (2020), pp. 1–16.

Licia Casaretto, Marta Picciulin and Anthony D. Hawkins, 'Mating behaviour by the haddock (*Melanogrammus aeglefinus*)', *Environmental Biology of Fishes* 98 (2015), pp. 913–23.

Licia Casaretto, Marta Picciulin, Kjell Olsen and Anthony D. Hawkins, 'Locating spawning haddock (*Melanogrammus aeglefinus*, Linnaeus, 1758) at sea by means of sound', *Fisheries Research* 154 (2014), pp. 127–34.

Anthony D. Hawkins, Licia Casaretto, Marta Picciulin and Kjell Olsen, 'Locating spawning haddock by means of sound', *Bioacoustics* 12 (2002), pp. 284–6.

'Haddock sounds', *Discovery of Sound in the Sea*, n.d., https://dosits.org/galleries/audio-gallery/fishes/haddock/.

Noam Pinsk, Avital Wagner, Lilian Cohen, Christopher J. H. Smalley, Colan E. Hughes, Gan Zhang, Mariela J. Pavan et al., 'Biogenic guanine crystals are solid solutions of guanine and other purine metabolites', *Journal of the American Chemical Society* 144 (2022), pp. 5180–9.

Arthur N. Popper, Anthony D. Hawkins, Olav Sand and Joseph A. Sisneros, 'Examining the hearing abilities of fishes', *Journal of the Acoustical Society of America* 146 (2019), pp. 948–55.

Arthur N. Popper and Anthony D. Hawkins, 'The importance of particle motion to fishes and invertebrates', *Journal of the Acoustical Society of America* 143 (2018), pp. 470–88.

Tanja Schulz-Mirbach, Friedrich Ladich, Martin Plath and Martin Heß, 'Enigmatic ear stones: what we know about the functional role and evolution of fish otoliths', *Biological Reviews* 94 (2019), pp. 457–82.

M. Ashokan, G. Latha and R. Ramesh, 'Analysis of shallow water ambient noise due to rain and derivation of rain parameters', *Applied Acoustics* 88 (2015), pp. 114–22.

Timothy A. C. Gordon, Andrew N. Radford, Isla K. Davidson, Kasey Barnes, Kieran McCloskey, Sophie L. Nedelec, Mark G. Meekan, Mark I. McCormick and Stephen D. Simpson, 'Acoustic enrichment can enhance fish community development on degraded coral reef habitat', *Nature Communications* 10 (2019), pp. 1–7.

Lanternfish

Robert S. Dietz, 'The sea's deep scattering layers', *Scientific American* 207, Aug. 1962, pp. 44–51.

Robert S. Dietz, 'Deep scattering layer in the Pacific and Antarctic Oceans', *Journal of Marine Research* 7 (1948), pp. 430–42.

Martin W. Johnson, 'Sound as a tool in marine ecology, from data on biological noises and the deep scattering layer', *Journal of Marine Research* 7 (1948), pp. 443–58.

Venecia Catul, Manguesh Gauns and P. K. Karuppasamy, 'A review on mesopelagic fishes belonging to family *Myctophidae*', *Reviews in Fish Biology and Fisheries* 21 (2011), pp. 339–54.

Elizabeth N. Shor, *Scripps Institution of Oceanography: Probing the Oceans 1936 to 1976*, (Tofua, 1978), https://escholarship.org/uc/item/6nb9 j3mt.

'Oceanographic research of deep scattering layer by sonar and hydrophone', n.d., *YouTube*, https://www.youtube.com/watch?v=EQLup b3aK8Q.

Kanchana Bandara, Øystein Varpe, Lishani Wijewardene, Vigdis Tverberg and Ketil Eiane, 'Two hundred years of zooplankton vertical migration research', *Biological Reviews* 96 (2021), pp. 1547–89.

Carl F. Eyring, Ralph J. Christensen and Russell W. Raitt, 'Reverberation in the sea', *Journal of the Acoustical Society of America* 20 (1948), pp. 462–75.

Whale earwax

Kathleen E. Hunt, Nadine S. Lysiak, Michael Moore and Rosalind M. Rolland, 'Multi-year longitudinal profiles of cortisol and corticosterone recovered from baleen of North Atlantic right whales (*Eubalaena glacialis*)', *General and Comparative Endocrinology* 254 (2017), pp. 50–9.

Ed Yong, 'The history of the oceans is locked in whale earwax', *The Atlantic*, 21 Nov. 2018.

Stephen J. Trumble, Stephanie A. Norman, Danielle D. Crain, Farzaneh Mansouri, Zach C. Winfield, Richard Sabin, Charles W. Potter, Christine M. Gabriele and Sascha Usenko, 'Baleen whale cortisol levels reveal a physiological response to 20th century whaling', *Nature Communications* 9 (2018), pp. 1–8.

Heard Island Feasibility Test

The Heard Island Feasibility Test (University of Washington, 2007), https://staff.washington.edu/dushaw/heard/index.shtml.

Victoria Kaharl, 'Sounding out the ocean's secrets', *Beyond Discovery* (National Academy of Sciences, 1999), http://www.nasonline.org/publications/beyond-discovery/sounding-out-oceans-secrets.pdf.

R. Monastersky, 'Climate test: hum heard "round the world", (world-wide sound wave experiment to test greenhouse effect)', *Science News* 139, 26 Jan. 1991.

Heard Island Science Daily, 5 Jan. 1991, https://staff.washington.edu/dushaw/heard/experiment/docs/heard_island_science_daily.pdf.

Arthur Baggeroer and Walter Munk, 'The Heard Island feasibility test', *Physics Today* 45 (1992), pp. 22–30.

Brian Kahn, 'A forgotten underwater sound experiment almost changed how we measure global warming', *Gizmodo*, 24 Nov. 2017, https://gizmodo.com/a-forgotten-underwater-sound-experiment-almost-changed-1820659353.

'History of the SOFAR channel', *Discovery of Sound in the Sea*, n.d., recording from channel, https://dosits.org/science/movement/sofar-channel/history-of-the-sofar-channel/.

Wenbo Wu, Zhongwen Zhan, Shirui Peng, Sidao Ni and Jörn Callies, 'Seismic ocean thermometry', *Science* 369 (2020), pp. 1510–15.

Clément de Boyer Montégut, Gurvan Madec, Albert S. Fischer, Alban Lazar and Daniele Iudicone, 'Mixed layer depth over the global ocean: an examination of profile data and a profile-based climatology', *Journal of Geophysical Research: Oceans* 109 (2004), https://doi.org/10.1029/2004JC002378.

Carl Wunsch, 'Advance in global ocean acoustics', *Science* 369 (2020), pp. 1433–4.

Chapter 5: Passengers

Tortoise

Justin Gerlach, Catharine Muir and Matthew D. Richmond, 'The first substantiated case of trans-oceanic tortoise dispersal', *Journal of Natural History* 40 (2006), pp. 2403–8.

Anthony S. Cheke, Miguel Pedrono, Roger Bour, Atholl Anderson, Christine Griffiths, John B. Iverson, Julian P. Hume and Martin Walsh, 'Giant tortoises spread to western Indian Ocean islands by sea drift in pre-Holocene times, not by later human agency – response to Wilmé et al. (2016a)', *Journal of Biogeography* 44 (2017), pp. 1426–9.

Dennis M. Hansen, Jeremy J. Austin, Rich H. Baxter, Erik J. de Boer, Wilfredo Falcón, Sietze J. Norder, Kenneth F. Rijsdijk, Christophe Thébaud, Nancy J. Bunbury and Ben H. Warren, 'Origins of endemic island tortoises in the western Indian Ocean: a critique of the human-translocation hypothesis', *Journal of Biogeography* 44 (2017), pp. 1430–5.

Nikos Poulakakis, Michael Russello, Dennis Geist and Adalgisa Caccone, 'Unravelling the peculiarities of island life: vicariance, dispersal and the diversification of the extinct and extant giant Galápagos tortoises', *Molecular Ecology* 21 (2012), pp. 160–73.

Adalgisa Caccone, Gabriele Gentile, James P. Gibbs, Thomas H. Fritts, Howard L. Snell, Jessica Betts and Jeffrey R. Powell, 'Phylogeography and history of giant Galápagos tortoises', *Evolution* 56 (2002), pp. 2052–66.

Fritz Haber

Minoru Koide, Vern Hodge, Edward D. Goldberg and Kathe Bertine, 'Gold in sea water: a conservative view', *Applied Geochemistry* 3 (1988), pp. 237–41.

Ross R. Large, Daniel D. Gregory, Jeffrey A. Steadman, Andrew G. Tomkins, Elena Lounejeva, Leonid V. Danyushevsky, Jacqueline A. Halpin, Valeriy Maslennikov, Patrick J. Sack, Indrani Mukherjee, Ron Berry and Arthur Hickman, 'Gold in the oceans through time', *Earth and Planetary Science Letters* 428 (2015), pp. 139–50.

Forrest H. Nielsen, 'Evolutionary events culminating in specific minerals becoming essential for life', *European Journal of Nutrition* 39 (2000), pp. 62–6.

Drifting life

Ian A. Hatton, Ryan F. Heneghan, Yinon M. Bar-On and Eric D. Galbraith, 'The global ocean size spectrum from bacteria to whales', *Science Advances* 7 (2021), eabh3732, https://doi.org/10.1126/sciadv.abh373.

Alexander von Humboldt, *Cosmos: A Sketch of the Physical Description of the Universe*, vol. 1 (Quality Classics, 2010), vol. 187703.

Blue whale faeces

Sam Kean, *Caesar's Last Breath: The Epic Story of the Air Around Us* (Random House, 2017).

Victor Smetacek, 'A whale of an appetite revealed by analysis of prey consumption', *Nature* 599 (2021), pp. 33–4.

Matthew S. Savoca, Max F. Czapanskiy, Shirel R. Kahane-Rapport, William T. Gough, James A. Fahlbusch, K. C. Bierlich, Paolo S. Segre et al., 'Baleen whale prey consumption based on high-resolution foraging measurements', *Nature* 599 (2021), pp. 85–90.

Lavenia Ratnarajah, Andrew R. Bowie, Delphine Lannuzel, Klaus M. Meiners and Stephen Nicol, 'The biogeochemical role of baleen whales and krill in Southern Ocean nutrient cycling', *PloS One* 9 (2014), e114067, https://doi.org/10.1371/journal.pone.0114067.

Jessica J. Williams, Yannis P. Papastamatiou, Jennifer E. Caselle, Darcy Bradley and David M. P. Jacoby, 'Mobile marine predators: an understudied source of nutrients to coral reefs in an unfished atoll', *Proceedings of the Royal Society B: Biological Sciences* 285 (2018), 20172456, https://doi.org/10.1098/rspb.2017.2456.

Pablo Alba-González, Xosé Antón Álvarez-Salgado, Antonio Cobelo-García, Joeri Kaal and Eva Teira, 'Faeces of marine birds and mammals as substrates for microbial plankton communities', *Marine Environmental Research* 174 (2022), 105560, https://doi.org/10.1016/j.marenvres.2022.105560.

Erica Sparaventi, Araceli Rodríguez-Romero, Andrés Barbosa, Laura Ramajo and Antonio Tovar-Sánchez, 'Trace elements in Antarctic penguins and the potential role of guano as source of recycled metals in the Southern Ocean', *Chemosphere* 285 (2021), 131423, https://doi.org/10.1016/j.chemosphere.2021.131423.

Lavenia Ratnarajah and Andrew R. Bowie, 'Nutrient cycling: are Antarctic krill a previously overlooked source in the marine iron cycle?', *Current Biology* 26 (2016), pp. R884–7.

Joe Roman, John Nevins, Mark Altabet, Heather Koopman and James McCarthy, 'Endangered right whales enhance primary productivity in the Bay of Fundy', *PLoS One* 11 (2016), e0156553, https://doi.org/10.1371/journal.pone.0156553.

Lavenia Ratnarajah, Jessica Melbourne-Thomas, Martin P. Marzloff, Delphine Lannuzel, Klaus M. Meiners, Fanny Chever, Stephen Nicol and Andrew R. Bowie, 'A preliminary model of iron fertilization by baleen whales and Antarctic krill in the Southern Ocean: sensitivity of primary productivity estimates to parameter uncertainty', *Ecological Modelling* 320 (2016), pp. 203–12.

Lavenia Ratnarajah, Delphine Lannuzel, Ashley T. Townsend, Klaus M. Meiners, Stephen Nicol, Ari S. Friedlaender and Andrew R. Bowie, 'Physical speciation and solubility of iron from baleen whale faecal material', *Marine Chemistry* 194 (2017), pp. 79–88.

E. J. Miller, J. M. Potts, M. J. Cox, B. S. Miller, S. Calderan, R. Leaper, P. A. Olson, Richard Lyell O'Driscoll and M. C. Double, 'The characteristics of krill swarms in relation to aggregating Antarctic blue whales', *Scientific Reports* 9 (2019), pp. 1–13.

London sewers

John Ashton and Janet Ubido, 'The healthy city and the ecological idea', *Social History of Medicine* 4 (1991), pp. 173–80.

Concrete

Aidan Reilly and Oliver Kinnane, 'Construction is a cause of global warming, but is concrete really the problem?', *Architects Journal*, 1 March 2019.

M. Garside, 'Cement production worldside from 1995 to 2021', *Statista*, 1 April 2022, https://www.statista.com/statistics/1087115/global-cement-production-volume/.

Colin R. Gagg, 'Cement and concrete as an engineering material: an historic appraisal and case study analysis', *Engineering Failure Analysis* 40 (2014), pp. 114–40.

Carbon

Pierre Friedlingstein, Matthew W. Jones, Michael O'Sullivan, Robbie M. Andrew, Dorothee C. E. Bakker, Judith Hauck, Corinne le Quéré et al., 'Global carbon budget 2021', *Earth System Science Data* 14 (2022), pp. 1917–2005.

Philip W. Boyd, Hervé Claustre, Marina Levy, David A. Siegel and Thomas Weber, 'Multi-faceted particle pumps drive carbon sequestration in the ocean', *Nature* 568 (2019), pp. 327–35. https://doi.org/10.5194/essd-14-1917-2022.

Hervé Claustre, Louis Legendre, Philip Boyd and Marina Levy, 'The oceans' biological carbon pumps: framework for a research observational community approach', *Frontiers in Marine Science* 8 (2021), 780052, https://doi.org/10.3389/fmars.2021.780052.

Christine Lehman, 'What is the "true" nature of diamond?', *Nuncius* 31 (2016), pp. 361–407.

Michael M. Woolfson, *The Origin and Evolution of the Solar System* (CRC Press, 2000).

Sophia E. Brumer, Christopher J. Zappa, Byron W. Blomquist, Christopher W. Fairall, Alejandro Cifuentes-Lorenzen, James B. Edson, Ian M. Brooks and Barry J. Huebert, 'Wave-related Reynolds number parameterizations of CO_2 and DMS transfer velocities', *Geophysical Research Letters* 44 (2017), pp. 9865–75.

Paul Falkowski, R. J. Scholes, E. E. A. Boyle, Josep Canadell, D. Canfield, James Elser, Nicolas Gruber et al., 'The global carbon cycle: a test of our knowledge of Earth as a system', *Science* 290 (2000), pp. 291–6.

B. P. V. Hunt, E. A. Pakhomov, G. W. Hosie, V. Siegel, Peter Ward and K. Bernard, 'Pteropods in southern ocean ecosystems', *Progress in Oceanography* 78 (2008), pp. 193–221.

Clifford C. Walters, 'The origin of petroleum', in Chang Samuel Hsu and Paul R. Robinson, eds, *Practical Advances in Petroleum Processing* (Springer, 2006), pp. 79–101.

Paul G. Falkowski, Edward A. Laws, Richard T. Barber and James W. Murray, 'Phytoplankton and their role in primary, new, and export production', in Michael J. R. Fasham, ed., *Ocean Biogeochemistry* (Springer, 2003), pp. 99–121.

Chapter 6: Voyagers

Ramisyllis multicaudata

Guillermo Ponz-Segrelles, Christopher J. Glasby, Conrad Helm, Patrick Beckers, Jörg U. Hammel, Rannyele P. Ribeiro and M. Teresa

Aguado, 'Integrative anatomical study of the branched annelid *Ramisyllis multicaudata* (Annelida, Syllidae)', *Journal of Morphology* 282 (2021), pp. 900–16.

Jennifer Frazer, 'One head, 1,000 rear ends: the tale of a deeply weird worm', *Scientific American*, Aug. 2021.

Christopher J. Glasby, Paul C. Schroeder and Maria Teresa Aguado, 'Branching out: a remarkable new branching syllid (Annelida) living in a Petrosia sponge (Porifera: Demospongiae)', *Zoological Journal of the Linnean Society* 164 (2012), pp. 481–97.

Heinz-Dieter Franke, 'Reproduction of the Syllidae (Annelida: polychaeta)', *Hydrobiologia* 402 (1999), pp. 39–55.

Penguins

Helen Phillips, Benoit Legresy and Nathan Bindoff, 'Explainer: how the Antarctic Circumpolar Current helps keep Antarctica frozen', *The Conversation*, 15 Nov. 2018, https://theconversation.com/explainer-how-the-antarctic-circumpolar-current-helps-keep-antarctica-frozen-106164.

Mati Kahru, Emanuele Di Lorenzo, Marlenne Manzano-Sarabia and B. Greg Mitchell, 'Spatial and temporal statistics of sea surface temperature and chlorophyll fronts in the California Current', *Journal of Plankton Research* 34 (2012), pp. 749–60.

Natalie M. Freeman, Nicole S. Lovenduski and Peter R. Gent, 'Temporal variability in the Antarctic polar front (2002–2014)', *Journal of Geophysical Research: Oceans* 121 (2016), pp. 7263–76.

Alejandro H. Orsi, Thomas Whitworth III and Worth D. Nowlin Jr, 'On the meridional extent and fronts of the Antarctic Circumpolar Current', *Deep Sea Research Part I: Oceanographic Research Papers* 42 (1995), pp. 641–73.

Donata Giglio and Gregory C. Johnson, 'Subantarctic and polar fronts of the Antarctic Circumpolar Current and Southern Ocean heat and freshwater content variability: a view from Argo', *Journal of Physical Oceanography* 46 (2016), pp. 749–68.

Yves Cherel, Keith A. Hobson, Christophe Guinet and Cecile Vanpe, 'Stable isotopes document seasonal changes in trophic niches and winter foraging individual specialization in diving predators from the Southern Ocean', *Journal of Animal Ecology* 76 (2007), pp. 826–36.

Kozue Shiomi, Katsufumi Sato, Yves Handrich and Charles A. Bost, *Diel shift of king penguin swim speeds in relation to light intensity changes*, Marine Ecology Progress Series no. 561 (Inter-Research Science Publisher, 2016), pp. 233–43, https://doi.org/10.3354/meps11930.

Andrew J. S. Meijers, Michael P. Meredith, Eugene J. Murphy, D. P. Chambers, Mark Belchier and Emma F. Young, 'The role of ocean dynamics in king penguin range estimation', *Nature Climate Change* 9 (2019), pp. 120–1.

Annette Scheffer, Philip N. Trathan, Johnnie G. Edmonston and Charles-André Bost, 'Combined influence of meso-scale circulation and bathymetry on the foraging behaviour of a diving predator, the king penguin (*Aptenodytes patagonicus*)', *Progress in Oceanography* 141 (2016), pp. 1–16.

L. G. Halsey, P. J. Butler, A. Fahlman, Charles-André Bost and Yves Handrich, *Changes in the foraging dive behaviour and energetics of king penguins through summer and autumn: a month by month analysis*, Marine Ecology Progress Series no. 401 (Inter-Research Science Publisher, 2010), pp. 279–89.

Jean-Benoît Charrassin and Charles-André Bost, *Utilization of the oceanic habitat by king penguins over the annual cycle*, Marine Ecology Progress Series no. 221 (Inter-Research Science Publisher, 2001), pp. 285–98.

Charles-André Bost, Cédric Cotté, Frédéric Bailleul, Yves Cherel, Jean-Benoît Charrassin, Christophe Guinet, David G. Ainley and Henri Weimerskirch, 'The importance of oceanographic fronts to marine birds and mammals of the southern oceans', *Journal of Marine Systems* 78 (2009), pp. 363–76.

Barbara A. Block, Ian D. Jonsen, Salvador J. Jorgensen, Arliss J. Winship, Scott A. Shaffer, Steven J. Bograd, Elliott Lee Hazen et al., 'Tracking apex marine predator movements in a dynamic ocean', *Nature* 475 (2011), pp. 86–90.

Igor M. Belkin, 'Remote sensing of ocean fronts in marine ecology and fisheries', *Remote Sensing* 13 (2021), art. 883.

Alberto Baudena, Enrico Ser-Giacomi, Donatella D'Onofrio, Xavier Capet, Cédric Cotté, Yves Cherel and Francesco D'Ovidio,

'Fine-scale structures as spots of increased fish concentration in the open ocean', *Scientific Reports* 11 (2021), pp. 1–13.

S. V. Prants, 'Marine life at Lagrangian fronts', *Progress in Oceanography* 204 (2022), 102790, https://doi.org/10.1016/j.pocean.2022.102790.

Young-Hyang Park, Isabelle Durand, Élodie Kestenare, Gilles Rougier, Meng Zhou, Francesco d'Ovidio, Cédric Cotté and Jae-Hak Lee, 'Polar Front around the Kerguelen Islands: an up-to-date determination and associated circulation of surface/subsurface waters', *Journal of Geophysical Research: Oceans* 119 (2014), pp. 6575–92.

Herring

Donna Heddle, '"Sharp tongues and sharp knives": the herring lassies', *Orkney News*, 20 Dec. 2020.

Minna Kajaste-McCormack, 'The Scottish Fisheries Museum: herring lasses and their silver darlings', *Art UK*, 20 Oct. 2021, https://artuk. org/discover/stories/the-scottish-fisheries-museum-herring-lasses-and-their-silver-darlings.

Mark Dickey-Collas, *The current state of knowledge on the ecology and interactions of North Sea herring within the North Sea ecosystem*, CVO Report no. CVO 04.028 (Stichting DLO-Centre for Fishery Research, 2004), https://library.wur.nl/WebQuery/wurpubs/fulltext/121078.

Pierre Helaouët and Grégory Beaugrand, *Macroecology of* Calanus finmarchicus *and* C. helgolandicus *in the North Atlantic Ocean and adjacent seas*, Marine Ecology Progress Series no. 345 (Inter-Research Science Publisher, 2007), pp. 147–65.

Kanchana Bandara, Øystein Varpe, Frédéric Maps, Rubao Ji, Ketil Eiane and Vigdis Tverberg, 'Timing of *Calanus finmarchicus* diapause in stochastic environments', *Ecological Modelling* 460 (2021), e109739, https://doi.org/10.1016/j.ecolmodel.2021.109739.

A. Corten, 'A possible adaptation of herring feeding migrations to a change in timing of the *Calanus finmarchicus* season in the eastern North Sea', *ICES Journal of Marine Science* 57 (2000), pp. 1270–2000.

Vimal Koul, Corinna Schrum, André Düsterhus and Johanna Baehr, 'Atlantic inflow to the North Sea modulated by the subpolar gyre in a historical simulation with MPI-ESM', *Journal of Geophysical Research: Oceans* 124 (2019), pp. 1807–26.

Cecilie Broms, Webjørn Melle and Stein Kaartvedt, 'Oceanic distribution and life cycle of *Calanus* species in the Norwegian Sea and adjacent waters', *Deep Sea Research Part II: Topical Studies in Oceanography* 56 (2009), pp. 1910–21.

Shuang Gao, Solfrid Sætre Hjøllo, Tone Falkenhaug, Espen Strand, Martin Edwards and Morten D. Skogen, 'Overwintering distribution, inflow patterns and sustainability of *Calanus finmarchicus* in the North Sea', *Progress in Oceanography* 194 (2021), 102567, https://doi.org/10.1016/j.pocean.2021.102567.

Margaret H. King, 'A partnership of equals: women in Scottish east coast fishing communities', *Folk Life* 31 (1992), pp. 17–35.

Tuna

Bruce B. Collette (2020) 'The future of bluefin tunas: ecology, fisheries management, and conservation', *Reviews in Fisheries Science & Aquaculture* 28, 2020, pp. 136–7, https://doi.org/10.1080/23308249.2019.1665237.

Peter Gaube, Caren Barcelo, Dennis J. McGillicuddy Jr, Andrés Domingo, Philip Miller, Bruno Giffoni, Neca Marcovaldi and Yonat Swimmer, 'The use of mesoscale eddies by juvenile loggerhead sea turtles (*Caretta caretta*) in the southwestern Atlantic', *PloS One* 12 (2017), e0172839, https://doi.org/10.1371/journal.pone.0172839.

Barbara A. Block, Steven L. H. Teo, Andreas Walli, Andre Boustany, Michael J. W. Stokesbury, Charles J. Farwell, Kevin C. Weng, Heidi Dewar and Thomas D. Williams, 'Electronic tagging and population structure of Atlantic bluefin tuna', *Nature* 434 (2005), pp. 1121–7.

David E. Richardson, Katrin E. Marancik, Jeffrey R. Guyon, Molly E. Lutcavage, Benjamin Galuardi, Chi Hin Lam, Harvey J. Walsh, Sharon Wildes, Douglas A. Yates and Jonathan A. Hare, 'Discovery of a spawning ground reveals diverse migration strategies in Atlantic bluefin tuna (*Thunnus thynnus*)', *Proceedings of the National Academy of Sciences* 113 (2016), pp. 3299–3304.

P. Reglero, A. Ortega, R. Balbín, F. J. Abascal, A. Medina, E. Blanco, F. de la Gándara, D. Alvarez-Berastegui, M. Hidalgo, L. Rasmuson, F. Alemany and Ø. Fiksen, 'Atlantic bluefin tuna spawn at suboptimal temperatures for their offspring', *Proceedings of the Royal Society*

B: Biological Sciences 285 (2018), e20171405, https://doi.org/10.1098/rspb.2017.1405.

Walter J. Golet, Nicholas R. Record, Sigrid Lehuta, Molly Lutcavage, Benjamin Galuardi, Andrew B. Cooper and Andrew J. Pershing, *The paradox of the pelagics: why bluefin tuna can go hungry in a sea of plenty*, Marine Ecology Progress Series no. 527 (Inter-Research Science Publisher, 2015), pp. 181–92.

Ango C. Hsu, Andre M. Boustany, Jason J. Roberts, Jui-Han Chang and Patrick N. Halpin, 'Tuna and swordfish catch in the US northwest Atlantic longline fishery in relation to mesoscale eddies', *Fisheries Oceanography* 24 (2015), pp. 508–20.

Brynn Devine, Sheena Fennell, Daphne Themelis and Jonathan A. D. Fisher, 'Influence of anticyclonic, warm-core eddies on mesopelagic fish assemblages in the northwest Atlantic Ocean', *Deep Sea Research Part I: Oceanographic Research Papers* 173 (2021), e103555, https://doi.org/10.1016/j.dsr.2021.103555.

Insha Ahmed Taray, Azmin Shakrine Mohd Rafie, Mohammad Zuber and Kamarul Arifin Ahmad, 'Hydrodynamics of bluefin tuna: a review', *Journal of Advanced Research in Fluid Mechanics and Thermal Sciences* 64 (2019), pp. 293–303.

Jay R. Rooker, Igaratza Fraile, Hui Liu, Noureddine Abid, Michael A. Dance, Tomoyuki Itoh, Ai Kimoto, Yohei Tsukahara, Enrique Rodriguez-Marin and Haritz Arrizabalaga, 'Wide-ranging temporal variation in transoceanic movement and population mixing of bluefin tuna in the North Atlantic Ocean', *Frontiers in Marine Science* 6 (2019), art. 398, https://doi.org/10.3389/fmars.2019.00398.

Dudley B. Chelton, Michael G. Schlax and Roger M. Samelson, 'Global observations of nonlinear mesoscale eddies', *Progress in Oceanography* 91 (2011), pp. 167–216.

Barbara A. Block, Steven L. H. Teo, Andreas Walli, Andre Boustany, Michael J. W. Stokesbury, Charles J. Farwell, Kevin C. Weng, Heidi Dewar and Thomas D. Williams, 'Electronic tagging and population structure of Atlantic bluefin tuna', *Nature* 434 (2005), pp. 1121–7.

Hōkūleʻa

Sam Low, *Hawaiki rising: Hōkūleʻa, Nainoa Thompson and the Hawaiian renaissance* (University of Hawaiʻi Press, 2018 [2013]).

Chapter 7: Future

Danielle Purkiss, Ayşe Lisa Allison, Fabiana Lorencatto, Susan Michie and Mark Miodownik, 'The big compost experiment: using citizen science to assess the impact and effectiveness of biodegradable and compostable plastics in UK home composting', *Frontiers in Sustainability*, 3 Nov. 2022, art. 132, https://doi.org/10.3389/frsus.2022.942724.

Sunke Schmidtko, Lothar Stramma and Martin Visbeck, 'Decline in global oceanic oxygen content during the past five decades', *Nature* 542 (2017), pp. 335–9.

Jozef Skákala, Jorn Bruggeman, David Ford, Sarah Wakelin, Anıl Akpınar, Tom Hull, Jan Kaiser et al., 'The impact of ocean biogeochemistry on physics and its consequences for modelling shelf seas', *Ocean Modelling* 172 (2022), 101976, https://doi.org/10.1016/j.ocemod.2022.101976.

Rebecca Loomis, Sarah R. Cooley, James R. Collins, Simon Engler and Lisa Suatoni, 'A code of conduct is imperative for ocean carbon dioxide removal research', *Frontiers in Marine Science* 9 (2022), https://doi.org/10.3389/fmars.2022.872800.

Norman G. Loeb, Gregory C. Johnson, Tyler J. Thorsen, John M. Lyman, Fred G. Rose and Seiji Kato, 'Satellite and ocean data reveal marked increase in Earth's heating rate', *Geophysical Research Letters* 48 (2021), e2021GL093047, https://doi.org/10.1029/2021GL093047.

I. Halevy and A. Bachan, 'The geologic history of seawater pH', *Science* 355 (2017), pp. 1069–71, https://doi.org/10.1126/science.aal415.

Chris Hadfield, *An Astronaut's Guide to Life on Earth* (Pan Macmillan, 2013).

Li Guancheng, Lijing Cheng, Jiang Zhu, Kevin E. Trenberth, Michael E. Mann and John P. Abraham, 'Increasing ocean stratification over the past half-century', *Nature Climate Change* 10 (2020), pp. 1116–23.

Andreas Oschlies, 'A committed fourfold increase in ocean oxygen loss', *Nature Communications* 12 (2021), pp. 1–8.

Lisa A. Levin, 'Manifestation, drivers, and emergence of open ocean deoxygenation', *Annual Review of Marine Science* 10 (2018), pp. 229–60.

Alison R. Taylor, Abdul Chrachri, Glen Wheeler, Helen Goddard and Colin Brownlee, 'A voltage-gated H+ channel underlying pH homeostasis in calcifying coccolithophores', *PLoS Biology* 9 (2011), e1001085, https://doi.org/10.1371/journal.pbio.1001085.

Fanny M. Monteiro, Lennart T. Bach, Colin Brownlee, Paul Bown, Rosalind E. M. Rickaby, Alex J. Poulton, Toby Tyrrell et al., 'Why marine phytoplankton calcify', *Science Advances* 2 (2016), e1501822, https://doi.org/10.1126/sciadv.1501822.

Wada Shigeki, Sylvain Agostini, Ben P. Harvey, Yuko Omori and Jason M. Hall-Spencer, 'Ocean acidification increases phytobenthic carbon fixation and export in a warm-temperate system', *Estuarine, Coastal and Shelf Science* 250 (2021), 107113, https://doi.org/10.1016/j.ecss.2020.107113.

N. Penny Holliday, Manfred Bersch, Barbara Berx, Léon Chafik, Stuart Cunningham, Cristian Florindo-López, Hjálmar Hátún et al., 'Ocean circulation causes the largest freshening event for 120 years in eastern subpolar North Atlantic', *Nature Communications* 11 (2020), pp. 1–15.

Peter Braesicke, Harald Elsner, Klaus Grosfeld, Julian Gutt, Stefan Hain, Hartmut H. Hellmer, Heike Herata et al., *World Ocean Review: Living with the Oceans 6: The Arctic and Antarctic – Extreme, Climatically Crucial and in Crisis* (Maribus, 2019).

Paul Voosen, 'Climate change spurs global speedup of ocean currents', *Science* 367 (2020), pp. 612–13, https://doi.org/10.1126/science.367.6478.612.

Sarah Stanley, 'Capturing how fast the Arctic Ocean is gaining fresh water', *Eos* 102, 8 Dec. 2021, https://doi.org/10.1029/2021EO210652.

Regin Winther Poulsen, 'Cutting the food chain? The controversial plan to turn zooplankton into fish oil', *Guardian*, 19 Jan. 2022, https://www.theguardian.com/environment/2022/jan/19/cutting-food-chain-faroe-islands-controversial-plan-to-turn-zooplankton-into-fish-oil-health-supplements.

Chloe Wayman and Helge Niemann, 'The fate of plastic in the ocean environment: a minireview', *Environmental Science: Processes & Impacts* 23 (2021), pp. 198–212.

Brian R. MacKenzie, Mark R. Payne, Jesper Boje, Jacob L. Høyer and Helle Siegstad, 'A cascade of warming impacts brings bluefin tuna to Greenland waters', *Global Change Biology* 20 (2014), pp. 2484–91.

Tiffany R. Anderson, Charles H. Fletcher, Matthew M. Barbee, Bradley M. Romine, Sam Lemmo and Jade Delevaux, 'Modeling multiple sea level rise stresses reveals up to twice the land at risk compared to strictly passive flooding methods', *Scientific Reports* 8 (2018), pp. 1–14.

Bradley M. Romine, Charles H. Fletcher, Matthew M. Barbee, Tiffany R. Anderson and L. Neil Frazer, 'Are beach erosion rates and sea-level rise related in Hawai'i?', *Global and Planetary Change* 108 (2013), pp. 149–57.

Index

Numbers in *italic* refer to pages with illustrations.

About the Author

Helen Czerski was born in Manchester. She is an associate professor in the Department of Mechanical Engineering at University College London. As a physicist, she studies the bubbles underneath breaking waves in the open ocean, seeking to understand their effects on weather and climate.

Helen has been a regular presenter of BBC TV science documentaries since 2011. She also hosts the Ocean Matters podcast, is part of the Cosmic Shambles Network, and is one of the presenters for the Fully Charged Show. She has been a science columnist for the *Wall Street Journal* since 2017 and she is also the author of the bestselling *Storm in a Teacup: The Physics of Everyday Life* and *Bubbles: A Ladybird Expert Book*.

She lives in London.

ALSO AVAILABLE FROM HELEN CZERSKI:

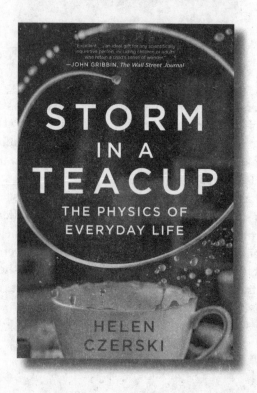

"A revelation. . . . [A]nyone who can write so appealingly about scientific complexities deserves the hats-off prize."
—Sarah Bakewell, *New York Times Book Review*

"An ideal gift for any scientifically inquisitive person."
—John Gribbin, *Wall Street Journal*

"[Czerski's] quest to enhance humanity's everyday scientific literacy is timely and imperative." —*Science*

W. W. NORTON & COMPANY
Independent Publishers Since 1923